Food Dehydration

BUTTERWORTH-HEINEMANN SERIES IN FOOD CONTROL

Editor-in-chief: Professor G. Campbell-Platt
University of Reading
UK

Food Dehydration

A Dictionary and Guide

J. G. Brennan
Department of Food Science and Technology
University of Reading

BUTTERWORTH
HEINEMANN

Butterworth-Heinemann Ltd
Linacre House, Jordan Hill, Oxford OX2 8DP

A member of the Reed Elsevier plc group

OXFORD LONDON BOSTON
MUNICH NEW DELHI SINGAPORE SYDNEY
TOKYO TORONTO WELLINGTON

First published 1994

British Library Cataloguing in Publication Data
Brennan, J. G.
Food Dehydration: Dictionary and Guide
I. Title
664

ISBN 0 7506 1130 8

Library of Congress Cataloguing in Publication Data
Brennan, J. G.
Food dehydration: a dictionary and guide/J. G. Brennan.
p. cm.
Includes bibliographical references.
ISBN 0 7506 1130 8
1. Food – Drying – Dictionaries. I. Title.
TP371.5.B74 94–3051
664'.0284–dc20 CIP

Composition by Genesis Typesetting, Rochester, Kent
Printed and bound in Great Britain

Contents

Preface

Drying is the oldest method of food preservation practised by man. Today the dehydration section of the food industry is large and extends to all countries throughout the world. Facilities range in size from simple sun driers to very large capacity, sophisticated drying installations. A very wide range of dehydrated foods is available including vegetables, fruits, meat, fish, cereal and milk products. Dehydrated soup and sauce mixes and complete meals made up of dried ingredients are also available. A great deal of research has been carried out to improve the quality and convenience of dehydrated foods. Many modern dried foods may be reconstituted quickly to yield products with good flavour and texture. Thus, they make an important contribution to the convenience food market.

This book is made up of four parts. Part 1 covers the historical background and general principles of food dehydration by means of heated air, by contact with a heated surface, by the application of radiant, microwave or dielectric energy and by freeze drying. In Part 2 information on drying phenomena, methods and equipment is presented in dictionary form. A guide to the procedures and conditions used for drying specific foods is given in Part 3. An extensive list of references on aspects of food dehydration makes up Part 4.

It is hoped that this work will be of interest to staff and students involved in undergraduate and postgraduate courses in food science, technology and engineering, and to researchers in the field of food dehydration. It should also be a quick and easy source of information for persons working in this section of the food industry.

J G Brennan

Part 1

Historical Background and General Principles

Historical Background

For thousands of years, man has sun dried foods to sustain him in off-season periods. The following are some recorded examples of the very early application of sun drying. As far back as 20,000 BC meat was cut into strips and sun dried in Russia. Around 10,000 BC salt was produced by sun drying seawater. American Indians made dried mashed potatoes about 3500 BC. The potatoes were frozen overnight and trampled to express out the juices. This process was repeated before the mash was dried. In ancient Egypt (2800–2300 BC) fruit such as apples, grapes and apricots were sun dried. Around 500 BC, dry-salted fish was produced.

Tea was dried in India in 300–400 AD. In the period 710–785 AD, large quantities of sun-dried foods were produced in Japan including fruits, vegetables, fish and shellfish, meat and poultry. The dried products were stored in warehouses. Around 900 AD, fish was sun dried in Norway. The Mongolian army used sun dried powdered milk in 1240 AD. Around 1650 AD colonists in North America dried boiled Indian corn over fires.

In 1780 AD, the first patent on vegetable drying was taken out in America. The vegetables were boiled in salt water, and kept for 20–30 hours. The quality was poor. In 1795 AD, in France, sliced vegetables were dried in air at 40°C, pressed, and sealed in foil. Enzyme activity occurred and vitamin C was destroyed.

Dried vegetables were shipped to the British troops during the Crimean war (1854–1856) and were also used by the Union troops in the American Civil War (1861–1865). In 1865 AD, a patent for producing dried egg was taken out. In 1872, Samuel Percy took out a patent on "Improvements in Drying and Concentrating Liquid Substances by Atomizing", i.e. spray drying/concentrating. Dried vegetables, produced in Canada, were used by British troops during the Boer war (1899–1902).

Developments accelerated in the 20th century. In 1901, Robert Stauff patented a spray drier for blood and milk. This featured an upward-spraying nozzle and perforated plate air disperser. In 1902, Just Hatmaker developed a drum drier. Merrel Soul, an American company, purchased the Stauff patent in 1905 and developed a spray drier for producing milk powder. This was a box-type, horizontal-concurrent drier which was operated on a batch principle. In 1912, George Krauss developed the centrifugal spray drier and in 1913 Grey and Jensen developed a conical spray drier. This type of spray drier was used extensively for many years. Dried vegetables were supplied to British and American troops during World War I (1914–1918). During that period, there was considerable expansion in vegetable-drying facilities in Europe, including cabinet, tunnel and conveyor driers. Research on vacuum drying of foods was also undertaken. C. E. Rogers introduced the continuous box-type, horizonal-concurrent, spray-drier in 1917. The jet-spray drier was developed by Coulter in 1940. Before and during World War II (1939–1945) a whole range of dried products was developed, including drum-dried soup mixes and tomato flakes, spray-dried milk and egg products, vacuum-dried fruits, and air-dried onions and garlic. In the period 1939–1945 a huge expansion in drying facilities occurred in Europe and America and dried foods were used extensively by the troops of all the parties to the conflict. In 1945, Flosdorf first used vacuum freeze drying for foods. Considerable research into freeze drying was undertaken in the UK in the early 1950s which led to the development of the accelerated freeze drying (AFD) method. Freeze-drying plants for meat, fish and vegetables were set up in many countries, notably Ireland. However, mainly because of the costliness of the process, interest in freeze drying of such products waned over the next decade.

The first instantized milk was introduced by Peebles in 1954. The BIRS drier was introduced in 1962 but it did not receive widespread application. In 1965, the first patent on instant coffee was taken out. Since then considerable advances have been made in improving the organoleptic quality and reconstitution properties of instant beverages. In 1960, the first freeze-dried instant coffee was produced. The production of freeze-dried instant coffee grew rapidly in the late 1970s and 1980s. Pneumatic drying of small particulate foods came into use in the early 1960s. This was followed by the introduction of fluidized-bed driers. These have found increasing application to food dehydration in the last twenty years. They are also used as agglomerators for spray-dried powders. The spouted bed drier, a variation on the fluidized bed, has also found some useful applications, particularly for larger-sized particles, greater than 5 mm. A novel application for the spouted-bed drier is for drying liquid foods. The liquid is sprayed onto a spouted bed of inert solid particles. This application is still under investigation. The toroidal-bed drier is another variation on the fluidized bed. This was first introduced in the 1980s and is still being evaluated.

The advantages of multistage drying became recognized in the 1970s and it is now widely practised. Conveyor, fluidized-bed and pneumatic drying may be carried out in two or more stages. Powder from spray driers may receive secondary drying in fluidized beds. Spray-drying systems featuring multistage drying in one unit were introduced in the 1980s and are finding increasing application.

The need for energy conservation became a priority in the 1980s and has led to improvements in the thermal efficiency of drying systems and in methods of recovering heat from the exhaust air from drying chambers. The use of microwave heating for the purpose of drying has so far found only limited application. However, this is being researched and is likely to find wider application in the near future.

Hand-in-hand with developments in drying equipment and techniques over the last fifty years, our understanding of drying processes has increased. A great deal of research into the mechanisms of heat and mass transfer during drying has been undertaken and, in recent years, models for the prediction of drying time and temperature/moisture distribution within food pieces during drying, has been developed (Van Arsdel et al, 1973a; Hayashi, 1989; Dalgleish, 1990; Masters, 1991).

General Principles of Dehydration

Many authors use the term 'drying' to describe the removal of moisture by exposure to the sun and the term 'dehydration' to moisture removal by the application of other heating methods. In this text, no such distinction is made. Both terms are used interchangeably to describe the unit operation in which nearly all the water normally present in a foodstuff is removed by evaporation or sublimation as a result of the application of heat. Thus mechanical de-watering methods, such as filtration, centrifugation or expression are not included. Osmotic drying and azeotropic drying are treated briefly even though they do not fall within the definition given above.

Usually, the main objective of dehydrating food is to prolong its shelf life beyond that of the fresh material. This is achieved by reducing the water activity (a_w) of the food to a value which will inhibit the growth and development of pathogenic and spoilage microorganisms, significantly reducing enzyme activity and the rate at which undesirable chemical reactions occur. The influence of a_w on such changes is discussed under Water activity and food quality (page 129). By this adjustment of a_w, and the use of appropriate packaging, the shelf life of the food can be extended without the need for refrigerated storage. The removal of most of the water from the food reduces the weight to be carried per unit food value. This can lead to substantial savings in the costs of handling and transporting the dried product as compared with the fresh material. A reduction in volume of the dried material, as compared with the fresh, can lead to savings in the cost of storage and transport. The maximum reduction in bulk is attained when dilute liquid foods are dried to powders, particularly if the powder is then compressed into blocks or tablets. On the other hand, little or no change in volume occurs when solid pieces of food are freeze dried. In between these extremes, varying degrees of shrinkage occur, depending on the food, the method of drying and the drying conditions.

Drying can also bring about undesirable changes in foods. The size and shape of solid food pieces change during drying, due to the shrinkage discussed above. When reconstituted, they may not return to their original shape and size. Colour changes may also occur due to the removal of water or as a result of exposure to high temperatures during drying. Again, the colour of the reconstituted product may differ from that of the fresh material. The texture of the reconstituted material may be less acceptable than that of the fresh because of changes in structure due to shrinkage and/or excessive exposure to heat. The capacity of dried food pieces to reabsorb water may be limited which would also contribute to their poor texture. In the case of food powders, it is usually desirable that they reconstitute rapidly and completely in hot or cold liquid, as appropriate. The extent to which this occurs depends on the drying method and conditions (see under Reconstitutability of dried food powders, page 91). Changes in flavour may also occur as a result of drying. These may be due to the loss of volatile flavour compounds during drying and/or to the development of an undesirable cooked flavour because of exposure to high temperatures. The extent of these changes depends on the drying method. Freeze-dried products generally exhibit the least changes in flavour. Spray drying and other rapid drying methods bring about moderate changes. Drying techniques in which the food is exposed to relatively high temperatures, e.g. drum drying, and/or which entail relatively long drying times, e.g. in cabinet driers, are likely to bring about important changes in flavour.

Changes in the nutritional quality of foods may occur as a result of drying. Considerable losses of water-soluble nutrients may occur during the preparation of the food prior to dehydration, i.e. during cleaning, peeling, blanching or cooking. Similar losses would be encountered when preparing foods prior to freezing or canning. During the drying operation itself, the loss of water- and lipid-soluble nutrients will depend on the drying method and conditions. Exposure of the food to a high temperature at a moisture content intermediate between that of the fresh material and the dried product is likely to lead to high losses. Such conditions should be minimized by careful selection of the drying method and conditions and good control of the drying operation. In general, drying should result in high retention of nutrients with the exceptions of vitamins C and A. Vitamin C losses in drying are usually somewhat higher than in canning and much higher than in freezing. Vitamin A losses in drying can be very much higher than in canning or freezing. Vitamin B losses in drying are relatively low, comparable with freezing and lower than in canning.

As a result of a considerable amount of research in recent years, many modern dehydrated foods have good organoleptic characteristics and reconstitute rapidly. Consequently, they constitute a significant component of the convenience food market.

Dehydration is a simultaneous heat-and-mass transfer operation. The necessary sensible and latent heat of evaporation, or sublimation, must be supplied to the food, while water or water vapour must move within the food to the evaporating surface and the water vapour must transfer from that surface to the surrounding atmosphere. The mechanisms whereby heat is transferred to food provide a convenient way of classifying the many drying methods used today. On this basis, there are three categories of drying methods as follows:

In group 1 heated air is the drying medium. The food is placed in a current of heated air. Most of the heat is supplied to the food by convection from the air. Such methods are also known as convective or convection drying methods.

In group 2 the food is placed in contact with a heated surface, usually a metal surface. Most of the heat is transferred to the food by conduction from the hot surface. Such methods are also known as conductive or conduction drying methods.

In group 3 the food is exposed to radiant heat, and radiation is the main mechanism of heat transfer. These methods are also known as radiative drying methods. Sun drying fits into this category.

In addition to these three main categories, the use of microwave and dielectric energy for the purposes of drying should be considered, as well as freeze drying which involves a freezing and a drying stage.

Drying in heated air

(i) Solids

During the drying of a wet solid in heated air, the air supplies the necessary sensible and latent heat and also acts as a carrier for the water vapour formed, moving it away from the drying surface and permitting further evaporation to occur. Consider a wet solid in the form of a thin slab positioned in a current of heated air flowing parallel to one of its large faces. Assume that drying takes place from this large face only. The slab consists of an inert solid, wetted with pure water, and the temperature, humidity and velocity of the air remain constant. Assume that all the heat is transferred by convection from the air. If the moisture content of the material is monitored throughout drying and the data presented in the form of curves as shown in *Figure 1.1*, it can be seen that the drying cycle can be considered to consist of a number of stages or periods as follows:

Period A-B. This represents a '*settling down*' or equilibration period during which the solid surface conditions come into equilibrium with the drying air. The length of this period is usually small compared to the overall drying time.

Period B-C. During this period the rate of drying remains constant. Hence it is known as the *constant rate period*. During this period the surface of the solid is saturated with water. As water evaporates from the surface it is replaced with water which migrates from within the solid to the surface. The rate of evaporation of water from the surface balances the rate of heat transfer to the surface, from the air, and so a state of equilibrium exists at the surface. Throughout this period the surface temperature remains constant at a value which corresponds to the wet-bulb temperature of the drying air. This is understandable if one compares the conditions at the surface to those which prevail at the wick of a wet-bulb thermometer. This state of equilibrium persists as long as the movement of water to the surface is sufficient to maintain it in a saturated condition. Water evaporates into the air stream as a result of a water-vapour pressure gradient between the surface of the solid and the main stream of the air. The rate of mass transfer ($-\mathrm{d}w/\mathrm{d}t$) may be described by an expression such as:

$$\frac{\mathrm{d}w}{\mathrm{d}t} = -K_g A(p_s - p_a) \qquad \text{(I)}$$

where K_g = mass transfer coefficient; A = drying area; p_s = water vapour pressure at the surface of the solid (i.e. the vapour pressure of water at surface temperature, since the surface is saturated); p_a = water vapour pressure in the main stream of the air.

Equation (I) may also be written as:

$$\frac{\mathrm{d}w}{\mathrm{d}t} = -K_g{}'A(H_s - H_a) \qquad \text{(II)}$$

where $K_g{}'$ = mass transfer coefficient; H_s = absolute humidity at the surface of the solid (i.e. saturation humidity of the air at surface temperature); H_a = absolute humidity in the main stream of the air.

The rate of heat transfer to the surface of the solid ($\mathrm{d}Q/\mathrm{d}t$) may be described by an expression such as:

$$\frac{\mathrm{d}Q}{\mathrm{d}t} = h_c A(\theta_a - \theta_s) \qquad \text{(III)}$$

where h_c = heat transfer coefficient for convection heating: θ_a = dry-bulb temperature of the air; θ_s =

temperature at the surface of the solid (i.e. the wet-bulb temperature of the air).

Since a state of equilibrium exists at the surface of the solid, and if sensible heat changes are neglected, the rates of mass and heat transfer may be related as follows:

$$\frac{dw}{dt} L_s = -\frac{dQ}{dt} \qquad \text{(IV)}$$

where L_s = latent heat of evaporation at θ_s.

Thus, the rate of mass transfer (i.e. the rate of drying) may be expressed in terms of a heat transfer coefficient and temperature difference as follows:

$$\frac{dw}{dt} = -\frac{h_c A}{L_s}(\theta_a - \theta_s) \qquad \text{(V)}$$

the drying rate may also be expressed in terms of the rate of change of moisture content thus:

$$\frac{dW}{dt} = -\frac{h_c A'}{L_s}(\theta_a - \theta_s) \qquad \text{(VI)}$$

where $-dW/dt$ = the rate of change of moisture content (dwb); A' = effective drying surface area per unit mass of dry solids.

If the thickness (depth) of the slab is l and the bulk density of the material ρ_s, the rate of change of moisture content may be expressed thus:

$$\frac{dW}{dt} = -\frac{h_c}{\rho_s L_s l}(\theta_a - \theta_s) \qquad \text{(VII)}$$

If W_o is the moisture content of the wet material at the start of the constant rate period (dwb) and W_c its moisture content at the end of that period (dwb) then the constant-rate drying time t_c is:

$$t_c = \frac{(W_o - W_c)\,\rho_s L_s l}{h_c(\theta_a - \theta_s)} \qquad \text{(VIII)}$$

Thus, the factors which control the rate of drying during the constant-rate period are the drying surface area, the difference in temperature or humidity between the air and the drying surface and the mass or heat transfer coefficients. The velocity of the air and the dimensions of the system also influence drying rate by affecting the transfer coefficients. For example, the following relationship often holds:

$$h_c = \frac{aG^n}{D_c^{\,m}} \qquad \text{(IX)}$$

where G = the mass velocity of the air; D_c = characteristic dimension of the system; a, n and m are constants. D_c has been represented as the equivalent diameter of the flow channel (cross-sectional area × 4 ÷ perimeter) or the length of the drying surface parallel to the direction of flow of the air. For most tray-drying calculations the equivalent diameter is used. Values of n in the range 0.35–0.80 have been reported in the literature. Where no specific data are available, a value of 0.80 is often used in calculations. Alternative expressions for h_c are available to use in situations where the air does not flow parallel to the drying surface and for through-flow of air. When a significant proportion of the heat of drying is supplied by conduction, e.g. from the metal tray on which the food is placed, and/or by radiation, e.g. from the walls of the drying chamber, an overall heat transfer coefficient, taking this into account, must be used in the above equations. In such circumstances, the surface temperature may remain constant during the constant rate period of drying, but its value will be between the wet-bulb temperature of the air and the boiling point of the water. In most practical drying situations, some heat transfer by conduction and radiation will occur in addition to convection. Drying under constant rate conditions can be advantageous when heat sensitive foods are being dried, as high rates of evaporation may be accomplished at relatively low product temperatures. Some solid foods do exhibit constant-rate drying but the length of that period is usually only a small proportion of the total drying time. In the case of many foods no constant rate period of drying is evident.

As drying continues, a point is reached at which the rate of migration of moisture to the surface is no longer adequate to maintain the surface in a saturated condition (point C in *Figure 1.1*). From this point on, the rate of drying is no longer constant but falls progressively throughout the rest of the drying cycle. Point C is known as the critical point, the moisture content at that point – W_c, the critical moisture content and the drying period beyond that point – C-D, the falling-rate period.

In period C-D, the *falling-rate period*, the temperature at the surface of the solid rises as drying proceeds and approaches a value corresponding to the dry-bulb temperature of the air as the material approaches dryness. Many authors have reported the occurrence of two or more falling rate periods, i.e. points of inflexion in the falling-rate curve (*Figure 1.1(c)*). Attempts have been made to explain such curves in terms of what is happening within the solid. One such explanation is as follows: just beyond the critical point the surface begins to dry out but moisture is still evaporating from the surface. At some point E (*Figure 1.1(c)*) the plane of evaporation moves down into the solid. The vapour arising from this plane has to pass through a layer of dry solid which further reduces the rate of drying. This behaviour could account for a two-stage falling-rate period, but there is little experimental evidence to confirm this. Other explanations relate to the mechanism of moisture movement within the solid (see below). Usually, in food dehydration operations, a large proportion of the drying takes place under falling rate conditions.

Very many mathematical models have been proposed to represent drying under falling-rate conditions. These can be put into two categories: (a) those that relate to the mechanisms of moisture movement within the solid and (b) those that are empirical and are obtained by fitting expressions to drying curves constructed from experimental data.

(a) Several modes of transfer of moisture within the solid have been proposed. These include liquid diffusion resulting from concentration gradients; vapour diffusion due to partial pressure gradients; liquid movement caused by capillary forces; diffusion in liquid layers adsorbed at solid interfaces; vapour flow as a result of differences in total pressure; flow caused by pressure gradients brought about by shrinkage; movement by a vaporization-condensation mechanism. The mechanism which has received the widest acceptance is diffusion due to concentration gradients. Such diffusion may be represented by Fick's second law:

$$\frac{dW}{dt} = D\frac{d^2W}{dl^2} \qquad \text{(X)}$$

where W = moisture content (dwb); t = time; l = distance; D = liquid diffusivity. A well-known solution to this equation for a slab-shaped solid, drying from one large face only is:

$$\frac{W - W_e}{W_c - W_e} = \frac{8}{\pi^2}\left[\exp\left\{-Dt\left(\frac{\pi}{2l}\right)^2\right\} + \frac{1}{9}\exp\left\{-9Dt\left(\frac{\pi}{2l}\right)^2\right\}\right] \qquad \text{(XI)}$$

where W = average moisture content at time t (dwb); W_e = equilibrium moisture content (dwb); W_c = moisture content at the start of the falling-rate period (dwb), i.e. critical moisture content; l = depth of slab. For large values of t equation (XI) may be reduced to:

$$\frac{W - W_e}{W_c - W_e} = \frac{8}{\pi^2}\left[\exp\left\{-Dt\left(\frac{\pi}{2l}\right)^2\right\}\right] \qquad \text{(XII)}$$

This expression holds for values of $(W - W_e)/(W_c - W_e)$ less than 0.6.

Direct application of equation (XII) assumes that D remains constant throughout the falling-rate period. There is considerable evidence that this is not the case and that D varies with moisture content. Many authors who have reported two or more falling-rate stages have found that the diffusion equation could be applied to each stage provided that a different value of D was used. As far back as 1958, Jason found this behaviour when drying fish muscle under laboratory conditions. He presented the results during the falling-rate period in the form of a curve shown in *Figure 1.2*. In this figure, the difference between the weight at time t, W_t, and the equilibrium weight, W_e, is plotted on a logarithmic scale as a function of time, t. The results are seen to fall on two straight lines LL and MM, over most of the curve. The diffusion equation could be fitted to both of these lines if a lower value of D was used for line MM as compared with LL. More recently, in a project in which this author participated (Gutierrez-Lopez, 1989), a food model comprised of glucose syrup, agar, glucose and sucrose with an initial moisture content of 0.27 (dwb), was dried under carefully controlled conditions. Some of the data are shown in *Figure 1.3(a)*. Three falling-rate stages were detected. When the diffusion equation was applied to each stage three values of D were calculated, decreasing as drying proceeded. Similar experiments

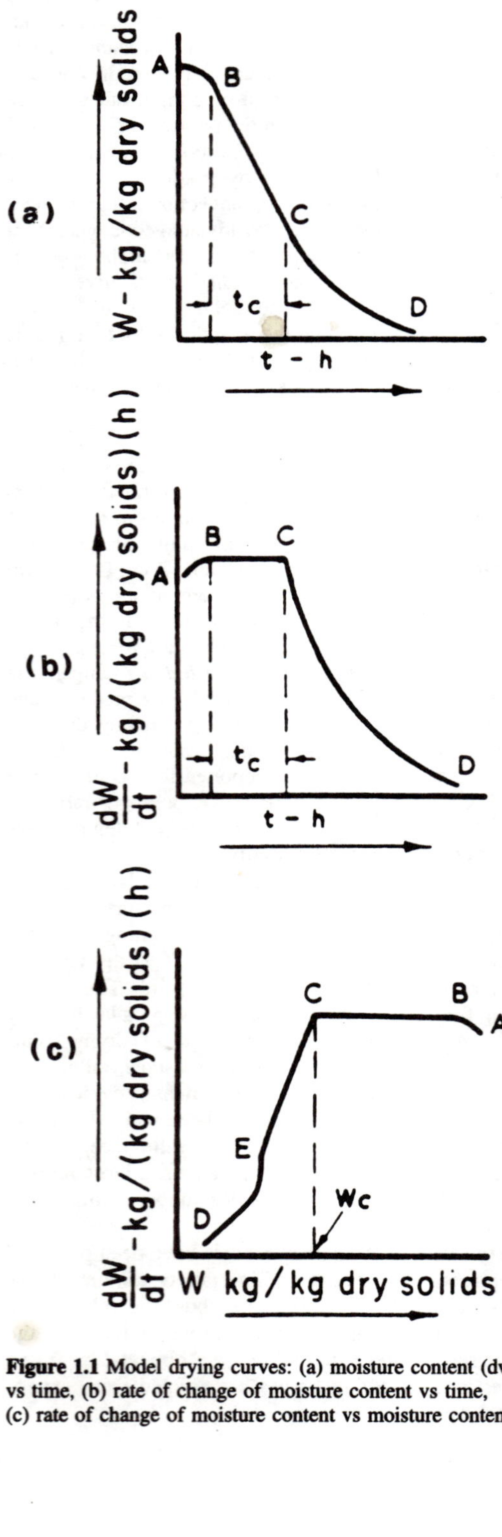

Figure 1.1 Model drying curves: (a) moisture content (dwb) vs time, (b) rate of change of moisture content vs time, (c) rate of change of moisture content vs moisture content

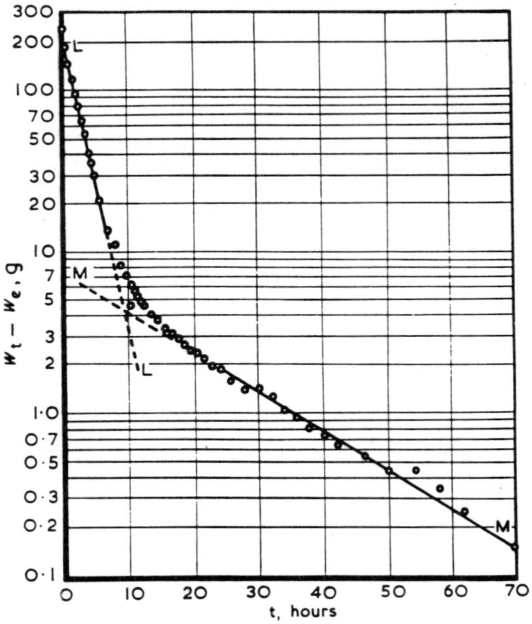

Figure 1.2 Amount of free water remaining in a fish fillet piece as a function of time (Jason, 1958)

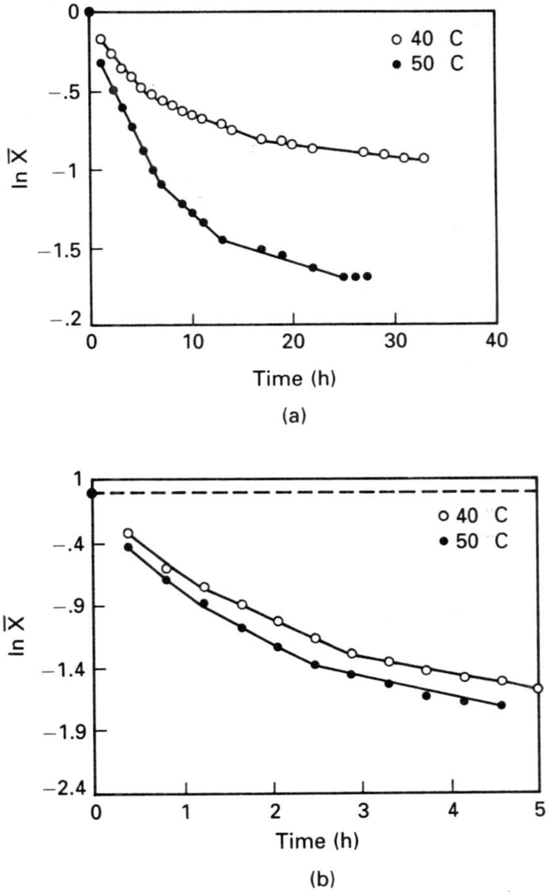

Figure 1.3 Dimensionless moisture content, X, as a function of time. $X = (W - W_e)/W_c{-}W_e)$ for (a) a food model and (b) pasta (Gutierrez-Lopez, 1989)

were carried out with pasta, with similar results (*Figure 1.3(b)*). This researcher obtained values of *D* for the food model and pasta by differential scanning calorimetry (DSC), which agreed well with those calculated from data obtained from laboratory drying experiments. More recently still, data collected by Wang (1992) when drying potato, exhibited a two-stage falling-rate pattern.

If a relationship between *D* and moisture content is known it can be incorporated into equation (XI) which then becomes a non-linear differential equation. One method of determining *D* at different moisture contents was reported by Saravacos (1967). A term known as the half equilibrium time was defined as the time required to reach a moisture content halfway between the moisture content at the start of the falling-rate period and equilibrium moisture content. Equation (XI) may be written in a more general form as:

$$0.5 = 1 - \frac{8}{\pi^2} \sum_{n=0}^{\infty} \frac{1}{(2n+1)^2} \exp\left\{-(2n+1)^2 Dt\left(\frac{\pi}{2l}\right)^2\right\} \qquad \text{(XIII)}$$

If the first term only of this equation is used, it reduces to:

$$D = \frac{0.194 l^2}{t_{(0.5)}} \qquad \text{(XIV)}$$

where $t_{(0.5)}$ is the half-equilibrium time. Approximate values of *D* can be obtained by applying equation (XIV) at various humidities or moisture contents in which the diffusivity is assumed to remain constant.

In diffusion equations, the drying time is proportional to l^2. In the literature, the drying time for food materials is generally reported as being proportional to l^n, with values of n ranging from 1.40 to 1.99. For example, Jason (1958) reports a value of 1.80 for fish muscle, Gutierrez-Lopez (1989) 1.94 to 1.98 for a model food and Wang and Brennan (1992) 1.78 to 1.88 for potato. The existence of external mass transfer resistances is one reason suggested in the literature for values of n of less than 2. This is most likely when air velocity is low. When high air velocity is used such resistances are less likely to occur. In such circumstances Vaccarezza and Chirife (1978) suggested that a heat effect may be responsible for low values of n and they developed a model for the calculation of material temperature as a function of drying time. Wang and Brennan (1992) used this model to calculate a corrected value of n for potato, which was 1.93.

Diffusivity varies with temperature. The usual relationship is an Arrhenius type model of the form:

$$\overline{D}_e = D_0 \exp\left(-\frac{Q}{RT}\right) \quad \text{(XV)}$$

where D_e = average effective diffusivity; D_0 = temperature independent constant; Q = energy of activation for diffusion; R = gas constant; T = absolute temperature. An example of such a relationship is shown in *Figure 1.4* where the values of ln D for the three falling-rate stages in the drying of pasta are linearly related to $1/T$ (Gutierrez-Lopez, 1989).

(b) Numerous mathematical models to represent falling-rate drying have been proposed, which were largely derived from experimental data. One of the earliest, proposed by Lewis (1921), took the form:

$$\left(\frac{dW}{dt}\right)_f = -K_c(W - W_e) \quad \text{(XVI)}$$

where dW/dt = drying rate at moisture content W during the falling-rate period and K_c = drying constant, which came to be known as the mass transfer coefficient and was related to temperature by an expression of the form:

$$K_c = a \exp\left(\frac{-b}{T}\right) \quad \text{(XVII)}$$

where a and b = material constants. Equation (XVI) may be integrated to the form:

$$\frac{W - W_e}{W_c - W_e} = \exp(-K_c t) \quad \text{(XVIII)}$$

An expression of this type has been said to represent drying when capillary movement of moisture takes place within the solid. In this case K_c is related to the drying rate in the constant rate period as follows:

$$K_c = \frac{-\left(\frac{dW}{dt}\right)_c}{W_c - W_e} \quad \text{(XIX)}$$

Thus by combining equations (VII) and (XVIII) the drying time in the falling-rate period, from an initial moisture content W_c (dwb) to a final moisture content W (dwb) is given by:

$$t = \frac{\rho_s L_s l\ (W_c - W_e)}{h_c(\theta_a - \theta_s)} \ln \frac{(W_c - W_e)}{(W - W_e)} \quad \text{(XX)}$$

There is not much experimental evidence to show that such an expression does depict capillary movement of moisture.

Equation (XVIII) above was used to describe the drying of agricultural materials but it did not apply to the whole falling-rate curve. The introduction of an empirical exponent n to give the expression:

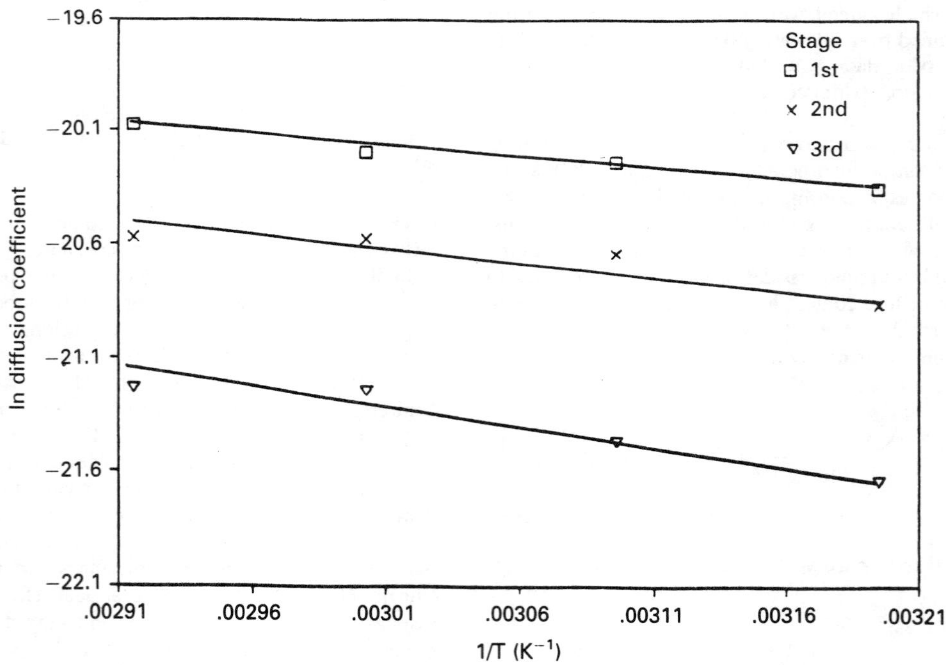

Figure 1.4 Liquid diffusivity, D, as a function of absolute temperature, T, for the three stages of drying of pasta (Gutierrez-Lopez, 1989)

$$\frac{W - W_e}{W_c - W_e} = \exp(-K_c t^n) \quad \text{(XXI)}$$

is said to have widened its applicability. This last equation gave good results when applied to drying shelled corn and soya beans (Sharaf-Elden et al, 1979). It was later applied successfully to experimental data from thin-layer drying of sunflower seeds and in-shell pecans.

Many other empirical expressions to represent falling-rate drying have been reported. Some of these were reviewed by Sharaf-Elden et al (1979). In general, such equations are usually applicable only under conditions close to those used when obtaining the experimental data. Many are specific to a particular food material or closely related materials. Within these limitations, they can be useful for predicting drying times.

Alvarez and Legues (1986) developed a model for the drying of seedless grapes which had both empirical and analytical elements. They defined an effective diffusion coefficient, D_e, as follows:

$$D_e = D_0(1 - F_0)^b \quad \text{(XXII)}$$

where D_0 and b are constants and F_0 is a dimensionless number accounting for variations of diffusivity with time thus:

$$F_0 = \frac{D_0 t}{l^2} \quad \text{(XXIII)}$$

The full expression proposed was as follows:

$$\frac{W - W_e}{W_c - W_e} = \frac{6}{\pi^2} \sum_{n=1}^{\infty} \frac{1}{n^2} \exp\left\{-\frac{n^2\pi^2}{(1 + b)} (1 + F_0)^{(1 + b)} - 1\right\} \quad \text{(XXIV)}$$

This model was simplified by taking only the first exponential term of the equation to give:

$$\frac{W - W_e}{W_c - W_e} = \exp\left\{\frac{\pi^2}{(1 + b)} (1 + F_0)^{(1 + b)} - 1\right\} \quad \text{(XXV)}$$

Equations (XXIV) and (XXV) fitted well to experimental data obtained for seedless grapes.

Hot air drying systems for solids

Various systems for drying solids in heated air are discussed elsewhere in this text. These include:

Kiln drier
Cabinet drier
Conveyor drier
Bin drier
Fluidized-bed drier (including spouted-bed and toroidal-bed)
Pneumatic drier
Rotary drier

(ii) Liquids

Spray drying is by far the most common method used for drying food liquids in heated air. The liquid is converted into a fine mist or spray (atomized) which is brought into contact with heated air in a drying chamber. Very rapid drying takes place and the spray is converted into a powder. Drying times are short, less than 20", and evaporative cooling maintains a relatively low product temperature throughout most of the drying cycle. If the powder is removed quickly from the drying chamber heat damage should be limited.

If it is assumed that drying takes place under constant rate conditions, the time, t, for a spherical droplet to dry from an initial moisture content W_o (dwb) to a final moisture content W_d (dwb) may be expressed thus:

$$t = \frac{r^2 \rho_1 L_s (W_o - W_d)}{3h_c(\theta_a - \theta_s)(1 + W_o)} \quad \text{(XXVI)}$$

where r = the radius of the droplet and ρ_1 = the density of the liquid.

In practice not all the moisture is removed under constant rate conditions. Much more detailed accounts of drying behaviour of droplets in spray driers have been published (Kerkhof and Schoeber, 1974; Masters, 1991).

The principles of spray drying are discussed elsewhere in this text and so also are air/powder separators, atomization and spray-drying chambers.

Drying by direct contact with a heated surface

If a wet material is placed in contact with a heated surface the necessary sensible and latent heat of evaporation is transferred to the material by conduction and drying can take place. The pattern of drying is similar to that of hot air drying in that drying takes place mainly in two stages. During the initial constant rate period the material temperature is close to its boiling point at the prevailing pressure. During this period the rates of drying will be higher than those attainable in air at the same temperature as the heated surface. When the rate of movement of moisture to the evaporating surface falls below the rate of evaporation, the falling rate period commences and the temperature of the material rises towards that of the heated surface. Assuming that drying takes place from one large face only and that shrinkage is negligible, the rate of drying may be expressed in terms of an overall heat transfer coefficient and temperature gradient as follows:

$$\frac{dw}{dt} = -\frac{dQ}{dtL_e} = -\frac{UA(\theta_w - \theta_e)}{L_e} \quad \text{(XXVII)}$$

where dw/dt = rate of change of weight (drying rate); dQ/dt rate of heat transfer by conduction; L_e = latent heat of evaporation at θ_e; U = overall heat transfer coefficient; A = drying area; θ_w = temperature of the heated surface (wall temperature); θ_e = evaporating temperature.

As drying proceeds $(\theta_w - \theta_e)$ decreases. If K_c is an overall heat transfer coefficient for the complete drying cycle (allowing for the decrease in $(\theta_w - \theta_e)$) then dw/dt may be written as:

$$\frac{dw}{dt} = -\frac{K_c A(\theta_w - \theta_e)}{L_e} \qquad \text{(XXVIII)}$$

The overall drying rate for a complete cycle may be expressed as:

$$\frac{dw}{dt} = \frac{(W_o - W_f)M}{t} \qquad \text{(XXIX)}$$

where W_o = initial moisture content of the material (dwb); W_f = final moisture content of the material (dwb); M = the mass of dry solid on heated surface; t = total drying time.

From the two above equations comes the relationship:

$$\frac{(W_o - W_f)M}{t} = \frac{K_c A(\theta_w - \theta_e)}{L_e} \qquad \text{(XXX)}$$

from which t can be calculated.

If drying is carried out at atmospheric pressure than θ_e will be in excess of 100°C. In order to achieve reasonable drying times and to dry to low moisture contents, θ_w needs to be appreciably higher than this. Towards the end of drying, the material temperature may be quite high and heat damage may occur. To limit such damage, when drying at atmospheric pressure, the material may be applied in a thin layer onto the heated surface resulting in short drying times. The drum drier operates in this way. Alternatively, drying may be carried out under reduced pressure so that relatively low values of θ_e prevail and hence low values of θ_w may be used. Such low temperatures are used in vacuum-shelf and vacuum-band driers. Drum driers and vacuum driers are discussed elsewhere in this text.

Drying by the application of radiant (infrared) heat

In *Figure 1.5* the types of electromagnetic radiation are presented. Infrared radiation, which is emitted by hot objects, occupies the wavelength range 0.7 μm to 300 μm. The rate of emission of infrared energy from a heated surface, dQ/dt, is given by:

$$\frac{dQ}{dt} = \sigma A T_1^4 \sigma \qquad \text{(XXXI)}$$

where A = the area from which the radiation is emitted; T_1 = the absolute temperature of the surface; σ = Stefan-Boltzmann constant (= 5.7×10^{-8} J s^{-1} m^{-2} K^{-4}); ϵ = the emissivity of the surface. Substances that are good emitters of radiation are also good absorbers of it. The rate of absorption of radiation by a surface, dQ'/dt, is given by:

$$\frac{dQ'}{dt} = \alpha A T_1^4 \sigma \qquad \text{(XXXII)}$$

where α = absorptivity of the surface. The absorptivity value of a surface is numerically equal to the emissivity value. Thus, for a surface exposed to infrared radiation the net rate of heat transfer by radiation to the surface is given by:

$$\frac{dQ''}{dt} = \epsilon A(T_1^4 - T_2^4)\sigma \qquad \text{(XXXIII)}$$

where T_1 and T_2 = the absolute temperatures of the emitting surface and the absorbing surface, respectively.

Complex relationships exist between the physical, thermal and optical properties of foods and their influence on the absorption of infrared radiation. Each of the major components of food, protein, fat and carbohydrate exhibits its own characteristic absorption pattern. In addition to this, the absorption characteristics of water

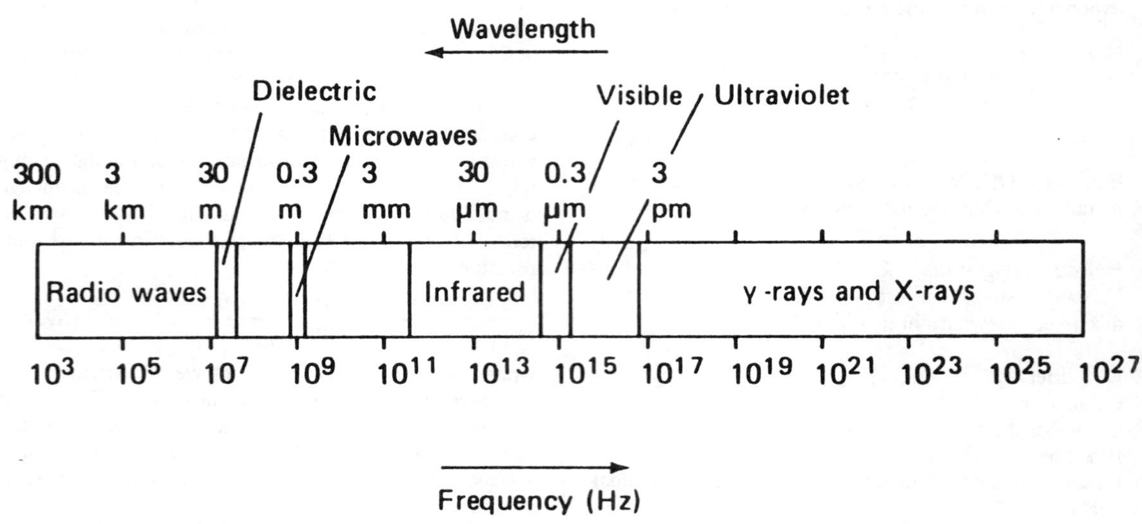

Figure 1.5 Electromagnetic radiation spectrum (Lewis, 1987)

in the liquid, vapour or frozen state affects the overall absorption by the food. In general, shorter wavelength radiation penetrates further than the longer waves. However, shorter wavelength radiation is more readily reflected. Consequently, it is difficult to predict the optimum wavelength to promote maximum transfer of heat to a given material.

It is, therefore, difficult to promote uniform heating of foods exposed to infrared radiation and to control the heating rate. Infrared heating is generally not used for removing water in bulk from foods. However, it has been applied to removing small amounts of moisture from granular materials such as breadcrumbs, spices and starches. Infrared driers are further discussed elsewhere in the text. Infrared heating is also used in vacuum-cabinet and band driers and in freeze driers. These driers are discussed elsewhere in this text. Radiant heat emitted from hot surfaces also plays a part in supplying the heat of evaporation in hot-air driers, even though the main mechanism of heat transfer is convection.

Approximately 48% of solar energy falls within the infrared range of frequencies. Solar drying is widely practised where sufficient sunlight is available. The applications for direct and indirect solar drying are discussed elsewhere in this text.

Microwave and dielectric heating in food dehydration

The position of microwave and dielectric radiation in the electromagnetic spectrum can be seen in *Figure 1.5*. Dielectric radiation is at a lower frequency (1–100 MHz) than microwave radiation (300 MHz–300 GHz). The phenomena of dielectric heating and microwave heating are essentially the same. Both are radio waves. The differences are in the frequencies used which determine the extent of the energy penetration. The higher frequency, shorter wavelength microwaves penetrate further than those in the dielectric range. By international agreement, the frequencies used for microwave heating are 915 MHz (896 MHz in Europe) and 2450 MHz. When food is located in the path of radio waves, distortions and deformations of the molecular structure occur and the applied energy is converted into heat. The power, P ($W\,cm^{-3}$), absorbed by the food is given by:

$$P = 55.61 \times 10^{-14} f E_f^2 \epsilon'' \qquad \text{(XXXIV)}$$

where f = the frequency of the radiation (Hz); E_f = the field strength ($V\,cm^{-1}$); ϵ'' = dielectric loss factor of the food. This factor is an important property of the food which will affect its heating rate. The loss factor for a given food will vary with its moisture content, temperature and whether it is frozen or not. It also depends on the frequency of the radiation.

The depth of penetration of radio waves into the food, D, may be expressed thus:

$$D = \frac{\lambda_o}{2\pi(\epsilon'')^{1/2}} \qquad \text{(XXXV)}$$

where λ_o = wavelength in free space.

To date the use of dielectric/microwave heating as the major source of energy for dehydration has been limited to removing small amounts of moisture from low-moisture products such as biscuits and cereals. It may also be employed in vacuum-cabinet and band driers and in freeze driers (Fellows, 1988; Lewis, 1989; Brennan et al, 1990). Microwave drying is further discussed elsewhere in this text.

Freeze drying

This method of drying involves freezing the material and subsequently subliming the ice from the frozen state to give a dried product. Sublimation occurs when the water vapour pressure in the immediate surroundings of the frozen material is less than that at the ice front within the material. In commercial operations, this water vapour pressure gradient is achieved by placing the frozen food in a vacuum cabinet and reducing the pressure in the cabinet to levels of the order of 13.5–270 $N\,m^{-2}$ (0.1–2.0 torr). The main components of a batch freeze drier are shown in *Figure 1.6*. The function of the condenser is to remove the water vapour formed by sublimation from the atmosphere so as to maintain the low water vapour pressure in the cabinet. The vacuum system removes the non-condensible gases from the chamber. Heat may be applied from above or below the sample or from both directions. Once sublimation has commenced, the main factors which affect the rate of drying are the rate at which water vapour moves through the dry layer and the rate at which heat is transferred to the ice front. Consider the case where a slab-shaped solid is being freeze dried from its upper surface only and where heat is supplied from above only, i.e. water vapour and heat move countercurrently through the dry layer (*Figure 1.7(a)*). The mass flow rate of water vapour through the dry layer, dw/dt, may be expressed as:

$$\frac{dw}{dt} = \frac{Ab(p_i - p_d)}{l} \qquad \text{(XXXVI)}$$

where A = drying area normal to the direction of flow of the vapour; b = the permeability of the dry layer to water

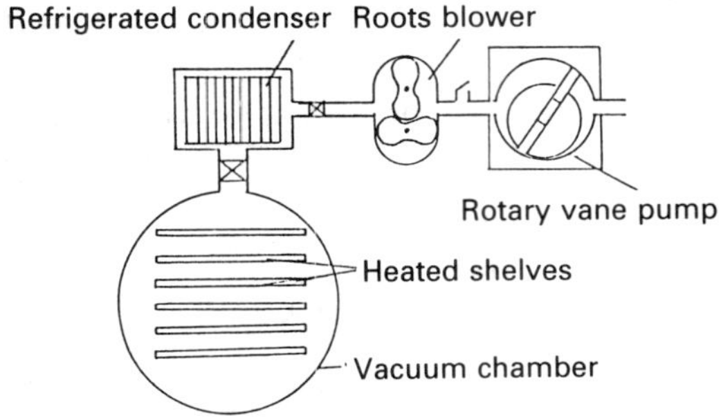

Figure 1.6 Main components of a batch freeze drier

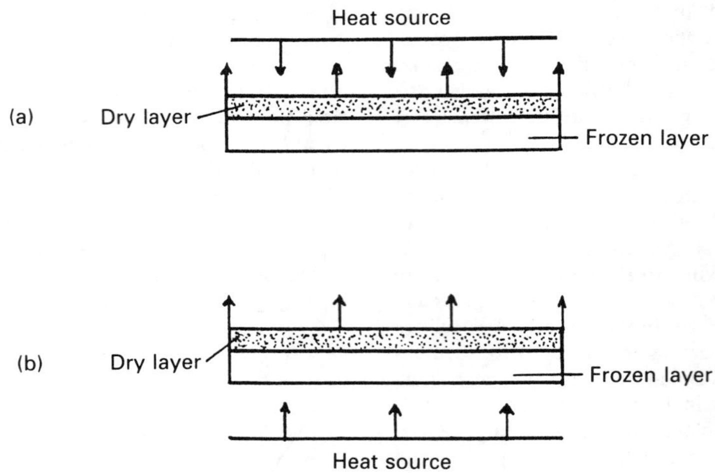

Figure 1.7 A slab-shaped solid being freeze dried: (a) with heat supplied through the dry layer, (b) with heat supplied through the frozen layer.

vapour; p_i = the water vapour pressure at the ice front at the prevailing temperature (θ_i, see below); p_d = the water vapour pressure at the top surface of the dry layer; l = the thickness of the dry layer.

The rate of heat transfer to the ice front through the dry layer, dQ/dt, may be expressed as:

$$\frac{dQ}{dt} = \frac{k_d A(\theta_d - \theta_i)}{l} \qquad \text{(XXXVII)}$$

where k_d = thermal conductivity of the dry layer; θ_d = temperature at the top surface of the dry layer; θ_i = temperature at the ice front.

An energy balance combining equations (XXXVI) and (XXXVII) gives:

$$\frac{k_d A(\theta_d - \theta_i)}{l} = -\frac{L_s Ab(p_i - p_d)}{l} \qquad \text{(XXXVIII)}$$

where L_s = heat of sublimation at θ_i.

The vapour pressure at the ice front, p_i, is given by:

$$p_i = p_d + \frac{k_d}{bL_s}(\theta_d - \theta_i) \qquad \text{(XXXIX)}$$

Note that l cancels out so that this relationship is independent of the extent of the drying. θ_i and p_i are related thermodynamically. If the dried surface temperature and chamber pressure are fixed, the ice surface temperature is also fixed.

If it is assumed that the ice front recedes uniformly into the slab, that the frozen layer is at its initial moisture content, W_o (dwb), and the dry layer at its final moisture content, W_f (dwb), the rate of change of weight of the slab, dw/dt, may be expressed as:

$$\frac{dw}{dt} = A\rho_s(W_o - W_f)\frac{dl}{dt} \qquad \text{(XL)}$$

where ρ_s = density of the dried solid.

Combining equations (XXXVIII) and (XL) we get the relationship:

$$\frac{dw}{dt} = -\frac{Ak_d(\theta_d - \theta_i)}{L_s l} = \frac{Ab(p_i - p_d)}{l}$$

$$= A\rho_s(W_o - W_f)\frac{dl}{dt} \qquad \text{(XLI)}$$

integrating within the limits $t = 0$, $l = 0$; $t = t_t$, $l = l_t$:

$$t_t = \frac{\rho_s(W_o - W_f)l_t^2}{b(p_i - p_d)2}$$

$$= \frac{L_s\rho_s(W_o - W_f)l_t^2}{k_d(\theta_d - \theta_i)2} \qquad \text{(XLII)}$$

where t_t = total drying time to W_f; l_t = total thickness of the slab.

If heat is applied through the frozen layer only while water vapour escapes through the dry layer, *Figure 1.7(b)*, then the equation becomes:

$$\frac{dw}{dt} = \frac{k_i A(\theta_p - \theta_i)}{L_s(l_t - l)} \qquad \text{(XLIII)}$$

where k_i = thermal conductivity of the frozen material; θ_p = temperature of the surface of the frozen material in contact with the heated plate. The vapour pressure at the ice front, p_i, now becomes:

$$p_i = p_d + \frac{k_i l}{bL_s(l_t - l)}(\theta_p - \theta_i) \qquad \text{(XLIV)}$$

As the ratio $l/(l_t - l)$ changes as drying proceeds so does p_i and hence θ_i. If p_i is expressed as a function of l then to calculate the total drying time, t_t, the following equation must be solved by analytical or numerical methods:

$$\int_{0}^{t_t} dt = \frac{\rho_s(W_o - W_f)}{b[f(l) - p_d]} \int_{0}^{l_t} l dl \quad \text{(XLV)}$$

Similar expressions may be derived to represent the case where heat is applied through the dry layer and the frozen layer simultaneously (Karel, 1974; Brennan et al, 1990).

The principles of freeze drying are discussed further elsewhere in this text as well as batch and continuous freeze drying equipment.

Part 2

Dictionary of Food Dehydration

A

Abrasive peeling
Most mechanical peelers remove skin from vegetables by means of a rubbing action (abrasive peelers). There are many different designs of such peelers. For relatively small-scale operations, the vegetables are placed in a cylindrical bowl rotating about a vertical axis. The inner surface of the bowl is coated with an abrasive material. As the vegetables rotate and tumble in the bowl, the peel is rubbed off by this abrasive coating. Water jets, directed into the bowl, help to remove the skin and transport the skin fragments out of the bowl. In another design, suitable for higher throughputs, an auger carries the vegetables through a circle of rotating rollers. The surfaces of the rollers are coated with an abrasive substance which removes the skin by rubbing. Water sprays clean the abrasive rollers not in contact with the vegetables. After abrasive peeling it is usually necessary to remove crevices, such as potato eyes, by hand. Otherwise, peeling losses would be large. The effectiveness of abrasive peelers and the peeling losses are influenced by the size and shape of the vegetables and the dwell time in contact with the abrasive surface.

In other types of mechanical peelers knives remove the skin. In one design, knife blades are fixed in the outside wall of a revolving cylinder in which the product is tumbled. There are many other, more specialized, designs of mechanical peelers to suit different vegetables (Luh and Woodroof, 1975; Arthey and Dennis, 1991). *See also* Flame peeling, Lye peeling, Hot water peeling, Steam peeling.

Absolute density
Mass per unit volume of a substance. Common units are kg m^{-3}. In the case of particulate matter, absolute density does not include the voids that exist between particles, only the density of the particles themselves. This is also known as solid or particulate density. *See also* Bulk density.

Absolute humidity
The water vapour content of moist air, usually expressed on a dry-weight basis, e.g. kilograms of water vapour per kilogram of dry air. Assuming ideal gas behaviour:

$$H = \frac{M_w p_v}{M_A(P - p_v)}$$

where H is the absolute humidity, M_w the molecular weight of water, M_A the molecular weight of air, P the total pressure of the system and p_v the partial pressure of water vapour in the system. Occasionally, absolute humidity is expressed on a volume basis, e.g. kilograms of water vapour per cubic metre of dry air. It is also known as specific humidity.

Absolute pressure
Total pressure above zero or perfect vacuum. Absolute pressure = gauge pressure + atmospheric pressure.

Absolute pressure gauge
An instrument which measures absolute pressure directly, as distinct from a *pressure gauge* which measures pressures above atmospheric pressure or a *vacuum gauge* which measures pressures below atmospheric pressure.

Absolute temperature
A temperature scale with its zero at absolute zero. In the Kelvin scale °K = °C + 273. In the Rankine scale °R = °F + 460.

Absolute units *see* SI units and/or British units.

Absolute viscosity *see* Dynamic viscosity.

Absolute zero
When a body is cooled its internal energy is reduced. The point at which its internal energy is zero, i.e. the

point at which all molecular motion ceases, is known as absolute zero or zero Kelvin, °K. 0°K = −273°C.

Absorbate
A gas or liquid which is absorbed by a substance.

Absorbent
A substance which absorbs a gas or liquid. In the context of this work the term may be used to describe a dehydrated material which absorbs water during reconstitution. It is also applied to substances such as silica gel and calcium chloride which readily take up water vapour from a gas. The term adsorbent may be more appropriate for this latter application.

Absorption
The process whereby a gas or liquid is absorbed by another substance. The mechanism involved may be physical or chemical and the absorbent may change as a result of the absorption, e.g. change from solid to liquid form. More specifically, this term is used to describe the unit operation in which a soluble component of a gas phase is transferred into a comparatively non-volatile liquid.

Acid dip
Immersion of some fruits in an acid dip prior to sulphuring, can improve the stability of the colour of the dried product. A 1% ascorbic acid and 0.25% malic acid dip has been used to retard enzymic browning in peaches. Some reports recommend an acid dip instead of sulphuring. However, the dried product would need to be stored at low temperature to inhibit browning (Salunkhe et al, 1991). *See* Peach.

Adiabatic
Describes a thermodynamic process which occurs without transfer of heat to or from any outside source. *See* Adiabatic saturation temperature.

Adiabatic cooling
A cooling process which occurs under adiabatic conditions. *See* Adiabatic saturation temperature.

Adiabatic cooling line *see* Adiabatic saturation temperature, Psychrometric chart.

Adiabatic saturation line *see* Adiabatic saturation temperature, Psychrometric chart.

Adiabatic saturation temperature
If a stream of moist air passes concurrently through a spray of water at temperature θ_s under adiabatic conditions, so that the air leaves the system in equilibrium with the water (i.e. saturated) and at temperature θ_s the temperature θ_s is known as the adiabatic saturation temperature. The line relating the temperature and humidity of air during adiabatic cooling is known as the adiabatic cooling line or adiabatic saturation line. Such lines are often found on psychrometric charts. *See also* Wet-bulb temperature. For air-water vapour mixtures adiabatic saturation temperature and wet-bulb temperature are approximately equal.

Adsorption
The process whereby gases, liquids or solutes in solution are held on solid surfaces. In the context of this work the adsorption of water on the surface of solid food materials is the main interest. At low values of water activity water molecules are strongly bound to specific sites on the solid. Polar sites such as hydroxyl groups in polysaccharides, carbonyl groups and amino groups in proteins are held by hydrogen bonding, ion-dipole bonds or other strong interactions. This usually occurs at water activities below 0.35. Water bound in this way is generally regarded as not being available as a solvent to support chemical reactions. This strongly bound water is often referred to as the monomolecular or monolayer moisture content. Foods are generally most stable at moisture contents corresponding to the monomolecular value. At values of water activity above the monomolecular region water may still be bound but less strongly, perhaps in multiple layers. *See also* Bound water, Monomolecular moisture content, Sorption, Sorption isotherm, Water activity.

Agglomeration
Dried food powders consisting of very fine particles 50 μm and below are difficult to handle, may give rise to a fire and/or explosion hazard and have poor reconstitution characteristics. The factors influencing the reconstitution properties of food powders are discussed elsewhere in this text. The most common solution to poor reconstitutability is to deliberately form agglomerates of fine particles. When agglomerated particles come into contact with water they are wetted as they touch the water surface, the water penetrates rapidly into the pores of the granule structure, the granules disintegrate into discrete particles which disperse and dissolve in the water. Such agglomerated powders are said to be instant. A degree of agglomeration may be brought about by recycling the fines from the cyclone(s) of a spray drier back into the wet zone of the drying chamber (*Figure 2.49*). If the powder is discharged from a spray drier at a higher-than-usual moisture content, 5–8% (wwb), on to a vibrated fluidized-bed drier some agglomeration may occur (*Figure 2.82*). Agglomeration occurs in the fluidized bed of the integrated fluidized-bed spray drier and on the moving belts in a Filtermat drier. However, the most effective method of agglomeration is to rewet the surface of the dried particles with steam, humid air, a fine mist of water or the original liquid feed, then cause the wetted particles to come into contact so that they adhere together in clusters. Finally, the agglomerated powders are redried back down to a stable

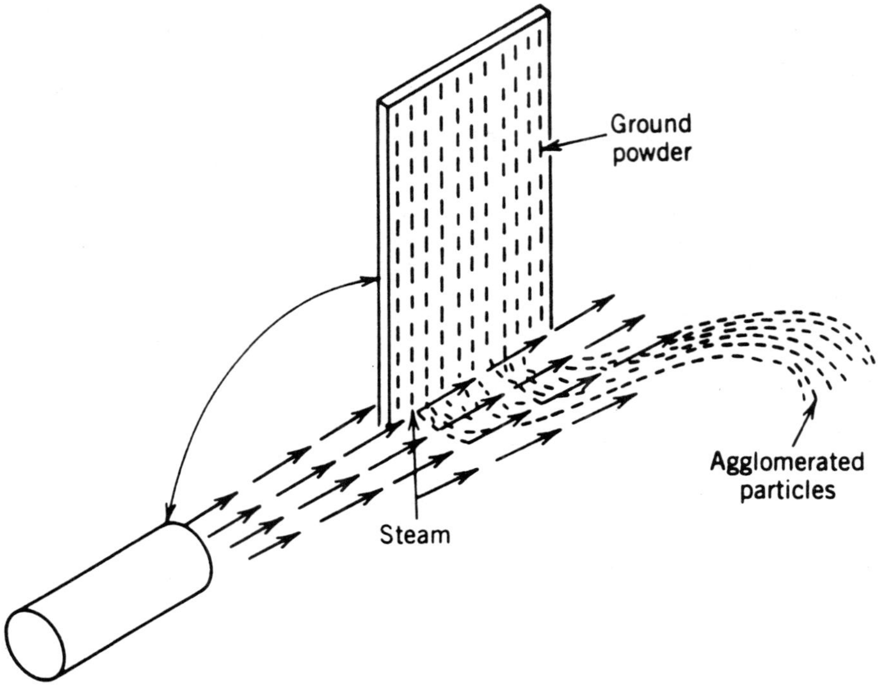

Figure 2.1 Agglomeration through a curtain of steam in feed port (Zemelman and Kettunen, 1992)

moisture content. This is followed by cooling and size classification of the agglomerates, with fine particles being recycled back into the rewetting zone. Various designs of rewet chamber are available. Some rely on the product contacting the wall of the chamber to increase agglomerate strength. In another design, steam flows through a curtain of the dry powder in a feed port before it enters a drying chamber (*Figure 2.1*). Agglomeration may be carried out in a vibrated fluidized bed (*Figure 2.2*). The powder is fed into the bed by means of a screw feeder. The rewetting liquid is applied through two-fluid nozzles. The bed consists of three sections. In the first, the bed is fluidized with heated air. Through a combination of high temperature, which may make the power particles sticky, and rewetting, agglomeration takes place. The second zone is also supplied with heated air to redry the material back down to a stable moisture content. In the third stage, the agglomerates are cooled to facilitate handling and packaging. The cooled material is then size classified (Brennan et al, 1990; Masters, 1991; Anon, 1992; Zemelman and Kettunen, 1992). *See also* Reconstitutability of dried food powders, Vibrated fluidized-bed drier.

Air-ballasting of vacuum pumps *see* Gas-ballasting of vacuum pumps.

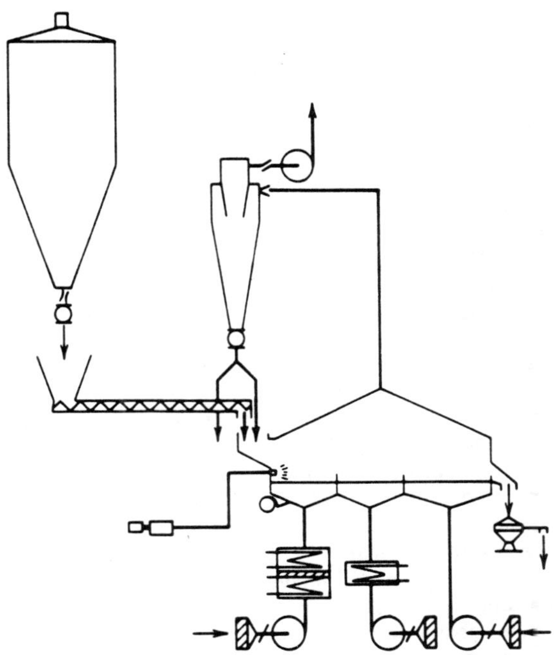

Figure 2.2 Fluidized-bed agglomerator (Zemelman and Kettunen, 1992)

Alkaline dip

Immersion of some fruits, which are to be dried whole, in an alkaline solution prior to drying can speed up the rate of drying. This process forms fine cracks in the skin of the fruit. Grapes and prunes are commonly treated in this way. A solution of sodium hydroxide or sodium carbonate, 5% or less, at a temperature in the range 93–100°C, may be used. Immersion times are of the order of 2–3 seconds. Immersion of grapes in cold solutions of sodium hydroxide or sodium carbonate together with olive oil can also accelerate drying, by causing dissociation of the wax platelets on the skin (Woodroof and Luh, 1975; Salunkhe et al, 1991). *See* Grape, Prune.

Air broom

A device used for cooling and dislodging powder deposits from the inner wall of a spray-drying chamber. It is particularly useful if the powder is tacky when hot. A perforated pipe rotates inside the chamber close to the inner wall. Cool air is directed on to the wall of the chamber. This disturbs any powder which has accumulated on the wall and at the same time cools it, thus making it easier to handle and transport.

Air heating

Air entering the drying chamber of a drier may be heated indirectly or directly. In indirect heating, there is no contact between the products of combustion from the fuel burner and the air being heated. Thus, there is no problem of contamination of the dried food with the combustion products. Indirect heating is widely used for food drying purposes. Steam heating is the most common form of indirect heating used. It is best suited to air temperatures below 250°C so that very high steam pressures are not required. To reach an air temperature of 250°C, the steam would need to be at 240°C. This would necessitate a steam pressure of 4 MN m^{-2}. The most common type of steam/air heater is the extended fin tube exchanger, which can have an efficiency of 98–99%. Indirect heating with fuel oils is also practised. Air temperatures up to 400°C are attainable by this means. Heat exchangers have efficiencies up to 85% depending on the heat load. Indirect heating with gas is also used. Heaters of this type perform very much like indirect fuel oil heaters. Thermal fluids may also be used to heat air indirectly for drying purposes. These special oils are circulated through a boiler and heat exchanger which heats the air. This type of heating may be used instead of or in combination with steam heating when air temperatures above 250°C are required. Electric air heating for drying is normally too costly to use on an industrial scale. It may be used in small laboratory driers. Direct heating of drying air, using natural gas, has become more common in recent years. This method of heating is more thermally efficient than indirect heating and savings in energy can be realized. Air temperatures up to 800°C may be achieved by direct heating. However, there are two main disadvantages associated with direct heating. Moisture of combustion increases the humidity of the air and so reduces its water-holding capacity. Thus the outlet temperature from a drier heated directly must be maintained at a higher value, as compared with one heated indirectly, to obtain the same product moisture content. This requirement offsets some of the savings mentioned above. The second problem is the possible contamination of the dried food with undesirable products of combustion. In a well-operated conventional gas burner, the main products of combustion are water vapour and carbon dioxide. However, other less acceptable products may be present in the air such as carbon monoxide, unburnt hydrocarbons and polycyclic aromatic hydrocarbons formed during combustion. Most concern has been expressed over the possible formation of nitrogen oxides, which could lead to an increase of nitrates and nitrites in the food. This would be undesirable in baby foods, for example. There is also the possibility of the formation of N-nitrosamines, in particular N-nitrosodimethylamine, which are considered to be carcinogenic. These compounds have been detected in spray-dried powders, dried by air directly heated with natural gas. This problem has led to the development of a number of so-called low-NO_x (low-NOX) burners, which are designed to reduce the contamination of the air with these undesirable products of combustion. There is some evidence in the literature that the use of such burners does reduce the level of contamination of powders as compared with conventional burners. However, this evidence is not conclusive and so it is advisable to check for the presence of such contaminants in powders produced using directly heated air (Kelly and Slattery, 1985; Jansen and Elgersma, 1985; Kelly et al, 1989). *See* Heat exchanger.

Air/powder separator

A device for separating powder from an air or other gas stream. Such separators are used to recover fine dried powder from the exhaust air leaving a spray-drying chamber. They may also be used for air/solids separation in fluidized-bed or pneumatic driers. There are four main types of such separators, i.e. cyclones, bag filters, wet scrubbers and electrostatic precipitators. These devices may be used singly or in various combinations. *See also* Bag filter, Cyclone, Electrostatic precipitator, Fluidized-bed drier, Pneumatic drier, Spray drying, Wet scrubber.

Ampere (A)

The basic SI unit of electric current. It is that constant current which, if maintained in two parallel rectilinear conductors of infinite length, of negligible circular cross-section, and placed at a distance of 1 metre apart in a vacuum, would produce a force between these conductors equal to 2×10^{-7} newton per metre length.

Angstrom (A)
A metric unit of length equivalent to 10^{-10} m.

Anti-caking agent (ACE)
The flow and handling properties of dried powders may be improved by the addition to them of small quantities of finely divided, free-flowing solids. These solids may be of organic origin, such as cornstarch (cornflour), powdered wood, seaweed and peat, or inorganic including fine synthetic silicates, metallic oxides, carbonates, phosphates or kaolin. In the case of food powders, inorganic materials are most commonly used. The particle size of the agent should be considerably smaller than that of the food powder. They act in one of two ways. If the food powder particles are relatively small, the particles of the agent come between them, reducing the incidence of contact between them. If the food particles are relatively large, they become coated with the fine particles of the agent. This coating prevents them from sticking together. Very small amounts of anti-caking agent, 1–2%, can improve the flow properties of powders. Larger quantities are required to prevent caking. It is important to comply with relevant regulations relating to the use of such additives. They are often used in food powders which are to be dispensed from vending machines, i.e. instant coffee, tea and whiteners.

Antioxidant
Dehydrated foods, particularly those containing polyunsaturated fats, are prone to oxidation, which can cause undesirable changes in odour, taste, colour and texture. Oxidation is discussed elsewhere in this text under water activity and food quality. Antioxidants may be used to inhibit oxidation in dried foods. The most commonly used antioxidants for this purpose are butylated hydroxyanisole (BHA) and butylated hydroxytoluene (BHT). These are described as free radical scavengers. However, they operate in different ways, depending on the substrate and the surrounding conditions. The result is the formation of stable hydroperoxides which interfere with the autocatalytic oxidation reaction. Other compounds can inhibit oxidation by chelating trace heavy metals, which catalyse the reaction. These include citric acid and citrates. Commercial antioxidant preparations often consist of mixtures of radical scavengers and metal chelators which work together to inhibit oxidation. Most countries set legal limits on the type and amount of antioxidant which can be used in foods. It is important to comply with such legislation (Morse, 1974). *See* Water activity and Food quality.

Apron drier *see* Conveyor drier.

Aseptic spray drying system
Where necessary spray drying may be carried out under aseptic conditions. A well-designed installation is required. It must be airtight and operate under a slight positive pressure to prevent contamination from outside. All internal surfaces should be smooth and free of crevices so that the whole system it can be thoroughly cleaned and sterilized before each run. The drying air and atomizing air, if a two-fluid nozzle is used, are pre-filtered through conventional air filters, heated indirectly and then passed through HEPA (High Efficiency Particulate Air) filters to sterilize them before entering the drying chamber. The feed material is pre-sterilized by heating or filtration before being atomized into the drying chamber. Drying takes place under aseptic conditions. The air containing the powder passes through a cyclone from which the dry, sterile powder is discharged, usually through a pinch-valve, into a sterile cabinet or room where packaging takes place. The room is maintained in a sterile condition by a laminar flow of HEPA filtered air. The exhaust air from the cyclone is released to atmosphere. Aseptic spray drying is widely used in the pharmaceutical industry. It may find some applications in food dehydration as quality specifications become more stringent (Pakowski, 1987; Masters, 1991). *See also* High efficiency particulate air (HEPA) filter, Spray drying.

Aspiration *see* Dry cleaning of raw materials.

Atmospheric pressure
The pressure exerted by the earth's atmosphere. At sea level this is $101.325\,kN\,m^{-2} = 14.69\,lb\,in^{-2} = 760\,mmHg$. The term atmosphere is used as a unit of pressure. *See* SI units and British units.

Atomization
The process of converting a liquid or slurry into a spray for the purpose of spray drying. The formation of a spray of feed within a narrow, specified size range, is an essential requirement for successful spray drying. Too wide a droplet size range can lead to non-uniform drying and/or the build up of powder on the inner wall of the drying chamber. *See also* Centrifugal atomizer, Pressure nozzle, Spray drying, Spray forming devices, Two-fluid nozzle.

Atomizer
A device for converting a liquid or slurry into a spray for the purpose of spray drying. *See also* Spray drying, Spray-forming devices.

Azeotropic drying
This method of drying involves the addition to the food of a solvent which forms a low-boiling-point azeotrope with water. The azeotrope is removed under vacuum. The food particles and solvent are mixed together and the mixture transferred to a reactor in which it is stirred. Vacuum is applied to the reactor, and the azeotrope boils off. The temperature of boiling depends on the solvent used and the pressure in the chamber. Typically the azeotrope boils at 24°C at a pressure of $13.3\,kN\,m^{-2}$ or

19°C at a pressure of 3.3 kN m^{-2}, absolute. Residual solvent is removed in a vacuum oven. Ethyl alcohol and ethyl acetate have been used as solvents. The process can be applied to frozen foods. Ethanol and ethanol/water mixtures have also been evaluated as solvents. The food is immersed in the solvent at −30°C for sufficient time for the azeotrope to form. A blast of dry air is used to evaporate the azeotrope. The presence of the solvent in the food accelerates the water diffusion rate. The process can be applied to large pieces of food, including whole fruits and vegetables. Some foods dried in this way have very good organoleptic properties, comparable with freeze-dried products. In other foods colours and flavours may be lost. No report of the commercial application of azeotropic drying was found (Holdsworth, 1969; Salunkhe et al, 1991).

Axial-flow fan *see* Fan.

B

Bag filter

This is a device used to separate powder particles from an air stream. The most common application is for recovering fine particles from the exhaust air leaving a spray-drying chamber. However, it may also be used for the same purpose on other types of driers, e.g. fluidized-bed and pneumatic driers. A bag filter system consists of a number of filter bags contained in a chamber, known as the baghouse. The principle of such a filter is shown in *Figure 2.3*. The powder-laden air is directed into the baghouse. The air passes through the fabric of the filter and is exhausted to atmosphere at the top of the chamber. The powder is retained on the outer surface of the filter fabric. As the powder layer builds up, the resistance to the flow of air increases. It is therefore necessary to remove the powder layer intermittently or continuously. If the drier can be operated batchwise then the filter bags may be cleaned manually during shut-down periods. If continuous operation of the drier is necessary, a number of filter bag units, arranged in parallel, may be used. Each unit is isolated from the air stream in turn and cleaned by shaking the bags by mechanical means. Another cleaning technique is to direct jets of air into the bags at predetermined intervals to loosen and dislodge the powder. Another method involves the use of a blow-ring. This consists of a hollow ring which moves up and down outside each bag. Air is directed continuously on to the surface of the bag to loosen and remove the powder. Bag filters can remove quite small particles. A typical efficiency curve for such a filter is shown in *Figure 2.4*. More than 99% of particles with diameters of 5 μm can be removed. To ensure good performance of such filters bag cleaning must be thorough and the bags inspected frequently for damage which could lead to leaks. Bags may be made of natural or synthetic fabrics. The fabric must be capable of withstanding the temperature of the exhaust air from the drying chamber. Cotton and wool fabrics can tolerate temperatures up to 90°C. Synthetic fabrics made from polypropylene, nylon or polyester can withstand higher temperatures of 100°C, 110°C and 130°C respectively. It is important to maintain good hygienic conditions in

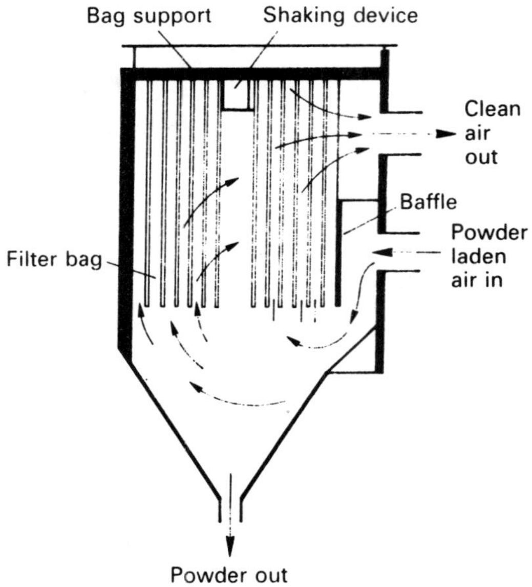

Figure 2.3 Principle of bag filter

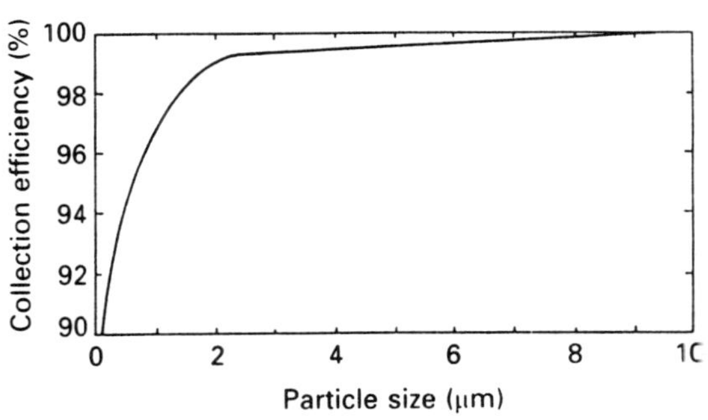

Figure 2.4 Efficiency curve for bag filter (Masters, 1991)

baghouses during shutdown. Bags must be removed and washed as necessary. Bag filters are not used for handling sticky or hygroscopic powders. Such filters may be used alone, or in combination with cyclones, to recover powder particles from the exhaust air from drying chambers. A cyclone may be used to remove the bulk of the larger particles while a bag filter recovers the very fine powder (Green, 1984; Masters, 1991). *See also* Cyclone, Electrostatic precipitator, Spray drying, Wet scrubber.

Band drier *see* Conveyor drier, Vacuum band drier.

Bar (bar) A metric unit of pressure equivalent to $10^5\,N\,m^{-2}$.

Barometric condenser
This is one type of condenser used to condense and remove water vapour from single- or multiple-effect evaporators. The vapour enters a chamber at the top of a tall column where it is contacted by a spray of cold water which condenses it. The column is full of water and is known as a barometric leg. The pressure of water in the column balances the atmospheric pressure to seal the system and maintain the vacuum. The temperature of the condensate-water mixture should be about 3°C below the temperature of the vapour in the chamber so that vapour will flow continuously into the chamber. The column must be tall enough to provide sufficient positive head at the base so that the condensate-cooling water mixture flows continuously out of the condenser at the same rate as it enters. A typical height for a barometric leg is in the range 7.5–11.0 m. A single effect evaporator equipped with a barometric condenser is shown in *Figure 2.5*, (Brennan et al., 1990; Toledo, 1991). *See also* Evaporator, Steam ejector, Vacuum pump.

Barometric leg
The tall column full of water through which the condensate from a condenser, used to condense vapour from an evaporator, vacuum drier or freeze drier, is discharged. *See* Barometric condenser.

Batch freeze-drier *See* Cabinet freeze drier.

Belt drier *See* Conveyor drier.

Belt trough drier
A conveyor drier designed to promote uniform drying of relatively large particulate foods. A sectional view of such a drier is shown in *Figure 2.6*. An endless wire mesh belt is supported loosely on two horizontal rollers. It hangs down and is partly supported by a grid, through which the heated air is directed, and a third roller to form a trough. The belt moves slowly in a clockwise direction. Material at the bottom of the bed moves up the slope, into the trough, where it turns over, and then back down the slope. The entire belt assembly is tilted towards one end so that the wet material follows a helical pathway to the lower end of the assembly. Fresh material is fed in at the raised end and dry product discharged over a weir at the lower end. The velocity of the drying air is sufficient to support, but not fluidize, the particles. The continuous movement of the particles exposes new surfaces to the drying air and promotes uniform drying. This drier was originally developed to be used in the drying stage of the dehydrofreezing process, in which uniform drying is particularly important. However, it has also been used to

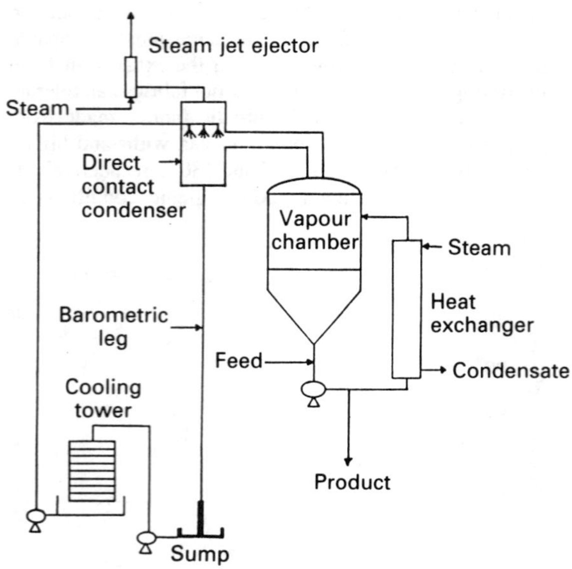

Figure 2.5 Barometric leg (condenser) (Toledo, 1991)

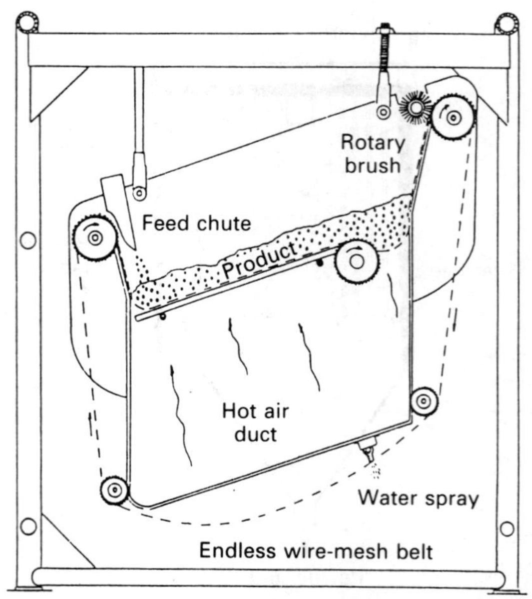

Figure 2.6 Belt trough dryer, (Van Arsdel et al, 1973a)

dry a wide range of food materials (Van Arsdel et al, 1973b; Salunkhe et al, 1991).

Belt washer *see* under Wet cleaning of raw materials.

BHA
Butylated hydroxyanisole, an antioxidant used in dried foods. *See under* Antioxidant, Water activity and food quality.

BHT
Butylated hydroxytoluene, an antioxidant used in dried foods. *See under* Antioxidant, Water activity and food quality.

Bin drier
A deep-bed through-flow drier of simple construction, mainly used to 'finish off' particulate foods partly dried in continuous tunnel or conveyor driers. It consists of a container of round or rectangular cross-section fitted with a false bottom. The partly dried food particles are loaded into the bin to a height of up to 2 m. Relatively dry air at a temperature in the range 40–70°C is directed up through the bed of particles at a flow rate of 0.5 $m^3 s^{-1}$ per square metre of bed area, or less. Very gentle drying takes place over an extended period of time, up to 36 h. Large stationary bins are fitted with their own fans and heaters. Smaller portable bins are fitted with wheels and can be 'plugged in' to a stationary air supply system. Typically, diced vegetables would be dried down to 10–15% moisture (wwb) in a tunnel or conveyor drier and then transferred to a bin drier. During their period in the bin, moisture migration between particles occurs in addition to the final drying, yielding a product of uniform moisture content (Van Arsdel et al, 1973a; Brennan et al, 1990).

BIRS process
This drying process was introduced in the late 1950s/early 1960s and a number of plants were constructed throughout Europe. The drying chamber consisted of a tall cylindrical tower, 72 m high and 16 m in diameter. Air with a relative humidity of not more than 3% and a temperature not more than 50°C entered at the bottom of the chamber and rose up the tower at a velocity in the range 0.045–0.90 m s^{-1}. The air was filtered and dehumidified by chemical means and/or refrigeration before entering the tower. The feed was introduced at the top of the tower through various types of distributor and fell gently down the tower, countercurrent to the air, taking about 90 s to reach the bottom from whence it was removed. Product moisture content range was 1.0–4.0% (wwb) and the air exhausted from the top of the chamber had a relative humidity in the range 80–90%. It was claimed that only 1.2–1.8 kg of steam was required to evaporate 1.0 kg water. A wide range of dried foods were produced in this way, including milk products, vegetable and fruit powders, cheese, gelatine and yeast. The organoleptic properties and handling characteristics of these products were claimed to be very good. However, the process proved to be uneconomic and no further plants have been constructed since that period (Ziemba, 1962; Anderson, 1962; Hayashi, 1989).

Blanching
An important heat process used in the preparation, for dehydration, of most vegetables and some fruits. Its main purpose is to inactivate enzymes or destroy enzyme substrates such as peroxides. If not inactivated, enzymes may cause discoloration, softening and undesirable flavour development during subsequent processing and storage of the dried product. Blanching involves rapidly heating the food to a predetermined temperature, holding it at that temperature for a predetermined time and rapidly cooling it. In addition to the inactivation of enzymes, blanching assists in cleaning the food, causes a reduction in its bacterial load and influences the texture of the product. On the other hand, blanching may result in the destruction of heat-sensitive vitamins and the loss of water-soluble nutrients through leaching. The effluent from blanchers contributes significantly to the waste disposal load from a food plant. Blanching may be carried out in water or steam. *Water blanching* may be effected on a batch principle by immersing the food in a pan of water held at an appropriate temperature, usually in the range 75–100°C, for the required time, usually in the range 20 seconds to 15 minutes, followed by cooling in cold water. Continuous water blanchers are of two types. In one, the food is passed at a controlled rate through a perforated drum rotating in a tank of water held at the appropriate temperature. In the other, the food is suspended in water, heated to the appropriate temperature and pumped through a coiled holding-tube, before water cooling (pipe blancher). In general, leaching losses are greater in water blanching, as compared with steam blanching. However, improvements in the design of water blanchers have reduced this difference. Good plant hygiene is necessary to avoid the build-up of thermophilic microorganisms in water blanchers. *Steam blanchers* make use of saturated steam at atmospheric or at low positive pressure, up to 150 kN m^{-2}. The food is conveyed through a steam chamber on a wire mesh belt, or by means of a helical screw. The correct blanching time is obtained by controlling the speed of the belt or screw. After blanching, the product is discharged through an outlet lock to a washer and cooler. This washing after blanching is necessary because the washing effect of steam blanching is less than water blanching. *Individual quick blanching (IQB)* is a technique designed to promote uniform blanching of food pieces with reduced leaching losses and effluent disposal problems. The food pieces are deposited as a single layer on a moving mesh belt which carries them through a steam cabinet. The pieces are steam heated for a sufficient time to inactivate the enzymes halfway to the centre of each piece. The pieces are then redistributed on to a second slower-moving belt, in a deeper layer, where

thermal equilibration occurs so that each piece is adequately blanched. The pieces then pass on to a cooling section. *Vacuum steam blanching* is carried out in an end-over-end rotating conical vessel equipped with a heating jacket. The food pieces are loaded into the vessel. It is rotated slowly and the pressure within it is reduced. The vacuum is broken with saturated steam and the pressure in the vessel is increased to atmospheric pressure. The vessel is held at atmospheric pressure for c. 1 minute. The food pieces are then vacuum cooled, by again reducing the pressure in the vessel. Vacuum blanching is relatively quick and can result in an improved texture in frozen or dehydrated vegetables. The use of microwaves, and mixtures of steam and flue gases for blanching, have been investigated but have not been used commercially. The effectiveness of a blanching process is usually evaluated by testing for the inactivation of indicator enzymes. The enzymes most commonly used as indicators are catalase and peroxidase. The latter is more heat resistant than the former (Brennan et al, 1990; Arthey and Dennis, 1991; Salunkhe et al, 1991). *See also* Water activity and food quality.

Blast freezing
A method of freezing in which the food is placed in contact with a cooled gas, usually air. Freezing takes place by heat transfer from the food to the cooled gas. *See also* under Freezing methods.

Bound moisture
Liquid water which is held within a solid structure by adsorption on to the solid surface, within capillaries or through solution effects and which exhibits a vapour pressure less than the vapour pressure of water at the same temperature. To remove bound water energy in excess of the latent heat of evaporation of water at the relevant temperature must be supplied to overcome this bonding. *See also* Adsorption, Sorption, Sorption isotherm, Unbound moisture, Water activity.

British units *see* Table 2.1.

Bulk density
Applies to particulate matter and is the mass per unit volume of such material, including voids that exist between the particles. It is usually measured by placing

Table 2.1 British (Imperial) units

Quantity	*Name*	*Symbol*	*Unit*
Mass	pound	lb	lb
	ounce	oz	0.0625 lb
	ton	ton	2240 lb
Length	yard	yd	yd
	inch	in	0.0278 yd
	foot	ft	0.3333 yd
	mile	mile	1760 yd
Time	second	s	s
Temperature	degree Fahrenheit	°F	°F
Area			ft^2
Volume			ft^3
	UK gallon	gal	gal*
Velocity			$ft\ s^{-1}$
Acceleration			$ft\ s^{-2}$
Density			$lb\ ft^{-3}$
Force	poundal	pdl	$lb\ ft\ s^{-2}$
	pound force	lbf	32.2 pdl
Pressure		psi	$lbf\ in^{-2}$
	inch of mercury	inHg	$0.4917\ lbf\ in^{-2}$
	foot of water	ftH_2O	$0.4336\ lbf\ in^{-2}$
Dynamic viscosity			$lb\ ft^{-1}\ s^{-1}$
Kinematic viscosity			$ft^2\ s^{-1}$
Energy			ft lbf
Power	horsepower	hp	$550\ ft\ lb\ s^{-1}$
Energy (heat)	British thermal unit	Btu	Btu†
Specific heat			$Btu\ lb^{-1}\ R^{-1}$
Thermal conductivity			$Btu\ h^{-1}\ ft^{-1}\ R^{-1}$
Specific energy			$Btu\ lb^{-1}$
Heat transfer coefficient			$Btu\ h^{-1}\ ft^{-2}\ R^{-1}$

* UK gallon is the space occupied by 10 pounds weight of water.
† Quantity of heat required to raise the temperature of 1 lb water by 1°F (or 1°R).

a known weight of the particulate material in a graduated container, tapping, shaking or vibrating the container in a specific way, and then measuring the volume occupied by the solid particles. Common units are kg m^{-3}. It is also known as apparent density. Bulk density will depend on the absolute density, geometry, size and surface characteristics of the solid particles and the method of measurement. *See also* Absolute density (Lewis, 1987).

Butylated hydroxyanisole (BHA)
An antioxidant used in dried foods. *See* Antioxidant, Water activity and food quality.

Butylated hydroxytoluene (BHT)
An antioxidant used in dried foods. *See* Antioxidant, Water activity and food quality.

C

Cabinet (compartment, tray) drier
A batch-operated drier consisting of an insulated cabinet fitted with a fan which pushes air through a heater and then either horizontally between trays of food material or vertically through the trays and layers of food. In some units both patterns of air flow may occur. Baffles located in the cabinet direct the air flow. Air velocities of $2–5\,m\,s^{-1}$ are used in crossflow driers, while air flow rates of $0.50–1.25\,m^3\,s^{-1}$ per square metre of tray area are used in throughflow systems. Dampers are fitted to control the fresh air intake and the outflow of exhaust air. By manipulation of these dampers the amount of air recirculated through the cabinet can be adjusted. Most cabinet driers are steam heated. However, cabinets equipped for other forms of direct and indirect heating are available. Small cabinets housing one or a small number of trays are useful in research and development projects. Larger cabinets, used for commercial drying, may contain trays measuring 2×1 m, stacked up to 25 high. The trays may be placed on fixed shelves in the cabinet or loaded on to trolleys which are wheeled into the cabinet. Such cabinets, used singly or in groups, can accommodate up to 20,000 kg day^{-1} of prepared fresh vegetables (Van Arsdel et al, 1973a; Williams-Gardner, 1976; Charm, 1979; Brennan et al, 1990).

Cabinet freeze drier
Much commercial freeze drying is carried out on a batch principle in cabinet freeze driers. Such a drier consists of a high-performance vacuum cabinet, usually cylindrical in shape, positioned horizontally with doors front and back. The cabinet is equipped with heated shelves. In early designs, these shelves were fixed and the trays of frozen food were placed on them. Heat was transferred both by conduction from below and by radiation from the undersides of the plates above. To decrease drying times, moving plates were introduced. The gap between the plates could be increased or decreased by pneumatic or hydraulic means. When the plates were closed together heat was conducted to both sides of the food and drying speeded up. However, the escape of the water vapour produced by sublimation was restricted and this caused some melting of the ice on the surface of the food. Drying was not uniform and the product was of poor quality. In the accelerated freeze drying (AFD) method, sheets of expanded metal mesh were placed in contact with both faces of the food pieces. As the plates were closed together, the mesh penetrated into the food bringing the heat source closer to the ice front, as drying proceeded. This shortened the drying time and the vapour could escape through the mesh, thus yielding a good quality product. This technique was mainly applied to meat and fish steaks. The use of mesh has now largely been discontinued and cabinets are equipped with fixed heater plates. The plates and trays are carefully machined to provide good thermal contact between them. Special designs of trays are available to suit different types of material. For example, compartmented trays or dishes, such as those shown in *Figure 2.7*, enable heating of granular material from three sides without restricting the escape of vapour. Trays of this type can also be used in tunnel freeze driers. In some freeze drier designs the trays are located between the heater plates out of contact with them. Heat is transferred by radiation from both above and below the trays. Microwave heating in freeze drying has been the subject of much investigation. Ice and liquid water have loss factors which are considerably higher than dry tissue. Thus, in frozen food exposed to microwaves, the ice will absorb energy much faster than dry tissue. This type of heating would appear to be very suited to freeze drying. However, controlling the heating can be a problem. Loss factor increases with rise in temperature. As the ice core warms up it absorbs more energy. If local melting of ice occurs, the water formed will absorb energy very rapidly. Sudden vaporization may occur which can cause explosive damage to the food. Ionization of the rarefied gases in a freeze drier cabinet can occur which may lead to plasma discharge and possible overheating of the food. This problem can be reduced by using high frequency (2450 MHz) microwaves. With the sophisticated methods of control now available, the use of

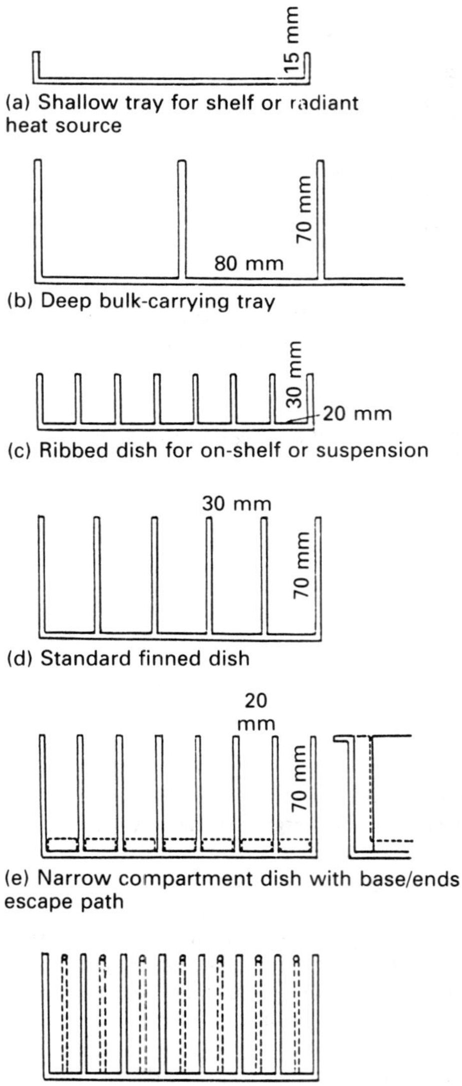

Figure 2.7 Different designs of trays or dishes for freeze drying (Dagleish, 1990)

microwaves for freeze drying is technically feasible. The main limitation on its application is cost.

The vacuum system for a batch freeze drier must be capable of reducing the pressure in the chamber rapidly so as to avoid melting of the prefrozen food when it is placed in the cabinet. It must be capable of maintaining the very low pressure required throughout the drying cycle. It has to cope with the water vapour produced by sublimation and the non-condensible gases, O_2, N_2 and CO_2. The most widely used arrangement is a refrigerated condenser backed up by a mechanical pumping system. The condenser, which may be a plate or coil, is located in the drying chamber or in another chamber connected to it via a wide-bore tube. The water vapour condenses as ice on the surface of the plate or coil and is thus removed from the atmosphere in the chamber, reducing the water vapour pressure and promoting sublimation of the ice. The temperature of the refrigerant in the coil or plate must be below the saturation temperature corresponding to the pressure in the chamber. The difference between these two temperatures must be sufficient to cause the vapour to flow to the refrigerated surface and heat to transfer through the wall of the condenser and any ice layer which has formed on its outer surface. This temperature difference is a measure of the efficiency, and hence the energy consumption, of the vapour removal system and should be kept to a minimum. Refrigerant temperature is usually in the range –50°C to –10°C. Build-up of ice on the condenser will reduce heat transfer and hence efficiency. A large condenser surface area must be provided if it is not to be defrosted during a drying cycle. One arrangement for defrosting is to have two condensers working in parallel. The condensers are located in separate chambers, both of which are connected to the drying chamber. At any one point in time, one of the condensers is being defrosted while the other is operating normally. The roles are reversed periodically throughout the drying cycle. The use of scraped surface condensers to remove the ice formed, or adsorption systems in which the water vapour is taken up by glycerol or glycol, have been investigated. However, such methods have not been employed commercially. The mechanical pumping system may consist of a single pump or two pumps in series. When a single pump is used it is usually a gas-ballasted rotary type. In the case of a two-stage pumping system, the first pump in the series may be a Roots pump (blower), or an oil-sealed rotary pump. The second pump in the series is usually an oil-sealed rotary pump and is gas-ballasted.

As an alternative to the combination of refrigerated condenser and vacuum pump(s), a steam ejector system may be used to create and maintain the vacuum in a freeze-drying chamber. This normally consists of four or more stages. A four-stage system is represented in *Figure 2.8*. Such a system can cope with the water vapour and non-condensible gases. Another possible combination is a refrigerated condenser in series with a small capacity steam ejector system to handle the non-condensible gases. Steam ejectors are generally less costly to purchase and maintain as compared with refrigerated condensers and pumps. However, they do require very large volumes of water, both for steam generation and cooling in the condensers. A typical cycle of operation of a cabinet freeze drier is as follows: the food is prefrozen on trays of appropriate design; these are loaded into the cabinet, between the heater plates; the cabinet is closed and pumped down rapidly to a pressure of $135\,Nm^{-2}$ or lower in about 10 minutes; when the water vapour pressure in the chamber reaches a level less

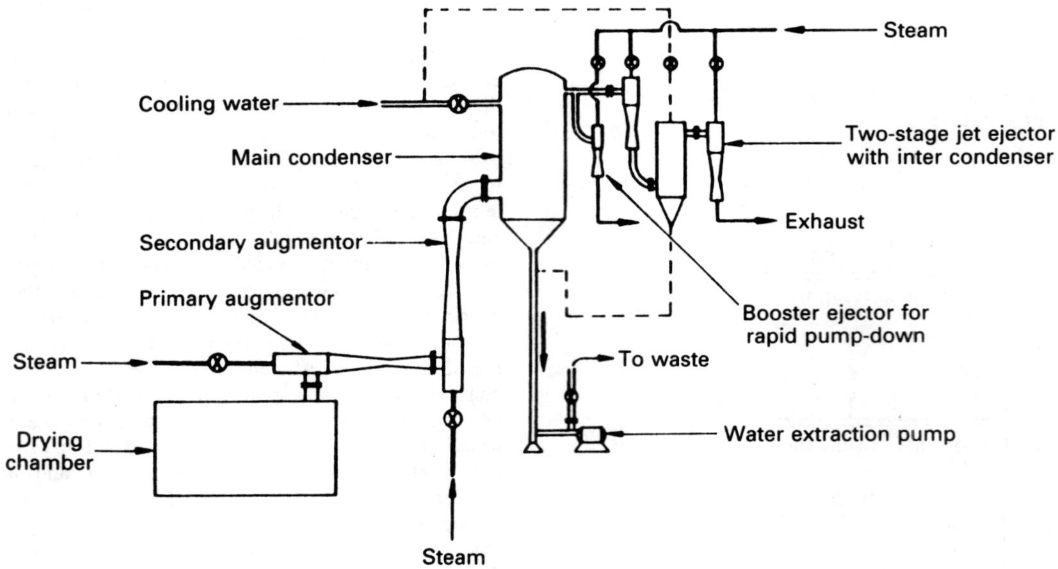

Figure 2.8 Four-stage steam ejector for freeze drier (Mellor, 1978)

than the vapour pressure of ice, sublimation will commence and the temperature of the food will fall slightly; the plates are heated rapidly to an appropriate temperature, typically in the range 70–150°C, and held at that temperature for a prescribed period of time, 30–120 minutes; the heating is stopped and the plates cooled slowly to a final temperature suited to the product, typically 40–80°C. Typical temperature changes during a cycle are shown in *Figure 2.9* (Cotson and Smith, 1963; Lorentzen, 1974 & 1975; Mellor, 1978; Brennan et al, 1990; Dalgleish, 1990). *See also* Freeze

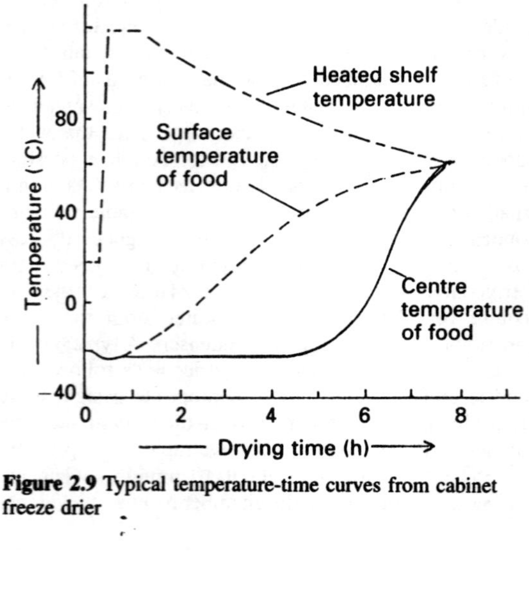

Figure 2.9 Typical temperature-time curves from cabinet freeze drier

drying, Multicabinet freeze drier, Steam ejector, Tunnel freeze drier, Vacuum pump, Vacuum-spray freeze drier.

Calorie (cal)
A metric unit of energy equivalent to 4.187 J.

Candela (cd)
The basic SI unit of luminous intensity. It is the luminous intensity, in the perpendicular direction, of a surface 1/600,000 metre squared of a black body at the freezing temperature of platinum under a pressure of 101,325 newtons per metre squared.

Capillary flow
When a capillary tube is immersed in a liquid vertically, with one end touching the liquid, the liquid will rise in the tube to some height above the liquid surface, under the influence of surface tension forces. The relationship between the height reached by the liquid, h, the surface tension of the liquid, γ, the contact angle between the tangent to the surface of the liquid at the point of contact with the inner surface of the tube, and the surface itself, θ, the density of the liquid, ρ, and the radius of the capillary tube, r, is:

$$h = \frac{2 \cos \theta\gamma}{\rho g r}$$

where g is acceleration due to gravity (*Figure 2.10*). The movement of moisture to the surface during the drying of a bed of solid particles, or in some cases a piece of solid material, has been said to occur as a

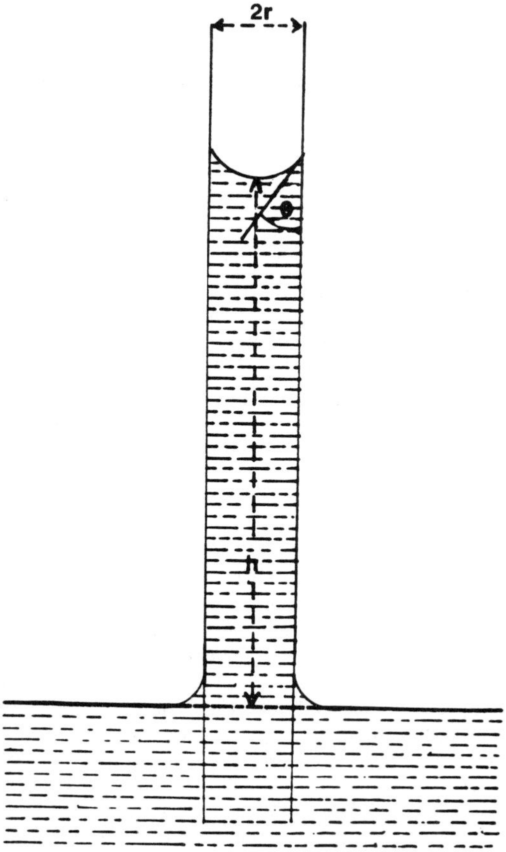

Figure 2.10 Representation of liquid in capillary tube

result of capillary flow. *See* under General Principles of Dehydration. It also plays a part in the reconstitution of dried material, in particular, freeze-dried and agglomerated products. *See* Reconstitutability of dried food powders.

Cascading rotary drier *see* Rotary drier.

Centrifugal atomizer
One type of device for converting a liquid or slurry into a spray for the purpose of spray drying. It usually consists of a disc, wheel or bowl fixed to the end of a rotating shaft. The liquid feed is introduced near the centre of rotation of the disc, accelerated to the peripheral velocity of the disc and spun off, in the form of a spray, from the edge of the disc, into the drying chamber. There are many designs of rotating head, including discs with holes at the periphery, wheels with vanes, rimless wheels with spokes and inverted bowls. Examples of atomizer designs are shown in *Figure 2.11*. They have diameters in the range 5 to 30 cm and rotate at speeds of 50,000 to 10,000 revolutions per minute. They are usually driven by electric motors, either directly or through belt or worm drives. Some small laboratory versions are driven by air turbines. Centrifugal atomizers can produce uniform sprays within a droplet diameter range of 1 to 600 μm. Many factors affect the size of droplet produced including the viscosity, density and surface tension of the feed and the design and speed of rotation of the atomizer. Many equations to predict droplet size have been recommended. The following relationship, for an atomizer featuring vanes, is an example:

$$\frac{D}{R} = 0.4\left(\frac{M}{N_v b \rho N R^2}\right)^{0.6}\left(\frac{N_v \mu b}{M}\right)^{0.2}\left(\frac{\gamma b^3 \rho N_v^3}{M^2}\right)^{0.1}$$

where D is the Sauter mean diameter (m), R is the

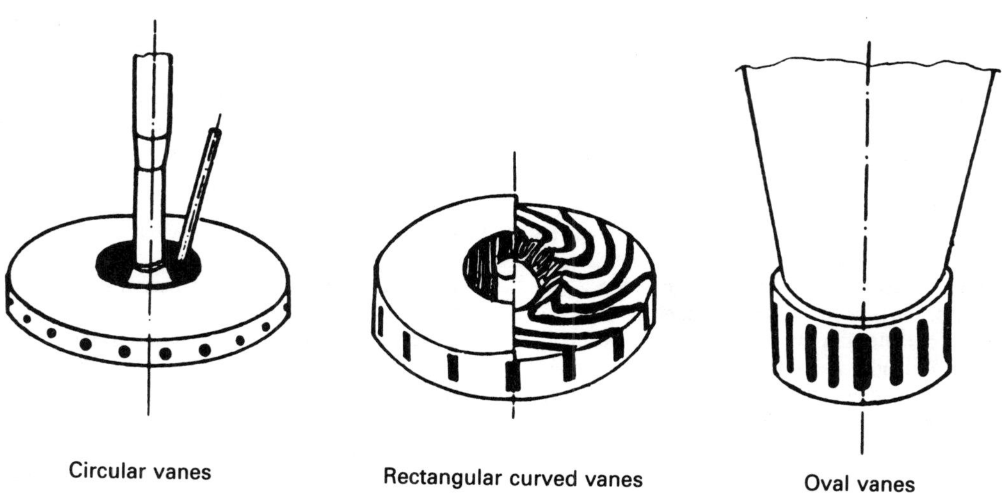

Figure 2.11 Types of centrifugal atomizer (Filkova and Mujumdar, 1987)

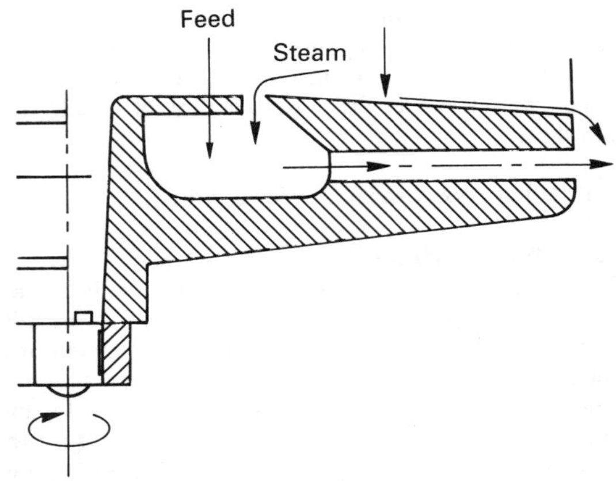

Figure 2.12 Centrifugal atomizer with steam injection (Anon, 1992)

diameter of the wheel (m), M is the mass flow rate of feed (kg s^{-1}), N is the rotational speed (revolutions per second), b is the height of the vanes (m), N_v is the number of vanes, ρ is the density of the feed (kg m^{-3}), γ is the surface tension of the feed (N m^{-1}), μ is the feed viscosity (P_s), (Masters, 1985). Centrifugal atomizers are not subject to blocking and abrasion by insoluble solid particles in the feed, to the same extent as pressure nozzles, and can handle viscous feeds at relatively low pumping pressures. Special designs of centrifugal atomizer are available by means of which steam can be injected (*Figure 2.12*) to replace the air atmosphere around the disc. This reduces the amount of air incorporated in the droplets so that the dried powder will consist of dense, void-free particles. By controlling the amount of steam injected, the bulk density of the powder can be adjusted. This type of atomizer is also known as the steam-swept atomizer (Van Arsdel et al, 1973; Coulson et al, 1978; Green, 1984; Masters, 1985; Filkova and Mujumdar, 1987; Pisecky, 1987; Brennan et al, 1990; Anon, 1992). *See also* Pressure nozzle, Spray drying, Two-fluid nozzle.

Centrifugal fan *see* Fan.

Centrifugal pump *see* under Pumping of foodstuffs.

Cleaning of raw materials
All raw food materials intended for dehydration must be cleaned to remove contaminants which might be a health hazard to the consumer or which might reduce the acceptability of the product to the consumer. In addition, the cleaning operation may play a role in reducing microbiological loads and undesirable chemical reactions which might adversely influence the effectiveness of the drying process and the quality of the product. An efficient cleaning operation should result in near complete separation of contaminants with minimum wastage of good material. It should cause minimum damage to the product and in particular leave the cleaned surface in good condition and avoid recontamination of the cleaned food. Contaminants in raw foods may be: *metal* such as nuts, bolts, metal fragments or parts; *mineral* such as soil, stones, glass fragments, grease, oil; *plant* such as twigs, foliage, stalks, pits, skins, husks, pieces of wood, rope, string; *animal* such as whole small animals or insects, body parts, excreta, hairs, feathers, insect eggs, blood; *chemical* such as pesticides, herbicides, fertilizers; *microbial*, i.e. rot, fungal growth, bacteria, yeasts. Both wet and dry cleaning methods are employed. The former may involve the use of screens, brushes, tumbling devices, inclined belts, aspirators and magnetic separators. Wet cleaning may make use of soak tanks, spray devices, flumes, filters and ultrasonic generators (Fellows, 1988; Brennan et al, 1990; Arthey and Dennis, 1991). *See* Dry cleaning of raw materials, Wet cleaning of raw materials.

Coacervation
A technique used to encapsulate flavouring materials. *See* Encapsulation.

Closed-cycle spray-drying system
In this type of spray-drying layout all the drying medium is recycled back through the heater and drying chamber. It is most commonly employed when gases other than air are used as the drying medium and/or when the solvent being removed from the feed is flammable. The drying medium leaving the drying chamber, containing the fine dried particles, passes though high efficiency cyclones or filters and then to a condenser/scrubber. Here it is cooled and washed with cool solvent, similar to that removed in the drying chamber but cooled. It is then returned to the drying chamber through the heater, which is of the indirect type. A proportion of the cooled solvent, equivalent to that evaporated in the drying chamber, is removed. The whole system is gas-tight and operates under a slight positive pressure. It is used when drying feeds that are highly susceptible to oxidation or which may form an explosive mixture of air and powder. In these cases, nitrogen gas would be used as the drying medium. It may also be used when toxic materials are being dried or if pollution of the atmosphere with strong, unpleasant odours or solvent vapours is likely (Masters, 1985). *See also* Air heating, Semi-closed-cycle spray-drying system, Spray drying.

Compartment drier *see* Cabinet drier.

Concurrent tunnel drier *see* Tunnel drier.

Concurrent spray drier *see* Spray-drying chamber.

Continuous freeze drier *see* Tunnel freeze drier.

Control of driers
The moisture content of the final product is usually the most critical variable that should be controlled in drying. While equipment for the continuous measurement of the

moisture content of solid materials is available, it has not yet found wide commercial application in food dehydration. Consequently, in most dehydration operations it is the drying conditions that are controlled, thus controlling the product moisture content indirectly. The number of variables which are controlled and the level of sophistication of the control systems will depend on the type of drier and the specifications for the product.

Hot-air driers
The important variables in such driers are air inlet and outlet temperatures, air flow rate/velocity and residence time of the product in the drier. In most cases air velocity and residence time are fixed for a given product and only the air temperatures are controlled during the drying operation. In simple batch driers, a thermostat controlling the air inlet temperature is all that may be required. The air outlet temperature may be continuously measured instrumentally and the signal from the instrument used to control the supply of fuel or the steam pressure to the air heater. For some applications control of air humidity may be desirable. The humidity of the outlet air is measured with a hygrometer and the signal from that instrument is used to operate dampers which control the amount of air which is recirculated, thus indirectly controlling the air humidity. In continuous hot-air driers it may be necessary to control the air temperature at different stages in the drying cycle, e.g. tunnel, conveyor and fluidized-bed driers. Spray driers are usually automatically controlled by one of two main methods. (1) The air inlet temperature is monitored instrumentally and the signal from the instrument used to regulate the fuel supply or steam pressure to the air heater. The outlet air temperature is monitored instrumentally and the signal from that instrument used to regulate the feed supply to the atomizer, so as to maintain the correct outlet temperature. This method is used with rotary atomizers. (2) The feed rate is maintained constant, the outlet air temperature is monitored instrumentally and the signal from the instrument used to regulate the fuel supply or steam pressure to the air heater. This method is mainly used with pressure nozzle atomizers. Fluctuations in outlet air temperature tend to be greater in this second method as compared to the first. The second method cannot compensate for variations in the solids content of the feed. Should a blockage develop in the feed pipe or atomizer or should the feed pump fail, the outlet air temperature will rise, perhaps to a dangerous level. A safety feature may be incorporated into the system which shuts down the air heater when a certain critical outlet air temperature is reached. In addition, water may be sprayed into the chamber through nozzles to cool the air or even flood the chamber to prevent a fire or explosion.

If a reliable on-line method of moisture content measurement is available this can be used to control the outlet air temperature. Such a system should be able to compensate for variations in the solids content of the feed and the humidity of the ambient air and result in a more uniform moisture content in the product as compared with the methods of control outlined above.

Drum driers
The important variables in drum driers are drum temperature, residence time and thickness of the feed layer on the drum surface. For a given product the thickness of the feed layer will be set by the feeding system used. The drum speed will be fixed to give the optimum residence time. The only variable which is controlled during the drying operation will be the drum temperature. This is achieved by controlling the pressure of the steam entering the drum.

Freeze driers
Since freeze drying is such an expensive operation it is important to optimize the drying conditions so as to minimize the drying time. This is achieved by controlling the heat input and the chamber pressure. There are two important constraints during freeze drying. The dry surface temperature must not rise so high as to cause heat damage to the dry material and the temperature at the ice front must not rise so high as to cause melting. Liapis and Litchfield (1979) carried out a quasi-steady-state analysis to ascertain the optimum control policy for a freeze-drying cycle. Their analysis suggested that at the beginning of the cycle both the heat supply and chamber pressure should be at their maximum values. When the temperature at the ice front approaches its limit, before melting occurs, the chamber pressure should be reduced rapidly to maintain the interface temperature at the maximum permitted value. The heat supply may need to be reduced as the dry surface temperature approaches its maximum value before heat damage occurs. Litchfield and Liapis (1982) studied the complete un-steady-state optimal control problem for two food materials, turkey meat and non-fat-reconstituted milk. In the case of the milk, the results were close to those from the quasi-steady-state analysis. In the case of the turkey meat, the results of the two analyses were similar at low chamber pressures, but some differences occurred at higher chamber pressures. They concluded that control policies arrived at by quasi-steady-state analysis may be useful guidelines but should be interpreted with caution. Only the complete un-steady-state analysis can yield the true optimal control policy (Mellor, 1978; Liapis, 1987). *See also* Freeze drying, Spray drying.

Conversion factors *See* Table 2.2.

Conveyor (apron, band, belt) drier
A type of hot-air drier in which the feed material is conveyed though the drying chamber on a moving perforated belt or band. The heated air may be directed concurrent or countercurrent to the feed material or in a cross-flow pattern. However, in most cases the air is directed through the belt and bed of feed material. Thus, the application of such driers is limited to food particles

Table 2.2 Conversion factors

Quantity	*British unit*	*Conversion factor*	*SI unit*
Mass	Pound (lb)	0.454	kg
Length	Foot (ft)	0.3048	m
Temperature	Degree Fahrenheit (°F)	0.556	K (or °C)
	or Rankine (R)*	0.556	K (or °C)
Area	Square foor (ft^2)	0.0929	m^2
Volume	Cubic foot (ft^3)	2.83×10^{-2}	m^3
	UK gallon (gal)	4.55×10^{-3}	m^3
Density	lb ft^{-3}	16.022	kg m^{-3}
Force	Poundal (pdl)	0.138	N
	Pound force (lbf)	4.448	N
Pressure	lbf in^{-2} (psi)	6.895	kN m^{-2}
	inHg	3.386	kN m^{-2}
	ftH_2O	2.989	kN m^{-2}
Dynamic viscosity	lb ft^{-1} s^{-1}	1.488	kg m^{-1} s^{-1}
Kinematic viscosity	ft^2 s^{-1}	0.093	m^2 s^{-1}
Energy	ft lbf	1.356	J
Power	hp	745.7	W
Energy (heat)	Btu	1.055	kJ
Specific heat	Btu lb^{-1}	2.326	kJ kg^{-1}
Thermal conductivity	Btu h^{-1} ft^{-1} R^{-1}	1.731	W m^{-1} K^{-1}
Specific energy	Btu lb^{-1}	2.326	kJ kg^{-1}
Heat transfer coefficient	Btu h^{-1} ft^{-2} R^{-1}	5.678	W m^{-2} K^{-1}

* 492 R = 32°F = 0°C = 273 K.

that form a porous bed. Particles which are thermoplastic and fuse together to form an impervious layer on the belt cannot be dried satisfactorily on this type of drier. The air may flow up or down through the bed of solids. Usually upward movement of air is used at the 'wet end' of the drier, where the particles are wet and dense, and downward movement at the 'dry end'. Thus the low density, dry particles are not disturbed or entrained in the exhaust air. In some designs of drier the air may flow up and down in alternate sections to improve the uniformity of drying, particularly when deep layers of feed are being dried. The conveyor is typically 2–3 m wide and from 3 to 50 m long, depending on the capacity required. It is usually made of hinged stainless steel plates, suitably perforated to suit the size of the particles being dried. The plates are typically 1 mm thick and 50–300 mm wide. Belts made from wire mesh or woven plastic may also be used. The depth of the bed of wet material is typically 75–150 mm and the air velocity between 0.25 and 2.50 m s^{-1}. Inlet air temperatures in the range 70–200°C are used depending on the heat sensitivity of the feed being dried. Particulate solids are deposited on the belt, in a uniform layer, by means of chutes, vibrating or moving belt feeders. Scrapers or other devices may be employed at the 'dry end' of the chamber to assist in removal of the dried product from the belt. Provision may be made for wet or dry cleaning of the belt as it returns to the 'wet end' of the chamber.

Multistage drying may be carried out by having two or more conveyors in series, the partly dried particles from one being redistributed on to the next for further drying. This arrangement can lead to better utilization of the belt area. If significant shrinkage of the food occurs in the early stages of drying on a single conveyor, then the loading on the belt in the latter stages of drying may be unnecessarily low. In a multistage arrangement, the particles may be redistributed, in a suitably thick layer, on to the next belt, thus making more economical use of all the belts. In addition, the particles will be mixed during transfer from one belt to the next, exposing new surfaces to the drying air, and leading to quicker and more uniform drying. The drying chamber may be compartmented so that air temperature and velocity may be set independently for each stage. In multistage drying, the conveyors may be arranged in series, horizontally or superimposed (*Figure 2.13*). The latter arrangement requires less floor space. For products that require slow, gentle drying, such as pasta, a number of belts may be arranged above each other in the drying chamber without partitions. The particles move from top to bottom of the chamber, transferring from one belt to the next. Air at relatively low flow rate (0.25–0.50 m s^{-1}) passes partly across and partly through the layers of food. The belt trough drier and the vacuum band drier, discussed elsewhere in this text, are types of conveyor drier.

Some feeds, which do not readily form a porous bed, may be converted into a form suitable for conveyor drying. Semi-solids and pastes may be extruded in strips or lumps on to the conveyor. In foam mat drying, the feed is aerated before being extruded on to a perforated belt for drying. Back-mixing some dried product with

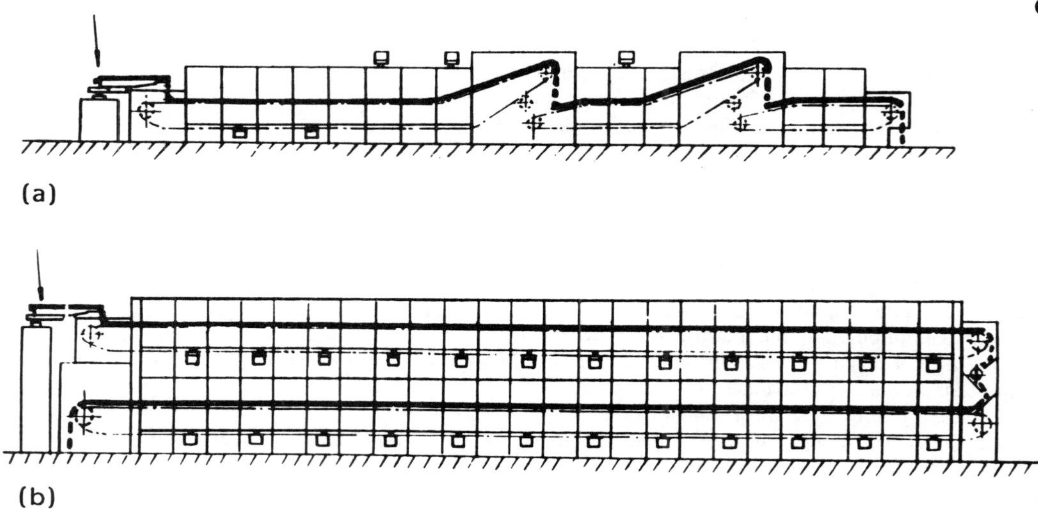

Figure 2.13 Multistage conveyor drier (Sturgeon, 1987)

the feed may make it suitable for through-flow drying. Some sticky materials may be rapidly pre-dried in a shallow layer and then redistributed on to a second belt in a thicker layer for further drying. In some instances, it may be more economical to partly dry a material on a conveyor drier, down to 10–15% moisture (wwb), and to finish the drying in a bin drier or some other type of batch drier, rather than drying to completion on the conveyor drier (Van Arsdel et al, 1973; Williams-Gardner, 1976; Sturgeon, 1987). *See also* Belt trough drier, Foam mat drying, Impingement drying, Vacuum band drier.

Countercurrent tunnel drier *see* Tunnel drier.

Countercurrent spray drier *see* Spray-drying chamber.

Cryogenic freezing
A freezing method in which the cooling medium is a subliming solid, usually solid carbon dioxide, or boiling liquid, usually liquid nitrogen. *See also* Freezing methods.

Cyclic-pressure freeze drying
A freeze drying technique in which the chamber pressure is varied on a predetermined cycle throughout the drying period. This results in an increase in the average effective thermal conductivity of the dried layer. This, in turn, can lead to more rapid heat transfer and shorter drying times as compared with drying under constant low pressure conditions (Mellor, 1978).

Cyclone separator
This device is widely used as an air/powder separator in drying plants. One of its main applications is for recovering fine particles from the exhaust air leaving a

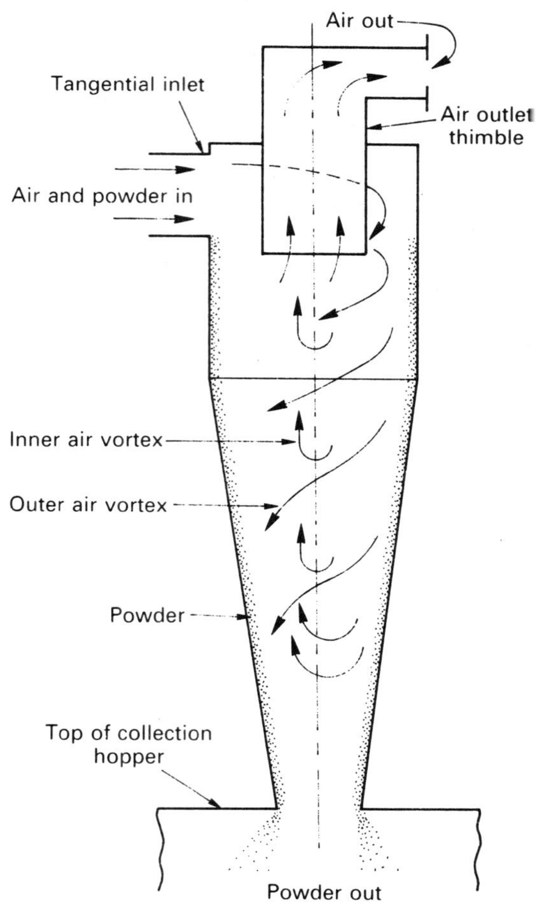

Figure 2.14 Principle of cyclone separator

spray-drying chamber. It may also be used in other types of driers where fine particle collection is necessary, e.g. fluidized bed and pneumatic driers and in pneumatic conveying systems. A simple cyclone separator is depicted in *Figure 2.14*. Air containing the solid particles enters a cylindrical chamber tangentially. The powder and air spiral downwards. The particles are thrown towards the wall of the chamber under the influence of centrifugal force and separated from the air. The powder is removed from the base of the chamber. The cleaned air spirals upwards at the centre of the chamber and exits through an outlet at the top. Powder particles that do not reach the wall during their dwell-time in the chamber are not separated out and leave in the air. Empirical expressions are available for calculating the diameter of the smallest particle that will be completely separated from the air. One of the most satisfactory relationships, based on Stokes law, is:

$$D_{min} = \left[\frac{9\mu B_c}{\pi N V_c \ (\rho_s - \rho)} \right]$$

where D_{min} is the diameter of the smallest particle to be completely removed, μ the viscosity of the air, B_c the width of rectangular inlet duct to cyclone, N number of turns the air makes in the cyclone, V_c the velocity of the air entering the cyclone, ρ_s and ρ the density of the solid and air respectively (Rosin, Rammler and Intelmann, 1932). Such equations are not very reliable. They are based on assumptions which may not be valid considering the complex flow patterns within the cyclone. Agglomeration of small particles may occur which aids their removal. Large particles may rebound off the cyclone wall and become entrained again in the air stream and so not separated out. A typical cyclone efficiency curve is shown in *Figure 2.15*. More than 85% of particles above 20 μm are removed from the air.

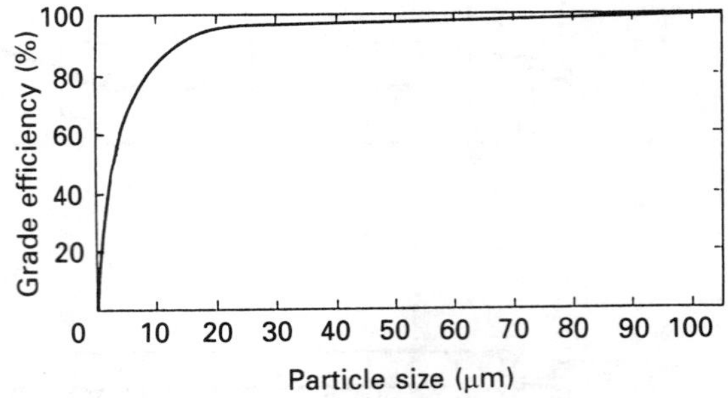

Figure 2.15 Typical cyclone efficiency curve (Masters, 1991)

However, only particles greater than 100 μm are completely removed. Two or more cyclones may be used where necessary. These may be arranged in series or parallel. In the first of these arrangements, the exhaust air from the first cyclone passes through the second. Larger particles are removed in the first unit and finer ones in the second. In the parallel arrangement the air from the drying chamber is divided and distributed into two cyclones. Four or more small cyclones can be arranged in parallel in a battery. Such an arrangement may be more effective in removing very fine particles. However, where possible, single cyclones are used as they lead to less handling and cleaning difficulties as compared to multiple systems. Other designs of cyclones, more complex than that depicted in *Figure 2.14*, are available. Cyclone separators may be used in combination with other air/powder separators, i.e. bag filters, electrostatic precipitators and wet scrubbers. Cyclone separators may also be used to separate liquid droplets from vapour, e.g. in evaporators (Green, 1984; Masters, 1985). *See also* Bag filter, Electrostatic precipitator, Spray drying, Wet scrubber.

D

Dehydrocanning
A method of preservation involving two stages. In the first stage, the food is partially dried. In the second stage, the partially dried food is filled into cans, sealed and heat processed in a retort. The principles are similar to those that apply in dehydrofreezing. This process is less widely used commercially as compared with dehydrofreezing. *See* Dehydrofreezing.

Dehydrofreezing
This is a two-stage method of preservation which involves partially drying the food and subsequently freezing the partially dried material. In dehydration of foods by heated air, a high proportion of the moisture is removed in the early stages of drying. Up to 90% of the moisture may be removed in the first half of the drying cycle. Most of the shrinkage of the tissues also occurs early in the drying cycle, particularly if a relatively low rate of drying prevails. Thus, significant reductions in weight and bulk can be achieved by partial dehydration. These can lead to savings in the costs of packaging, distributing and storing the frozen, semi-dried material, as compared with freshly frozen product. On the other hand, undesirable changes in colour and flavour occur during the later stages of drying, when the product temperature approaches the dry bulb temperature of the air. Such changes can be avoided by removing the product from the drier at an intermediate point in the cycle. However, the water activity of the partially dried material is, in most cases, not low enough to prevent microbial or chemical spoilage. It is therefore frozen to stabilize it, and treated as a frozen food. The extent of the drying necessary to obtain the best quality product depends on the food and the drying conditions. In many cases, a reduction of about 50% in the fresh weight of the food is brought about in the drying stage. There may be less thawing and drip losses from the frozen, semi-dried food, as compared with freshly frozen material. The freezing capacity of the plant may also be increased. The type of drier used in the first stage of the process should be one which promotes uniform drying. Cross-flow tunnel driers, conveyor driers, fluidized-bed driers and rotary driers have been used for this purpose. The belt-trough drier was designed with a view to producing a semi-dried product with a uniform moisture content. It is advisable to size grade the material prior to drying to assist in obtaining a uniform moisture content. Freezing may be achieved by any of the conventional methods. The method best suited to each particular product is used. Since this process is mainly applied to fruits and vegetables for catering purposes or for remanufacture, the partially dried material may be packaged in 2–25 kg quantities and frozen in freezing rooms or tunnels. As an alternative to hot-air drying, in the case of fruits, osmotic drying may be used in the first stage of dehydrofreezing (Luh and Woodruff, 1975; Salunkhe et al, 1991). *See also* Belt-trough drier, Conveyor drier, Dehydrocanning, Fluidized-bed drier, Osmotic drying, Tunnel drier.

Density
Mass per unit volume of a substance. Common units are kg m^{-3}. In the case of particulate solids, a distinction must be made between absolute density and bulk density. *See* Absolute density, Bulk density.

Dewpoint
The temperature at which a given air-water vapour mixture becomes saturated if cooled at constant absolute humidity. It is also known as saturation temperature.

Diaphragm pump *see* Pumping of foodstuffs.

Direct rotary drier *see* Rotary drier.

Direct-indirect rotary drier *see* Rotary drier.

Dispersibility of dried powders *see* Reconstitutability of dried powders.

Double-cone vacuum drier
This type of drier is used in the chemical and pharmaceutical industry but has very few applications in the drying of foods. It consists of a hollow vessel in the shape of a double-cone which rotates about a horizontal axis. The vessel is jacketed and heated by the introduction of steam or the circulation of hot water, or some other thermal fluid, through the jacket. It is connected to a vacuum system. It is suitable for drying heat-sensitive powders or granular materials which are tumbled within the rotating vessel. Heat is transferred to the particles mainly by conduction from the heated wall of the vessel. The feed is loaded into the vessel. The pressure within the vessel is reduced to the required level. The vessel is set in motion and heat is applied through the jacket. The pressures and temperatures used are similar to those used in vacuum shelf or vacuum band driers. It is not suitable for drying friable material which might break down as a result of the tumbling action (Anon, 1992). *See also* Vacuum band drier, Vacuum shelf drier.

Double drum drier *see* Drum drier.

Drum (film, roller) drier
A *single-drum drier* consists of a hollow metal cylinder rotating about a horizontal axis and heated internally by steam. The feed, which must be in the form of a solution, a suspension of solids or a paste, is applied as a thin film to the surface of the drum. This film dries as the drum rotates and the dry product is removed from the drum surface by a knife located ½ to ¾ of a revolution from the point of application of the feed. It is important that the film of feed on the drum surface be of uniform thickness. A number of techniques are used to apply the feed (*Figure 2.16*). The drum may dip into a trough containing the feed material (a). The contents of the trough may need to be continuously stirred. The feed effectively boils on to the drum surface. For feeds prone to frothing, this method may not be satisfactory, as the film formed may not be uniform. The feed in the trough may reach too high a temperature. This last effect can be reduced by using a shallow trough and continuously supplying it with cold feed. The feed may be picked up from the trough by a roller and may be cooled and then transferred to the drum surface by a second roller (b). The speed of rotation of the second roller may be adjusted to give a film of desired thickness. The temperature of the feed in the trough may be kept at a suitably low level. For feeds with a high starch content, a multi-roller system may be employed (c). The feed is introduced between the feed-box roller and the first of the applicator rollers. Gelatinization of the starch occurs and the feed takes up the form of a rotating cylinder. Each of the applicator rollers applies a layer to the drum surface until the required thickness is attained. Low-viscosity liquid feeds may be sprayed or splashed on to the drum surface. This method is not suitable for feeds that tend to froth.

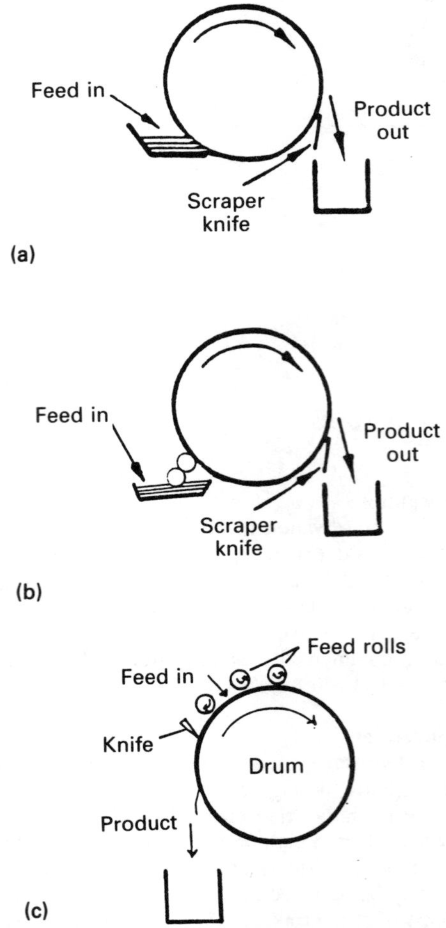

Figure 2.16 Single-drum driers with different feeding methods: (a) dip feed, (b) unheated roller feed, (c) multi-roller feed system for starchy products

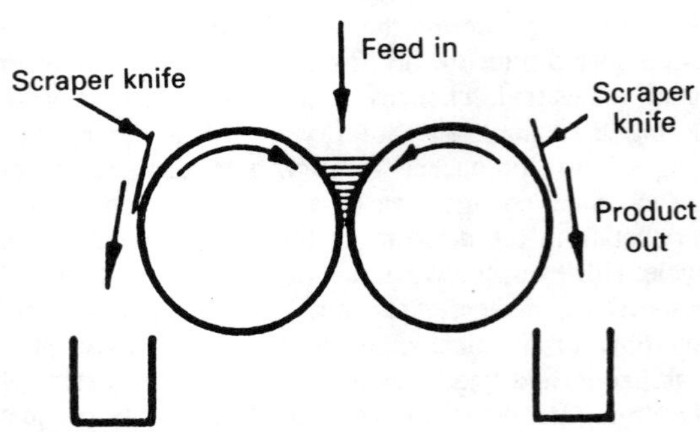

Figure 2.17 Double-drum drier

A *double-drum drier* consists of two drums situated close together and turning towards each other at the top (*Figure 2.17*). The feed is introduced into the trough formed between the drums. The clearance between the

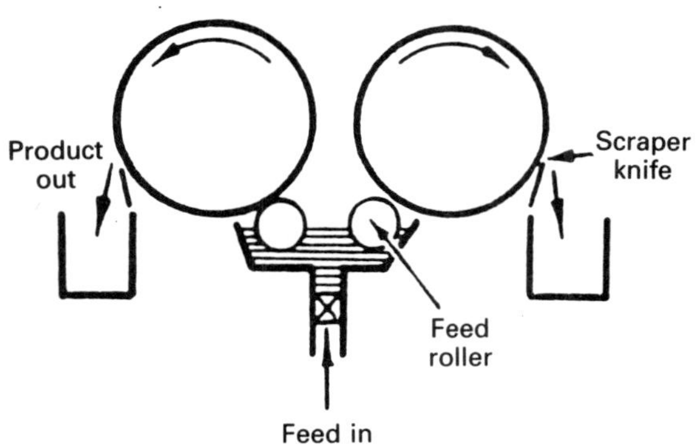

Figure 2.18 Twin-drum drier

drums is adjustable and influences the thickness of the film formed upon them. This design of drum drier is suitable for low viscosity liquids that are not very heat-sensitive. The liquid in the trough can reach quite a high temperature. The *twin-drum drier* also consists of two drums (*Figure 2.18*). However, there is a bigger gap between the drums as compared with the double-drum design, and they rotate away from each other at the top. The drums are fed from the same trough, either by dipping into it or by means of splashing devices. Twin-drum driers are suited to drying feeds containing crystalline solids or other relatively hard particles. Drums have diameters ranging from 0.15 to 1.5 m and lengths from 0.2 to 3.0 m. They rotate at speeds in the range 3–20 rpm. Drying times are in the range 2–30 s. Variable speed drives are usually fitted. Drums are made from cast iron, carbon steel, chrome-plated cast iron or stainless steel. The knives for product removal may be single blades extending the length of the drum with adjusting-screws at intervals to enable the pressure on the drum surface to be set. Alternatively, a number of blades, overlapping or butting on to each other, may cover the length of the drum. Each blade is adjustable. The drums are heated internally by pressurized steam. The steam pressure is selected to suit the heat sensitivity of the feed. Typically, pressures in the range 200–600 kN m^{-2} are used with a drum surface temperature reaching 110–165°C. It requires 1.1–1.3 kg of steam to evaporate 1 kg of water. Drum driers may be fitted with hoods, sometimes equipped with fans, through which the vapour is exhausted. They may be located inside vapour-tight or dust-tight enclosures for drying toxic, dusty or flammable materials. Many other special features have been added by various workers. A high-velocity stream of air may be directed counter-current to the drum rotation to increase the rate of drying and cool the film of feed on the drum by evaporation. For handling thermoplastic and sticky materials, a blast of chilled air may be directed at the product just before the knife. This can facilitate the removal of the product from the drum surface. In the case of highly hygroscopic products, low-humidity air may be introduced near the knife to facilitate removal and handling of the dry material.

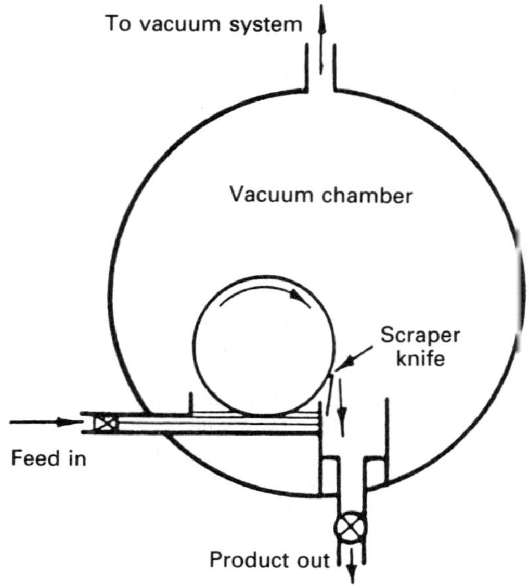

Figure 2.19 Vacuum-drum drier

When very heat-sensitive material is to be dried, a *vacuum-drum drier* (*Figure 2.19*) may be considered. Single- or double-drum driers may be enclosed within a vacuum-tight chamber, which is pumped down to a suitable pressure, so that drying can be carried out at a temperature appropriate for the material being dried. Chamber pressure may range from just below atmospheric, 100 kN m^{-2}, down to 5 kN m^{-2}, absolute. The drum(s) are heated internally with vacuum steam or hot water. Feeding arrangements similar to those used on atmospheric drum driers may be used. The dried product is removed from the drying chamber via a rotary valve or by a screw conveyor to two receiving vessels operated in parallel. While one receiver is under vacuum and receiving product, the other is being emptied. A vacuum is then drawn on the empty vessel which is then ready to receive product. The flow rate of feed and the adjustment of knives, spreader rollers or the nip between drums in a double-drum drier, are controlled from outside the vacuum chamber, through vacuum-tight stuffing boxes and glands. Vacuum-drum driers are expensive to purchase and maintain and are only used for very heat-sensitive food materials which can bear the relatively high drying costs. Such driers can be designed to operate under sterile conditions (Van Arsdel et al, 1973a; Williams-Gardner, 1976; Charm, 1981; Hall and Upadhyaya, 1986; Moore, 1987; Brennan, 1989).

Drum screen *see* Dry cleaning of raw materials.

Drum washer *see* Wet cleaning of raw materials.

Dry basis *see* Dry weight basis.

Dry cleaning of raw materials
Removal of contaminants under dry conditions gives rise to less microbiological and waste disposal problems than wet cleaning techniques. Dry cleaning may be used as the sole cleaning stage or as a preliminary step prior to wet cleaning. Dry cleaning methods include:

Screening
This involves separation on the basis of size. Screens may be of the flat-bed type. Two screens may be positioned one above the other. The apertures in the top screen are just large enough to let the food particles pass through, but prevent larger particulate contaminants from doing so. The apertures in the lower screen are just small enough to hold back the food particles but allow smaller particulate contaminants to pass through. Such screens may be stationary and inclined at an angle to the horizontal. The raw material flows down the top screen and separation is effected. Alternatively, the screens may be horizontal but are vibrating. In rotary-drum screens the raw material is introduced into a hollow rotating screen. The apertures in the screen are just large enough to let the food particles pass through, but retain larger particulate contaminants. The food particles may then be fed to a second rotating screen which retains the food particles but allows smaller particulate contaminants to pass through.

Brushing
The food items may be carried on conveyors beneath rotating brushes, or rubber fingers, which remove contaminants adhering to them.

Tumbling
Food particles may be tumbled in large, perforated cylinders to loosen dirt adhering to them. The contaminants are then removed by screening. Food particles may be carried on vibrating, perforated belts to loosen adhering material. Rotating abrasive discs may be used for a similar purpose.

Inclined belts
If spherical food particles, such as peas, are fed on to a moving inclined belt, the peas will roll down the belt while non-spherical particulate contaminants will be carried up on the belt and discharged from the top.

Aspiration (winnowing, air classification)
If food particles are introduced into a stream of air flowing at an appropriate velocity, light contaminants may be separated off (*Figure 2.20*). Dry cyclone separators may also be used for this purpose.

Magnetic separators
By passing the food particles over magnets, ferrous metal contaminants may be removed. The food particles may be cascaded or carried on belts over the magnets.

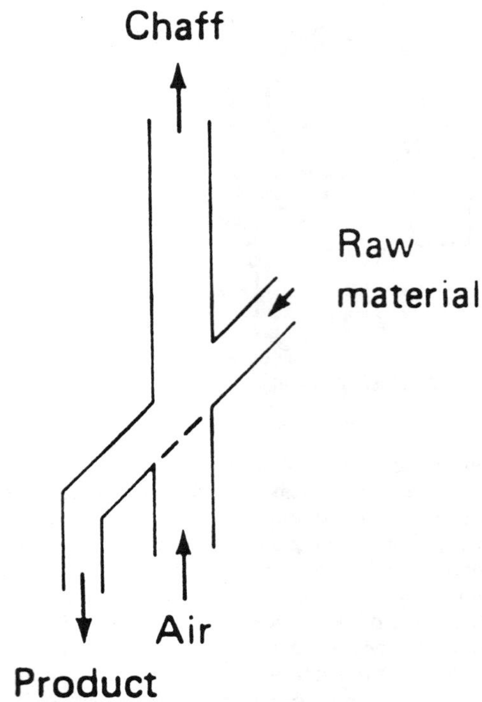

Figure 2.20 Principle of aspirator

Rotating magnetized drums or belts are also available. Both permanent and electromagnets are used.

Electronic metal detectors are available which will detect both ferrous and non-ferrous contaminants. X-ray detectors are used to locate stone, glass and metal contaminants in foods. Such detectors are used after the cleaning is complete or in later stages of processing.

When using dry cleaning techniques it is important to minimize the spread of dust, to prevent recontamination of the food, to safeguard the comfort and health of operatives and to reduce the risk of fire or explosion. The maximum concentration of dust in the atmosphere which is regarded as tolerable by personnel is 10 mg per litre. Higher concentrations can lead to explosion and fire. The fire and explosion hazards associated with fine powders (and dust) have been discussed elsewhere in this text, under safety. Good dust extraction equipment, dust-proofing of equipment, regular and thorough cleaning and spark-proofing of electrical equipment reduce the likelihood of explosion or fire. Explosion venting is the most common technique used to safeguard personnel and reduce damage to plant and buildings if an explosion does occur (Fellows, 1988; Brennan et al, 1990; Arthey and Dennis, 1991). *See* Cyclone separator, Safety aspects of drying, Wet cleaning of raw materials.

Dryeration
A method used for drying cereal grains which involves an initial drying stage, using heated air, followed by a

holding period and final drying in ambient air. This method relieves the stresses set up in the grain during the initial drying and reduces its brittleness compared to grain dried by conventional methods (Van Arsdel et al, 1973b). *See also* Cereal grains.

Dry weight basis (dwb)
A method of expressing the moisture content of a wet material based on unit mass of dry solids. If M is the mass of water and S the mass of dry solids in a sample, then the moisture content (dwb) may be expressed as M/S e.g. kg moisture per kg dry solids or as a percentage thus $(M/S) \times 100$. The relationship between moisture content, expressed as a percentage on a dry weight basis, M_{dwb}, and a wet weight basis, M_{wwb}, is:

$$M_{dwb} = \frac{M_{wwb}}{100 - M_{wwb}}$$

In the context of dehydration, moisture content is usually expressed on a dry weight basis, since the mass of dry solids does not change during drying. Hence weight/time curves will have the same shape as moisture content (dwb)/time curves. *See also* Wet weight basis.

Dynamic viscosity
For fluids which exhibit Newtonian flow, i.e. a linear relationship between shear stress and shear rate, dynamic viscosity is the ratio of shear stress to shear rate. Common units are $N\,s\,m^{-2}$ or Pl. Also known as absolute viscosity.

Dyne (dyn)
A metric unit of force equivalent to 10^{-5} N.

E

Economy
In the context of drying, this term is used to quantify the efficiency of utilization of fuel in a drier. It is usually expressed as mass of fuel required to evaporate unit mass of water, under specified conditions.

Electrical energy
When a potential difference is established, e.g. in a wire, and an electrical current flows through that difference, energy is dissipated as heat. The energy thus dissipated (E) when a current (I) flows through a potential difference (V) for a time t is given by the relationship:

$$E = V \times I \times t$$

Units of E are joules; V, volts; I, amperes and t, seconds.

Electrical energy supplied nationally is usually measured in kilowatt hours (kWh). 1 kWh is equivalent to 3.6 MJ.

Electrostatic precipitator
This is a device for removing powder particles from an air stream. The air containing the powder flows into a strong electrical field formed between two electrodes. A potential difference from 10 to 60 kV exists between the electrodes. The discharge electrode, on which the charge density is greater, gives off ions which combine with the solid particles. They obtain a negative polarity and move towards the other electrode which is earthed. When they contact this electrode the particles lose their charge. They are removed from the plate by a tapping device and fall into a hopper. The discharge electrode may be in the form of a wire located down the centre of a hollow cylinder which is earthed. A number of such cylinders would make up the precipitator. Alternatively, earthed plates may be mounted in series, parallel to each other, with the discharge electrodes suspended between them. They are best suited to removing fine powders and can recover up to 99% of all particles with diameters greater than 1 μm. They are expensive to purchase and install. They may be used in combination with other types of air/powder separators, e.g. the larger particles may be removed by a cyclone separator and the fines by a precipitator. Such precipitators are not widely used on food dehydration plant (Coulson et al, 1978; Green, 1985; Masters, 1991). *See also* Bag filters, Cyclone separators, Wet scrubbers.

Encapsulation
This is a process whereby a core material is encased or enclosed within an outer shell or coating. When relatively small capsules are being produced, less than 50 μm in diameter, the term microencapsulation is often used. The process has been applied to food flavouring materials and other food ingredients in recent years. Various techniques are used to form capsules including:

Spray drying
Where the core material is an oil, it is mixed with a polymeric substance such as gelatine, modified starch, vegetable gum, nongelling protein or dextrin. An emulsifying agent is added and the mixture homogenized to produce an oil-in-water emulsion. The emulsion is then spray dried. Both centrifugal and nozzle atomizers are used. During the atomization the oil moves to the centre of the droplets while the polymeric material forms a shell around it which subsequently dries in the chamber. The protective coating limits the loss of volatiles and oxidative changes in the core material during drying. This technique is widely used for food flavours. Some oxidation and volatile loss does occur, particularly in the case of citrus oils rich in terpenes. Flavours with excess volatiles and reduced content of oxygen-sensitive components have been developed for spray drying. Various designs of spray drier are used, including the integrated fluidized bed chamber. The dried capsules may be agglomerated to improve their handling and reconstitution properties. Water-soluble flavours and other food ingredients may be protected

during spray drying by mixing with hydrocolloidal material. However, they are not contained inside a well-defined shell but are uniformly dispersed within the dried particles.

Fluidized-bed drying
Solid particles may be encapsulated in a protective coating in a fluidized bed. The coating material may be a lipid or solutions of water-soluble or insoluble polymeric materials, similar to those used to encapsulate flavour compounds. A vigorously agitated fluidized bed or spouted bed is best. The coating material is sprayed on to the solid particles through nozzles. In the case of lipid coating materials, cool air is used to fluidize the particles and the coating material solidifies on to the surface of the particles. Where solutions of coating material are used, heated air is introduced into the bed and the solvent evaporates leaving a dry coating on the particles. On leaving the bed the coated product may be further dried and/or cooled prior to packaging.

Solvent dehydration
For encapsulating flavours, an emulsion is formed using similar materials and methods to those used in preparing them for spray drying. This emulsion is atomized directly into a polar solvent such as isopropanol, ethanol, glycerine or a polyglycol. These solvents extract moisture from the coating material. The capsules are separated from the solvent by filtration and dried under vacuum to a stable moisture content. This technique results in a well-protected core material with very good retention of volatiles. However, it is a relatively costly method as compared with spray drying.

Coacervation
A colloidal material such as gelatin is dispersed in water. The core material, such as citrus oil, is dispersed in the mixture by agitating it. The solubility of the gelatine in the water is reduced, e.g. by adding sodium sulphate, and two phases are formed, one rich in gelatine, the other much less so. The colloid-rich phase is in the form of an amorphous cloud. Under the proper conditions, the gelatin coalesces on to the surface of the oil droplets, forming small capsules. A hardening agent, such as glutaraldehyde, is added. This causes crosslinking in the coating material. The pH is adjusted and the capsules collected, washed and dried under vacuum. It is possible to encapsulate water-soluble materials in oil soluble coatings by dispersing them in non-polar solvents at high temperature. The coating material is dissolved in the solvent and subsequently brought out of solution by cooling the dispersion. The coating material deposits on the aqueous droplets forming capsules. Coacervation is particularly suited to producing very small capsules, below 1 μm in diameter. However, it is an expensive technique and has found very limited application in foods. Some specialized flavours are coated in this way.

Extrusion
A low-moisture melt of low DE malto-dextrin is prepared. An emulsifying agent and antioxidant is added to the core material and it is extruded through fine holes into a bath of cold isopropynol. The melt solidifies and is broken into small rods by agitation. These are recovered by filtration or centrifugation, mixed with anti-caking agent, and packaged. This process is very expensive and only used for high cost products.

Centrifugation
Core material is pumped through a central tube and coating material through an annular channel surrounding it. The tubes converge to a single orifice from which emerges one liquid column within another liquid column. The tubes are rotated around a vertical axis and, under the influence of centrifugal force, the column of liquid breaks up into droplets, consisting of core material covered with coating material. The droplets are collected in starch powder or liquid solvents. The starch absorbs excess liquid and coating material. The solvents may contain cross-linking agents which harden the coating.

Spray chilling
Small solid particles may be coated with lipid material, usually hydrogenated vegetable oil, by this method. The particles are mixed into the molten oil and the mixture is sprayed through a nozzle into a chamber where it contacts refrigerated air. The coating material solidifies, forming small spherical particles of encapsulated material. Vitamins and minerals are coated by this technique. Capsules produced in this way provide a controlled release of core material. The process is relatively expensive.

Applications for encapsulation include: flavours such as citrus oils, menthol, peppermint, oleoresins of spices and herbs, artificial flavouring materials; dry granular acids, coated with oil or malto-dextrin, for use in puddings and pie fillings and to directly acidify meat in 'fermented' sausage products; leavening agents for frozen doughs; vitamins; minerals and salt (Dziezak, 1988; Versic, 1988; Graves and Weiss, 1992). *See also* Agglomeration, Fluidized-bed drier, Spray drying.

Endothermic reaction or process
A reaction or process during which heat is absorbed. *See also* Enthalpy, Exothermic reaction or process.

Energy
The capacity to do work. *See* Kinetic energy, Potential energy, Thermal energy, Electrical energy.

Energy used in drying
Drying is one of the most energy-intensive unit operations applied to foods. It is required to supply the necessary sensible and latent heat to bring about the evaporation or sublimation of the water. However, the thermal efficiency of driers may range from 30% to 90%. Thus, there is considerable scope for conserving

energy in drying operations. This scope is greatest in driers employing heated air as the drying medium. In single-stage spray drying, for example, only about 55% of the energy supplied is used directly to evaporate water. Typically, 30% is lost in the exhaust air, 3–5% by radiation and the remainder in the dry product, and used in driving fans, atomizers and other items of ancillary equipment. The main factors which influence energy losses in hot-air drying are:

Preconcentration of the feed
In a single-stage spray drier up to 6000 kJ of energy may be required to evaporate 1 kg water. In a six-stage, multiple effect evaporator with mechanical recompression, this figure may be as low as 200 kJ. Thus, liquid foods and slurries should be preconcentrated to as high a solids content as possible before spray drying. The upper limit of solids concentration is usually determined by the viscosity of the concentrate. High viscosity may cause problems in pumping and, in particular, atomization of the feed in spray drying. The upper limits of viscosity are usually less for jet as compared with centrifugal atomizers.

Method of heating the air
Direct heating of the drying air by burning oil or gas in the air stream generally requires less energy than indirect heating with those fuels or steam heating. However, direct heating may lead to contamination of the food with the products of combustion and is usually confined to the use of natural gas as the fuel. Special low-NO_x burners are available which minimize such contamination. Well-designed and maintained burners and heat exchangers will reduce energy losses. In the case of steam heating the condensate may be returned to the boiler or used for other heating purposes. Electric heating of drying air is not usual except in small-scale equipment.

Radiation losses
Energy lost by radiation from the walls of drying chambers, ducting, etc. can be significant. The proportion of energy wasted in this way depends on the design of the drier, in particular its surface area to volume ratio. Such losses can be reduced by efficient insulation of all external surfaces. Insulation material should have a suitable cladding, often stainless steel, and this must be effectively sealed to avoid leakage of feed material into the insulation. In some spray-drying applications, the temperature of the inner surface of a well-insulated chamber may be so high as to cause the powder to stick to it. Reducing or removing the insulation may be necessary to avoid this problem. This can lead to excessive heat losses. One solution to this is to pass cool air through a jacket covering the outer surface of the chamber wall. The warm air leaving the jacket is then fed as part of the drying air into the drying chamber, thus conserving energy.

Operating conditions
The greater the difference between the air inlet and outlet temperatures from a drying chamber the more thermally efficient it is. The upper limit on inlet temperature is usually determined by the heat sensitivity of the feed material and the design of the drier. For example, a higher air inlet temperature may be used in a concurrent tunnel or spray drier as compared with a countercurrent design, because of the evaporative cooling which occurs at the wet end of the concurrent type. The lower limit for outlet temperature from a drying chamber is often determined by the moisture content required in the product. In some cases, the humidity of the outlet air may be the limiting factor. As this approaches saturation level the likelihood of condensation occurring in outlet ducts, fans, etc. is increased.

Recycling of air
Recycling part of the exhaust air from the drying chamber back through the air heater and into the chamber will result in energy savings. In some driers, under certain circumstances, up to 75% of the exhaust air may be recycled leading to energy savings of up to 20%. However, the more air that is recycled the higher the humidity of the incoming air to the drying chamber and hence the lower its water-holding capacity. Thus, these savings must be balanced against a reduction in evaporative capacity of the drier.

Recovery of heat from the exhaust air
By far the major mode of energy loss from hot-air driers is in the exhaust air. For example, outlet temperatures from spray driers are usually in the range 80–95°C, but can reach 100°C and above for certain applications. Very large quantities of air are exhausted from spray driers, at these temperatures, representing large energy losses. Some of this energy may be recovered by exchanging heat between the incoming and outgoing air streams. This may be accomplished directly using heat exchangers, usually finned tubular types, or thermal wheels. If the exhaust air contains particulate solids, for example air from a spray-drying chamber, fouling of the heat exchangers can be a problem. Provision must be made for the periodic cleaning of the heat transfer surfaces and/or solids separation equipment, such as cyclones or bag filters, should be introduced upstream of the heat exchanger. The recovery efficiency of heat exchangers is of the order of 60–70% while that of thermal wheels may be as high as 90%. Heat may be recovered indirectly from the exhaust air using liquid coupled systems. These consist of two liquid-gas heat exchangers. One of these is located in the exhaust air stream where heat is transferred to a thermal fluid from the air. The second is sited in the inlet air stream where heat is transferred from the thermal fluid to the air. The recovery efficiency of such exchangers is about 50%. Heat may be recovered from the exhaust air by means of a wet scrubber. A portion of the liquid feed to the drier is diverted to the scrubber where it washes particles from

the exhaust air and at the same time recovers some heat from that air. The warm solution from the scrubber is then fed to the drying chamber. The heat recovery efficiency of scrubbers is of the order of 60%. Heat pumps may also be used to recover energy from the exhaust air from driers. Both open-cycle and closed-cycle pumps can be used. This approach is not widely used in food dehydration plants.

Some energy is also lost in the product. The amount depends on the temperature at which the product is discharged from the drier. However, such losses are relatively small and normally cannot be recovered.

Multistage drying

In some hot-air drying systems, reduction in energy requirement may be brought about by drying in two or more stages. For example, in spray drying it is common practice to remove the product from the spray-drying chamber at a moisture content in the range 5–7% (wwb) and complete the drying in a fluidized-bed drier. The higher the moisture content of the product leaving the first stage, the lower the outlet air temperature from that stage may be and hence the higher the thermal efficiency of the chamber. The upper limit on the moisture content of the product leaving the first stage is usually dependent on the ease with which the partially dried powder can be removed from the chamber. Too high a moisture content can cause particles to stick to the chamber wall. Since most of the water is removed in this first stage, under efficient conditions, there is an overall saving in energy as compared with single-stage drying. The most common type of secondary drier used in two-stage spray drying is a vibrating fluidized-bed drier. The integrated fluidized bed drier (IFB) features two-stage drying in one chamber. The Filtermat drier is another multistage spray drier. In multistage drying, better control over product characteristics, especially reconstitution properties, is possible as compared to single-stage drying. Conveyor, fluidized-bed and pneumatic driers are also operated on a multi-stage principle. The energy required to evaporate a unit mass of water by drum-drying is of the same order as that used in air drying. Energy losses in drum-drying include that in the steam condensate leaving from within the drum, that in the vapour produced in drying and normally exhausted to atmosphere, that lost from the drum due to convection of air over the drum surface and that in the solids. Energy in the steam condensate may be used for other heating duties or recycled to the boiler. Since the vapour is exhausted at temperatures close to ambient and in a large volume of air it is not feasible to recover the energy from it. The losses due to convection and in the solid are relatively small and normally cannot be recovered.

The energy required to remove a unit mass of water by freeze drying is about twice that required by air drying. Energy is used to freeze the food initially and then to sublime the ice in the freeze drier. The heat required to sublime a unit mass of frozen water is the sum of the heat of fusion and the latent heat. Energy losses in freeze drying include that dissipated in the vacuum pumps, that lost to the environment from the condensers in the refrigeration system of the freezer and freeze drier, that lost by radiation and that in the product. Heat losses by radiation and in the product are relatively small. Heat losses from the vacuum pumps and refrigeration equipment are relatively large. Recovery of this energy is not easy (Cook and Lang, 1979; Flink, 1977; Strumillo, 1987; Strumillo and Lopez-Cacicedo, 1987; Anon, 1992). *See also* Air heating, Heat exchangers, Heat pumps, Insulation, Low-NO_x burner, Spray drying, Thermal efficiency, Thermal wheel, Wet scrubber.

Enthalpy

The sum of the internal energy and the product of the pressure and volume of a substance. Thus:

$$H = U + pV$$

where H is enthalpy, U internal energy, p pressure and V volume. A change in enthalpy H may be represented as:

$$\Delta H = \Delta U + \Delta(pV)$$

or

$$\Delta H = \Delta U + p\Delta V + V\Delta p$$

If the pressure is constant:

$$\Delta H = \Delta U + p\Delta V$$

From the first law of thermodynamics:

$$q = \Delta U + p\Delta V$$

so

$$q = \Delta H$$

where q is the quantity of heat liberated.

So, for a process that takes place at constant pressure, the change in enthalpy is equal to the amount of heat evolved or absorbed. If the enthalpy change is negative, heat is evolved and the process is exothermic. If the enthalpy change is positive, heat is absorbed and the process is endothermic. Very often enthalpy is expressed as a value above an arbitrarily chosen reference temperature. For water and steam enthalpy is taken as zero at the triple point of water 0.01°C (usually quoted as 0°C). For refrigerants the reference temperature is usually −40°C. Specific enthalpy is the enthalpy associated with unit mass of a substance. Common units for specific enthalpy are $kJ\,kg^{-1}$. Values for specific enthalpy of water and steam and refrigerants are published in thermodynamic tables and in the form of pressure-enthalpy or enthalpy-entropy diagrams. Thermodynamic tables for water and steam only are also known as steam tables (Haywood, 1968; Lewis, 1987).

Entropy

A measure of the degree of order or disorder in a system. The entropy of a substance in the liquid state is greater than in the solid state. The entropy of a substance in the gaseous state is greater than in the liquid state. When an object loses heat to its surroundings it decreases in entropy. The change in entropy may be expressed as:

$$\Delta S_1 = \frac{Q}{T_1}$$

where S_1 is the change in entropy, Q is the heat lost (Q is negative, sign convention) and T_1 is absolute temperature. The surroundings, however, increase in entropy, the change being expressed as:

$$\Delta S_2 = \frac{Q}{T_2}$$

where S_2 is the change in entropy and T_2 the absolute temperature of the surroundings. The total change in entropy is:

$$\Delta S_2 + \Delta S_1 = \frac{Q}{T_2} - \frac{Q}{T_1}$$

For a reversible process $T_1 = T_2$ and there is no change in entropy. For an irreversible or spontaneous process where $T_1 > T_2$ the total entropy change will be positive. So, in a spontaneous process, the total entropy always increases.

Entropy is usually expressed as a value above an arbitrarily chosen reference. For water and steam, entropy is taken as zero at the triple point of water 0.01°C (usually quoted as 0°C). For refrigerants −40°C is usually the reference. Specific entropy is the entropy associated with unit mass of a substance. Common units are kJ kg^{-1} K^{-1}. Values of specific entropy for water and steam and refrigerants are published in thermodynamic tables or in the form of entropy-enthalpy diagrams (Haywood, 1968; Lewis, 1987).

Environmental spray drier
A semi-closed-cycle spray-drying system in which toxic or odorous materials in the vented air are deactivated or deodorized in a special design of direct air heater. *See also* Semi-closed-cycle spray-drying system.

Equilibrium moisture content
The moisture content of a wet material which is in equilibrium with its surrounding atmosphere. When a wet material is introduced into an atmosphere of fixed temperature and humidity, it will gain or lose moisture until it comes into equilibrium with that atmosphere. Exposure to that atmosphere beyond that point will cause no further change in the moisture content of the material. When such a state of equilibrium exists, the moisture content of the material is known as its equilibrium moisture content at the temperature and humidity of the atmosphere. The relative humidity of the atmosphere which is in equilibrium with the wet material is known as the equilibrium relative humidity of the material at its equilibrium moisture content. Relationships between equilibrium relative humidities, or water activities, and equilibrium moisture contents, expressed graphically, are known as sorption isotherms.

In hot-air drying, the moisture content of the material being dried which is in equilibrium with the drying air, at a specified temperature and humidity, is the lowest value attainable under those conditions. Equilibrium moisture content, W_e, features in many drying equations. *See* General Principles of dehydration, Sorption isotherms, Water activity.

Equilibrium relative humidity
The relative humidity of the atmosphere which is in equilibrium with a wet material, with a specified moisture content and at a specified temperature. *See* Equilibrium moisture content.

Erg (erg)
A metric unit of energy equivalent to 10^{-7} J.

Evaporation (vaporization)
The conversion of a substance from a liquid to a gaseous state. This occurs initially at the surface of the liquid. In a boiling liquid, evaporation occurs throughout the bulk of the liquid. The latent heat of evaporation is the heat required to convert a unit mass of a liquid to a gas at a specified temperature.

Evaporation (concentration)
The unit operation in which the solids content of a solution is increased by boiling off the solvent. It is widely used in the food industry to: (i) reduce liquid volume to be stored, packaged and transported. For example, fruit juices are concentrated in their country of origin and transported in concentrated form to packaging plants in other countries where they are diluted prior to bottling or cartoning; (ii) pre-concentrate a liquid prior to further processing. For example, sugar solutions are pre-concentrated prior to crystallization; milk is pre-concentrated prior to spray drying, to reduce energy costs; (iii) extend the shelf life of a food liquid by reducing its water activity. For example, sweetened condensed milk is manufactured by evaporation of whole milk, with added sugar, to give a shelf stable product.

The temperature at which a solution boils depends on a number of factors including: (a) the external pressure. The lower the external pressure the lower the boiling temperature. When evaporating heat-sensitive liquid foods, it is common practice to do so under a pressure less than atmospheric, to reduce heat damage, i.e. vacuum evaporation; (b) the soluble solids content of the solution. The higher this is the higher the boiling temperature. Since the soluble solids content of a solution increases during evaporation, the boiling temperature will rise as evaporation proceeds; (c) the hydrostatic head. The higher this is the higher the boiling temperature. Thus in an evaporator, the boiling temperature of the solution will increase as the depth of the liquid increases. Factors (b) and (c) above have to be taken into account when designing an evaporator, to ensure that it will have the specified capacity and that overheating of the liquid does not occur (Brennan et al, 1990). *See also* Evaporator.

Evaporator (vacuum evaporator)
An evaporation system to concentrate a food liquid at a pressure less than atmospheric, is comprised of: (i) a heat exchanger, or calandria, to supply the necessary sensible and latent heat for evaporation. In the food industry, saturated steam is the usual heating medium; (ii) a separator in which the concentrated liquid is separated from the vapour; (iii) a condenser to condense the vapour, which is then removed from the system. This must be backed up by vacuum pumps or steam ejectors and extraction pumps or barometric legs to create and maintain the appropriate, sub-atmospheric pressure in the evaporator. There are many different designs of vacuum evaporator. The simplest is a covered, steam-jacketed pan, usually equipped with a stirrer connected to a separator and condenser. Such pans are used for boiling sugar solutions for confectionery or jam manufacture, concentrating tomato pulp and soup and sauce preparations. A group of evaporators have heat exchangers comprised of bundles of short tubes, arranged horizontally or vertically, surrounded by steam jackets, with disengagement spaces above. These rely on natural circulation of the liquid to promote heat transfer. In one such design, the heat exchanger is outside the vapour disengagement space as shown in *Figure 2.21*. To increase the rate of circulation of the liquid, a pump may be included in the circuit. This type of forced circulation evaporator is useful when concentrating viscous liquids. In another group of evaporators, the calandria consists of a bundle of long (3–15 m) narrow (25–50 mm diameter) tubes, arranged vertically, and steam-jacketed. When such an evaporator is operated correctly, a thin film of liquid travels up (climbing film evaporator) or down (falling film evaporator) the walls of the tubes. Rapid evaporation occurs and residence time in the heating zone is relatively short. A cyclone separates the entrained liquid from the vapour. This type of evaporator is best suited to concentrating very heat-sensitive, relatively low-viscosity liquids. In the plate evaporator, thin layers of the liquid rise and fall over the surfaces of vertical plates, heated by steam. Evaporation occurs at the plate surfaces. Vapour and liquid are separated in a centrifugal entrainment separator. In a mechanically wiped film evaporator, the calandria is in the form of a cylindrical or frustrum-shaped, jacketed shell, within which an axially mounted multi-bladed rotor is located. As the blades rotate, they continuously wipe the heat transfer surface, creating a thin, fast-moving layer of the liquid. Evaporation occurs in this thin film. Vapour and liquid are again separated centrifugally. This type of evaporator is useful for concentrating viscous liquids prone to foaming and/or scale formation. The evaporators described above may be operated singly on a batch or continuous basis. A number of evaporators operating in series may be used. To conserve energy, the vapour produced in the first evaporator is used as the heating medium in the second. This is possible if the boiling temperature of the liquid in the second evaporator is sufficiently low to maintain an adequate temperature difference for the transfer of heat. This is achieved by operating the second evaporator at a lower pressure than the first. This arrangement can be extended to involve more than two evaporators. Each evaporator is known as an effect and the procedure is known as *multiple-effect evaporation*. A triple-effect system is shown in *Figure 2.22*. Exhaust vapour from the final effect passes through a condenser and is removed by pumps or a barometric leg. Whereas a single-effect evaporator requires 1.10–1.30 kg steam to evaporate 1 kg water, a triple-effect system only requires 0.37–0.45 kg. In general, the more effects the better the steam economy. However, the capital cost of the plant rises rapidly as the number of effects increases. So for each situation, an economic balance must be achieved between reduced operating costs and increased capital costs. It should be noted that multiple-effect evaporation does not lead to a larger throughput than that obtainable with a single-effect system with a comparable heat exchange surface. Usually five to six effects are used,

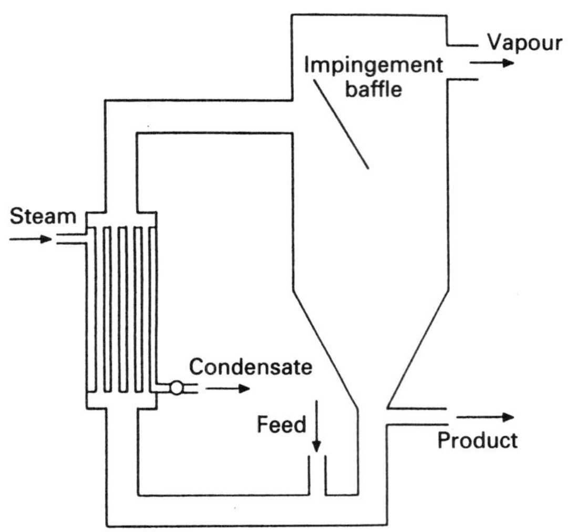

Figure 2.21 Natural circulation vacuum evaporator with external calandria (Brennan et al, 1990)

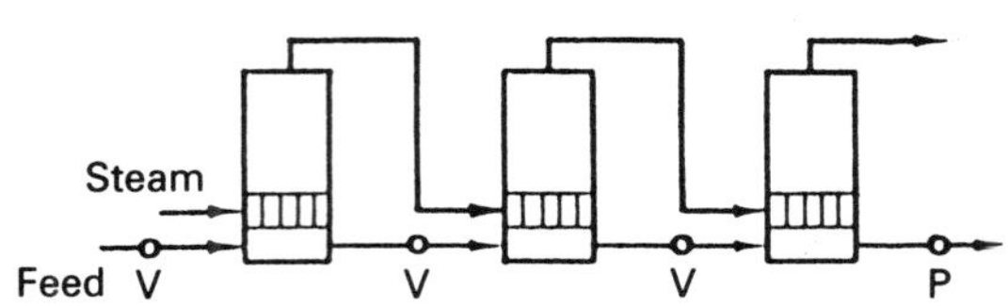

Figure 2.22 Triple-effect vacuum evaporator with forward feed (Brennan et al, 1990)

but, in certain circumstances, more than six may have economic advantages. In the triple-effect system depicted in *Figure 2.22*, the vapour and liquid travel concurrently through the effects. This is known as forward feeding. Other feeding arrangements are possible, each one with certain advantages and limitations. It is common practice to concentrate dilute food liquids by multiple-effect evaporation prior to drying them by drum drying, spray drying, vacuum drying or freeze drying. Steam usage per kg water evaporated is considerably less in multiple-effect evaporation than in drying by these methods (Brennan et al, 1990). *See also* Energy used in drying, Evaporation.

Exothermic reaction or process
A reaction or process during which heat is evolved. *See also* Enthalpy, Endothermic reaction or process.

Explosion hazard in drying *see* Safety aspects of drying.

Explosion puffing
A technique used to create a porous structure in partially dried, diced fruit or vegetable pieces, as a means of shortening the drying times. The diced material is dried to about 15–30% moisture content (wwb) by an appropriate hot-air drying method. The rate of drying is still reasonably high at this stage in the cycle. The pieces are then loaded into a puffing-gun. This is a cylindrical pressure vessel with a cover at one end, fitted with a quick release mechanism. The gun is sealed and pressurized by heating internally with superheated steam, or externally with a direct gas or oil flame. The temperature of the water in the food pieces rises above 100°C. The pressure in the gun is suddenly reduced by releasing the cover. The superheated water evaporates rapidly and the vapour formed flashes off. This creates a porous structure in the food pieces. These are then further dried to their final moisture content in the hot-air drier. Drying times may be reduced by a factor of two or three as compared with uninterrupted air drying. Because of the porous structure of the dried pieces, they reconstitute rapidly. The organoleptic quality of the reconstituted pieces may be better than conventionally dried material, because of the shorter drying times. Both batch and continuous explosion puffing systems have been described in the literature (Sullivan and Craig, 1984; Salunkhe et al, 1991). *See also* Foam mat drying, Vacuum puff drying.

F

Fan
This is a device for moving air or other gases. In so doing, it does not develop a pressure of more than 0.5 psi ($3.45\,kN\,m^{-2}$). Fans are used for a variety of duties including air ventilation and conditioning, supplying drafts to boilers and furnaces, conveying small particulate solids (pneumatic conveying) and supplying air for drying. There are two classes of fan – centrifugal and axial-flow. In a centrifugal fan, the air enters the fan housing from the side in a direction parallel to the axis of the rotor and exits from an outlet tangential to the circumference of the rotor. Thus the air direction changes through 90° on passing through the fan. In an axial-flow fan, the air passes straight through the fan in a direction parallel to the axis of the rotor (*Figure 2.23(d)*). Centrifugal fans may be further classified on the basis of the shape of the rotor blades into straight or radial blade (*Figure 2.23(a)*), forward-curved blade (*Figure 2.23(b)*), and backward-curved blade (*Figure 2.23(c)*) types. Each type has a characteristic performance curve (see *Figure 2.24*). The choice of a fan type for a particular duty will depend on the conditions, e.g. a backward-curved blade type is often preferred for supplying air to a spray drier where a large volume of air at low pressure is required. Axial-flow fans may also have different shaped impellers, i.e. discs, propellers and vanes. Both centrifugal and axial-flow fans are used to move air through drying systems. Cabinet and tunnel driers may be fitted with either type, conveyor driers usually feature axial-flow fans while spray driers, fluidized-bed and pneumatic driers are usually equipped with centrifugal fans. The ratio of the power developed by the fan to the power input through the shaft is known a fan efficiency. Values of fan efficiency range from 40–70% (Van Arsdel et al, 1973; Hall, 1979; Green, 1984; Masters, 1991).

Film drier *see* Drum drier.

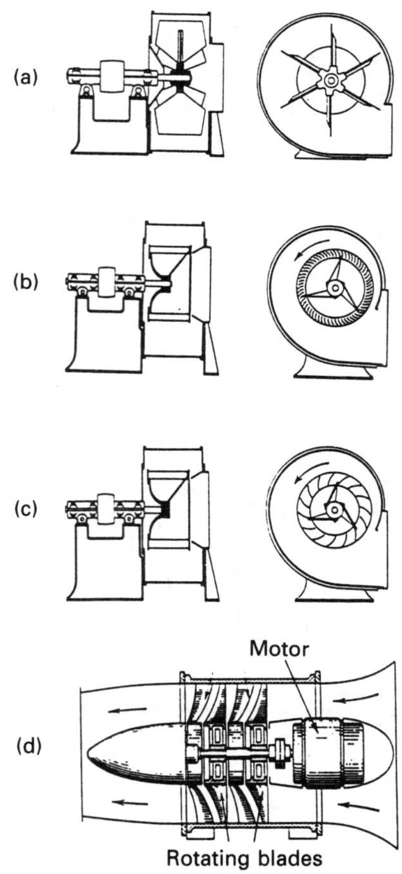

Figure 2.23 Centrifugal and axial-flow fans: (a) straight blade centrifugal fan, (b) forward-curved blade centrifugal fan, (c) backward-curved blade centrifugal fan, (d) two-stage axial-flow fan, adapted from Green (1984)

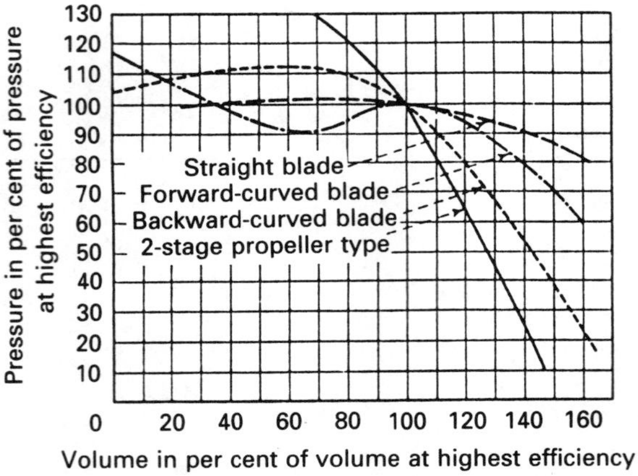

Figure 2.24 Performance curves for three types of centrifugal fan and a two-stage axial-flow fan (Green, 1984)

Filtermat drier

A spray-drying system which features multistage drying. The feed is dried down to 19–20% moisture (wet-weight basis) in the main drying chamber (*Figure 2.25*). This partially dried powder falls on to a moving perforated belt made of a polyester material. Secondary drying occurs as air is directed through the layer of powder and the belt. After a short holding period, the powder is subjected to a third drying stage followed by a cooling stage. The advantages claimed for this system as compared with a simple, single-stage spray drier include: energy savings, lower building height required, better control over product characteristics and flexibility in drying conditions. This makes the Filtermat drier suitable for a wide range of products including some that are difficult to dry by other means (Rheinlander, 1982; Filkova, 1987; Hansen, 1988). *See also* Energy used in drying, Spray drying.

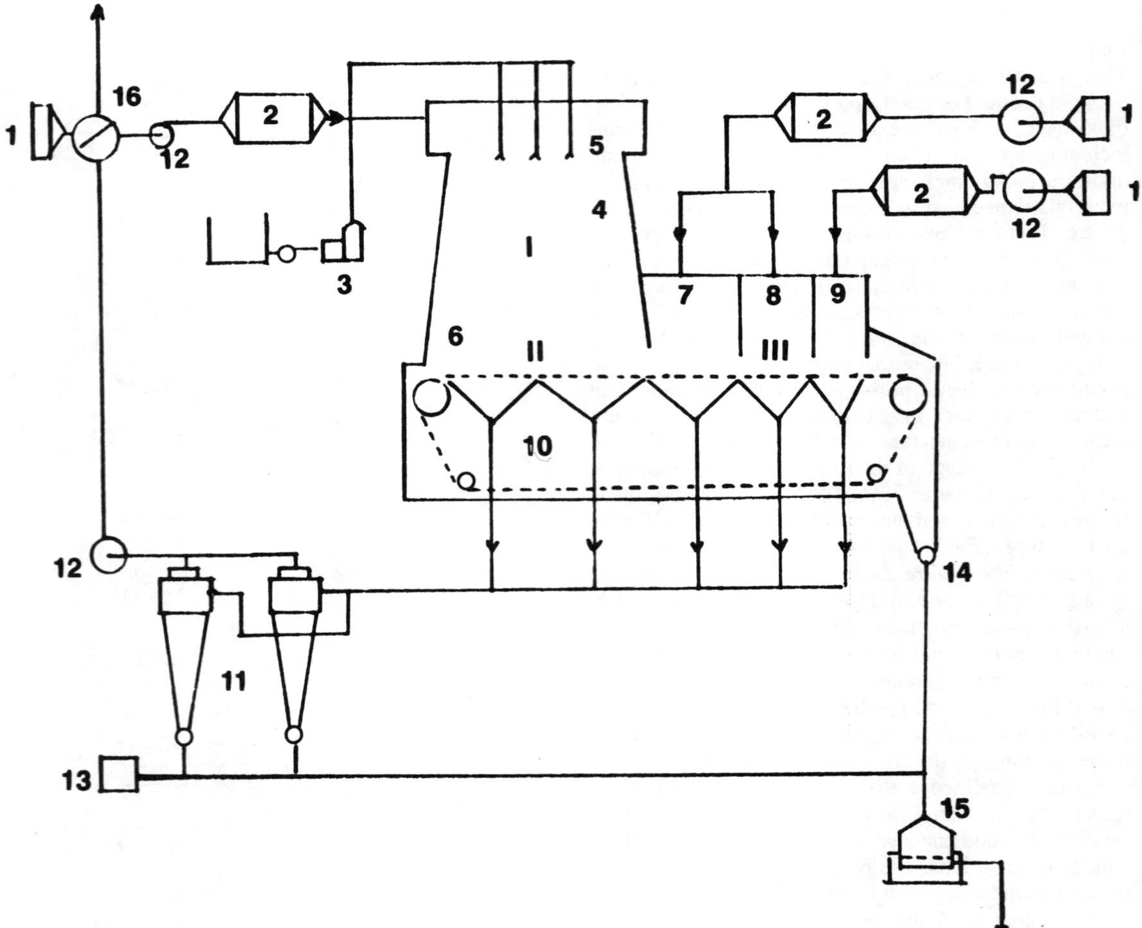

Figure 2.25 A three-stage spray-drying system (Filtermat), (1) air filter, (2) heater-cooler, (3) high-pressure pump, (4) nozzle system, (5) air distributor, (6) primary drying chamber, (7) retention chamber, (8) final drying chamber, (9) cooling chamber, (10) Filtermat belt assembly, (11) cyclones, (12) fan, (13) fines recovery system, (14) Filtermat powder discharge, (15) sifting system, (16) heat recovery system, (I) first drying stage, (II) second drying stage, (III) third drying stage (Brennan, 1989)

Fire hazard in drying *see* Safety aspects of drying.

Flame peeling
The skins of some vegetables have been removed in the past by the direct application of a gas flame. At temperatures up to 250°C for less than 1 minute, the skin was charred and subsequently removed by brushes or washers. This method of peeling was used for pimientos (chillis) and onions, but is seldom used now (Luh and Woodroof, 1975). *See also* Abrasive peeling, Hot water peeling, Lye peeling, Steam peeling.

Flash drier *see* Pneumatic drier.

Flat-sheet reverse osmosis or ultrafiltration membrane *see* Membrane concentration.

Flotation *see* Wet cleaning of raw materials.

Fluidized-bed drier
If air flows up through a bed of solid particles, supported by a perforated plate or grid, a pressure drop will develop across the bed due to the frictional force between the air and the particles. As the velocity of the air increases, the frictional force and hence the pressure drop will increase. When the frictional force overcomes the weight of the particles, the bed expands and begins to behave as a liquid. The bed is said to be fluidized at this stage. The velocity of the air at which fluidization first occurs is known as the incipient, or minimum fluidization, velocity. This velocity may be calculated for spherical particles by the relationship:

$$\nu_f = \frac{(\rho_s - \rho)g}{\mu} \frac{d^2\epsilon^3}{180(1 - \epsilon)}$$

where ν_f is the incipient velocity, ρ_s the density of the solid particles, ρ the density of the air at the operating temperature, g the acceleration due to gravity, μ the viscosity of the air at the operating temperature, d the diameter of the particles and ϵ the voidage of the bed.

As the air velocity is increased above the incipient value, bubbles of air pass up through the bed. These bubbles are small as they leave the perforated plate but coalesce to form larger bubbles as they move upwards through the bed, causing mixing of the particles. As the air velocity is further increased, the mixing becomes more vigorous and at one stage particles will leave the surface of the bed and then fall back on to it. At a still higher air velocity particles will be carried out in the exhaust air stream. The velocity at that point is known as the entrainment velocity. The entrainment velocity for spherical particles may be calculated from the relationship:

$$\nu_e = \sqrt{\left[\frac{4d(\rho_s - \rho)}{3C_d\rho}\right]}$$

where ν_e is the entrainment velocity and C_d the drag coefficient. The drag coefficient has a value of 0.44 when the Reynolds number is in the range 500–200 000.

If heated air, at a velocity between the incipient and entrainment values, is used to fluidize a bed of food particles containing moisture, then dehydration of the particles will occur. This is the principle of the fluidized-bed drier.

Not all food materials are suitable for fluidized-bed drying. The feed must be in particulate form, usually in the size range 20 μm–10 mm. The particles should be relatively uniform in size and shape, not very sticky or susceptible to damage due to abrasion. The optimum air velocity in a fluidized-bed drier will depend on the size and density of the feed particles and the extent to which fine particles are created in the bed due to abrasive action. It normally should be as high as possible to promote rapid mixing and high rates of heat transfer in the bed. The finer the particle size the lower the entrainment velocity. In order that high air velocities can be used, some loss of product by entrainment may be acceptable. The fines can be recovered from the exhaust air by means of cyclones or bag filters. Typical gas velocities for fluidizing particles with densities in the range 1000–2000 kg m^{-3} are given in Table 2.3. The support for the bed may be a simple perforated plate, but more complex designs incorporating jets or bubble caps are also used, especially for very fine powders. Driers may be operated under applied overhead pressure or reduced pressure by suitable siting of fans.

Table 2.3 Typical gas velocities for fluidizing particles with densities in the range 1000–2000 kg m^{-3}; adapted from Hovmand (1987)

Average particle size (μm)	*Velocity* (m s^{-1})
100–300	0.2–0.4
300–800	0.4–0.8
800–2000	0.8–1.2
2000–5000	1.2–3.0

Batch fluidized-bed driers are used for small-scale operations. In such driers, the temperature can be accurately controlled and good mixing can yield a product with uniform moisture content.

There are two main types of continuous fluidized-bed drier. In the plug-flow drier (*Figure 2.26(a)*), the length-to-width ratio of the path followed by the particles from feed point to exit is much greater than unity. The feed is introduced at the wet end of the drier and discharged over a weir at the dry end. The main advantage of this type of drier is that the distribution of residence time of the particles in the bed is relatively narrow and so the moisture content in the dried product is uniform. Its main drawback is that the range of feeds it can handle is limited. The feed to the drier must be readily fluidizable. In the well-mixed (or back-mixed) fluidized-bed drier (*Figure 2.26(b)*) the length-to-width ratio is unity or below. The particles are vigorously mixed and the bed

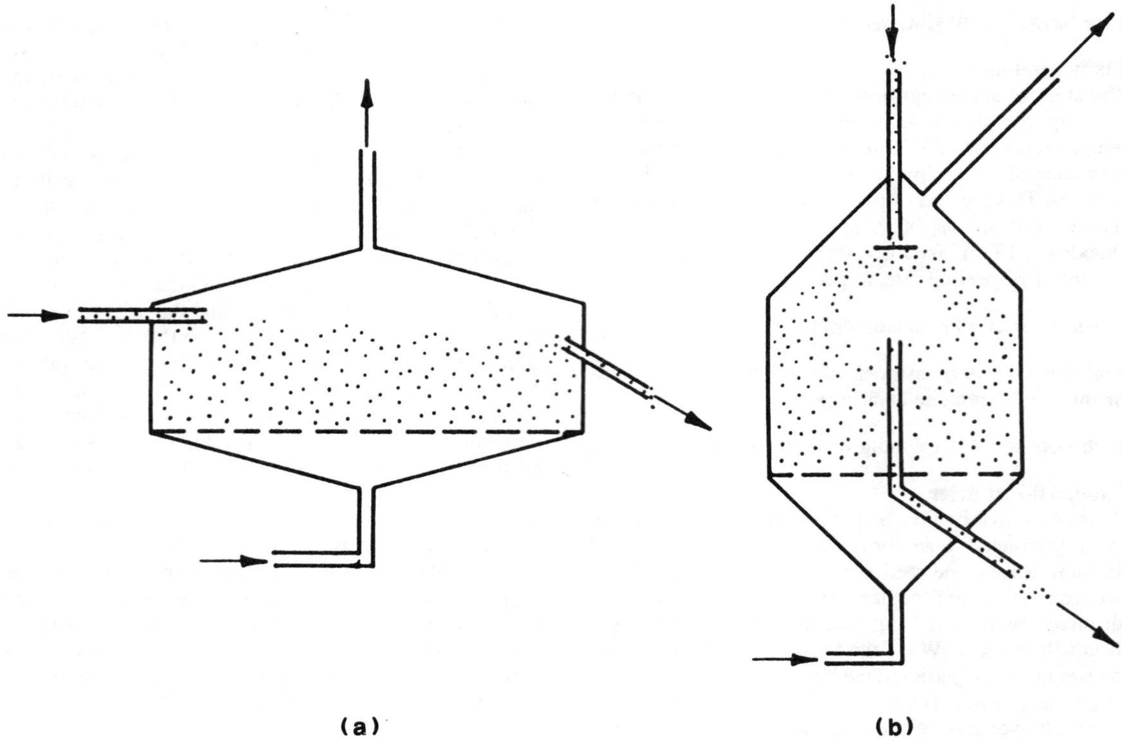

Figure 2.26 Types of fluidized-bed drier: (a) plug-flow type, (b) well-mixed type, adapted from Hovmand (1987)

will have a relatively uniform temperature. However, the distribution of residence time of the particles in this type of bed is wider than that in the plug-flow bed and so the moisture content of the product is less uniform. On the other hand, in the well-mixed bed the wet feed falls on to a bed of almost dry particles, is quickly dispersed and surface moisture rapidly evaporates. Thus, such a bed can handle feeds that are not so readily fluidizable. The well-mixed bed is best suited to removing surface moisture, whereas the plug-flow bed is good for removing bound moisture and so for drying to completion. For most food dehydration applications, the plug-flow design is used. Multistage fluidized-bed drying is also practised. In a two-stage system, the first stage may be a well-mixed bed, the second stage a plug-flow bed. Alternatively, two or more plug-flow beds may be used in series. This arrangement is most widely used in food dehydration. Up to six stages have been used. Such systems can lead to energy savings and better control over product quality as compared with single-stage drying.

Other designs of fluidized-bed driers are available including: the vibrated fluidized bed, the spin flash drier, the spouted bed and the toroidal bed. Further information on these driers will be found elsewhere in this text. Semi-closed-cycle fluidized-bed drying systems, including self-inertizing layouts, similar to semi-closed-cycle spray-drying systems, are available for drying materials that may form explosive powder/air mixtures (*Figure 2.27*). Closed-cycle layouts for handling flammable solvents are also in use. Fluidized beds may be built into spray-drying chambers to permit two-stage drying in one unit (Williams-Gardner, 1976; Charm, 1979; Hovmand, 1987; Boerson, 1988; Fellows, 1988; Brennan, 1989; Brennan et al, 1990). *See also* Integrated fluidized-bed spray drier, Closed-cycle spray-drying system, Semi-closed-cycle spray-drying system, Spin flash drier, Spouted bed drier, Toroidal bed drier, Vibrated fluidized-bed drier.

Foam mat drying

A technique developed to promote rapid drying of concentrated liquid foods, such as fruit juices, milk and coffee extract. The feed is converted into a stable foam by the addition of a foaming agent and the incorporation of air, or other gases, into the concentrate by batch or in-line mixing. The foam is spread on trays or moving mesh belts, in layers 2–3 mm thick, and dried in one or two stages in heated air. 1% or less of a foaming agent is added to the feed. Soya protein, albumin, fatty acid esters of sucrose and glycerol monostearate have been used as foaming agents. Moisture movement within the foam structure is rapid, leading to rapid drying. Drying times of the order of 1 h in air at 100°C have been

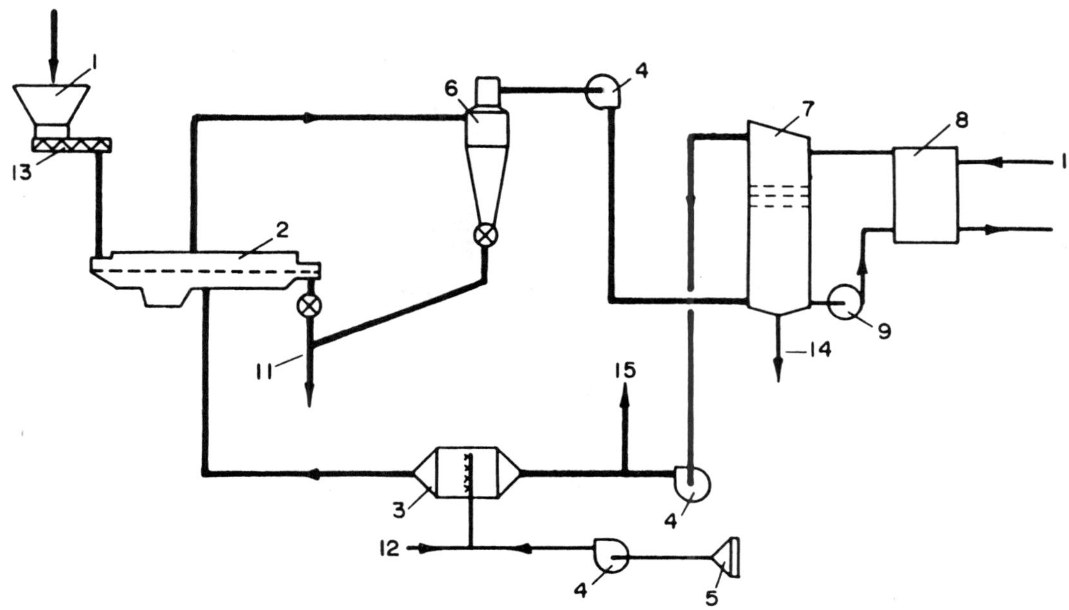

Figure 2.27 Self-inertizing fluidized-bed drying system: (1) product inlet, (2) vibro-fluidizer (vibrated fluidized-bed drier), (3) heater, (4) fan, (5) air filter, (6) cyclone, (7) scrubber-condenser, (8) heat exchanger, (9) pump, (10) cooling water inlet, (11) product outlet, (12) fuel gas inlet, (13) feeding device, (14) condensate outlet, (15) bleed-off (Hovmand, 1987)

reported. The porous nature of the dried product leads to its rapid reconstitution. To further accelerate drying, the foam may be extruded on to the tray or belt in strips. Another technique, known as cratering, has been used. The foam is spread on to a belt in a uniform layer and holes blown into it by controlled blasts of air directed up through the layer. Some throughflow of drying air then occurs. Nitrogen may be used as an alternative to air to form the foam (Van Arsdel et al, 1973; Charm, 1981; Salunkhe et al, 1991). *See also* Explosion puffing, Vacuum puff drying.

Force

An external agency which causes a change in the momentum of a body. Force acting on a body is directly proportional to the product of the mass of the body and its acceleration. Common units are $kg\,m\,s^{-2}$ or N.

Freeze concentration

This method of concentrating liquid foods involves the fractional crystallization of water to ice and removal of that ice to leave a more concentrated solution behind. Thus, it involves freezing the liquid followed by separation of the ice in a centrifuge, filter or wash column. The main advantage of this technique over vacuum evaporation is that relatively low temperatures prevail throughout the process and so the product does not suffer heat damage and loss of volatile aroma compounds is relatively low. However, the capital and running costs are relatively high, production rates are low and the degree of concentration attainable is limited to 40% or less. Suspended matter in the feed can act as heterogeneous nuclei and the concentrate may lack colour and in some cases flavour. It is important to limit the amount of soluble solids which leaves with the ice crystals, rather than the concentrate. The generation of large ice crystals, substantially free of inclusions of mother liquor, and the efficient separation of these from the concentrate, will minimize the loss of soluble solids. A representation of a typical freeze concentration system is shown in *Figure 2.28*. The liquid feed is frozen in a scraped-surface heat exchanger. The resultant slurry is fed to a mixing vessel fitted with paddle stirrers. Here the ice crystals increase in size. Separation of the ice from the concentrate may be effected by centrifugation, vacuum filtration, pressure filtration or washing. In a column the slurry enters at the bottom. The major portion of the concentrate drains through the crystals and is removed. A heater melts the ice crystals at the top of the column. Some of the melted water drains down through the bed of crystals, washing entrained soluble solids from the crystals. In another system, the bulk of the concentrate is removed by centrifugation or filtration and the crystals washed in a column to recover soluble solids. To conserve energy and to attain high production rates, two or more freezing and separation stages may be used. Freeze concentration has found only limited application in the food industry, including the concentration of fruit juices prior to freezing or freeze drying, the concentration of vinegar and alcohol in wines (Mellor,

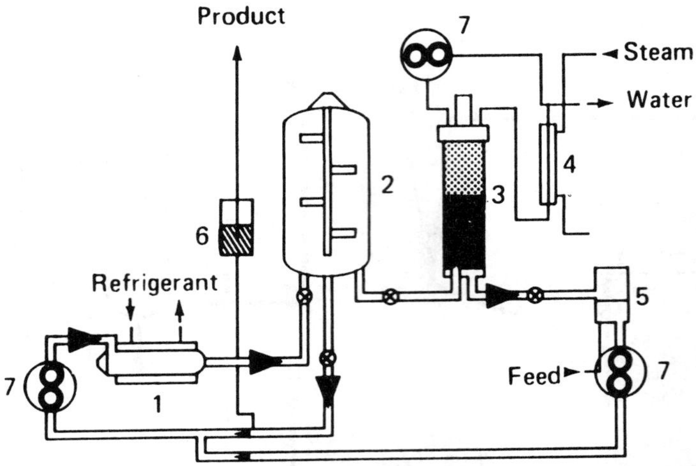

Figure 2.28 Freeze concentration plant: (1) scraped-surface heat exchanger, (2) mixing vessel, (3) wash column, (4) melting device, (5) storage tank, (6) expansion vessel, (7) pump (adapted from Kessler, 1986)

1978; Kessler, 1986; Fellows, 1988; Brennan et al, 1990). *See also* Evaporation.

Freeze drier *see* Cabinet freeze drier, Tunnel freeze drier, Vacuum-spray freeze drier.

Freeze drying (sublimation drying, lyophilization) A method of drying in which the food is first frozen and then the ice is caused to sublime to produce a dried product. Sublimation is brought about by creating and maintaining a water vapour pressure gradient between the atmosphere surrounding the frozen material and the ice front in that material. The vapour pressure of ice at −20°C is approximately 135 N m^{-2} (1 torr) absolute. Consequently, the water vapour pressure in the drying chamber must be reduced to this sort of level, or below, to promote sublimation. This reduction in water vapour pressure is usually achieved by reducing the total pressure in the drying chamber. Freeze-drying cabinets or tunnels are evacuated and maintained at pressures in the range 13.5–270.0 N m^{-2} (0.1–2.0 torr) absolute. Complete drying takes place in three stages. During the freezing stage, water is withdrawn from the hydrated components of the food by the formation of ice crystals or eutectic mixtures. The bulk of the water is removed from the food by the subsequent sublimation of these ice crystals. In the third stage, water strongly bound to the solid is removed, either in the freeze-drying chamber or in some other, less sophisticated, equipment.

During sublimation there is no movement of liquid water within the body of the food material and, consequently, no redistribution of soluble components. Little or no shrinkage of the tissues occurs so that solid food pieces retain their shape and size. The dried product has a light, porous structure which reconstitutes rapidly. The retention of volatile compounds, responsible for odour and taste, is high. The reasons for this are not fully understood. It may be due to the entrapment of volatiles in microregions within the solid or because the volatile compounds are very strongly adsorbed by the solid. Because relatively low temperatures prevail throughout the freeze drying of most foods, the development of heated or cooked flavours is not common. The organoleptic quality of freeze-dried foods is usually very good.

On the other hand, not all foods can be freeze dried satisfactorily. Foods whose structure is susceptible to damage during freezing will exhibit poor texture after reconstitution. Some denaturation of proteins may occur due to the concentration of solutes at the freezing stage. Freeze-dried foods readily adsorb water vapour and are also prone to oxidation because of their porous structure. They are also brittle and susceptible to mechanical damage. Consequently, sophisticated, relatively expensive packaging may be required, as compared with other forms of dried food. Overall freeze drying is relatively expensive, as compared with other methods of drying. A problem which can arise during the freeze drying of foods with a high sugar content is that of structural collapse. If the temperature of the dry material rises above its glass transition temperature its viscosity may be reduced to the extent that the structure will collapse. This may cause the material to partially melt and puff up. The resultant product is likely to be of poor quality.

Ideally, all of the moisture in a food to be freeze dried should be converted to ice. However, a proportion of the water present in foods is strongly bound to the solid and is effectively unfreezable. This water has to be removed at the end of the sublimation process, usually by vacuum drying in the freeze-drier chamber. With particulate foods it is possible to freeze in the drying chamber simply by reducing the pressure. If no heat is supplied to the food, evaporative cooling can cause it to freeze. Sublimation of the ice will commence when the water vapour pressure in the chamber is reduced to a level less than that of the ice. Controlled heating may then be applied. However, to evaporatively freeze the food, up to 20% of the moisture may have to be removed from the unfrozen food. This can result in such undesirable changes in the food that subsequent freeze drying is not warranted. In the vacuum-spray freeze drier, liquid food is sprayed into a large vacuum chamber, surrounded by a refrigerated coil. The droplets evaporatively freeze and subsequently dry, by sublimation, in the chamber.

However, food is usually pre-frozen prior to its introduction into the freeze-drying chamber. The method of freezing best suited to each food should be used. This is usually identified experimentally. Plate freezing, blast freezing, immersion freezing and two-phase freezing, usually with liquid nitrogen, have all been used. There is some evidence that the rate of freezing can influence the rate of drying. Slow freezing, leading to the formation of large ice crystals, may promote higher rates of drying as compared with quick freezing. However, other effects of freezing, such as structural changes, usually influence the choice of freezing method, rather than the drying rate. Frozen food temperatures range from −10 to −50°C. Freeze drying may be carried out on a batch principle using a single cabinet freeze drier, semi-continuously using a number of cabinets with shared vacuum and heating facilities or continuously using a tunnel freeze drier (Cotson and Smith, 1963; King, 1971; Van Arsdel et al, 1973; Goldblith et al, 1975; Gutcho, 1977; Mellor, 1978; Liapis, 1987; Brennan, 1989; Dalgleish, 1990; Brennan et al, 1990). *See also* Cabinet freeze drier, Glass transition, Multicabinet freeze drier, Tunnel freeze drier, Vacuum-spray freeze drier.

Freezing

A method of food preservation in which the temperature of the food is reduced to a level at which a large proportion of the water present in it is converted to ice. The frozen food must be held at that temperature, or below, throughout storage and distribution until it is thawed and prepared for consumption. The extension of the shelf life is a result of a combination of (i) the low water activity in the frozen product and (ii) the low temperature at which the frozen food is held during its life. The reduction in water activity arises from the immobilization of the water when converted to ice and the increase in soluble solids content of the water which remains unfrozen. This combination of low water activity and temperature has the effect of inhibiting the growth and development of spoilage microorganisms, the activity of enzymes and the rate at which chemical reactions occur. Thus, the rate at which undesirable changes in the organoleptic and nutritional quality of the frozen food occur is greatly reduced as compared with

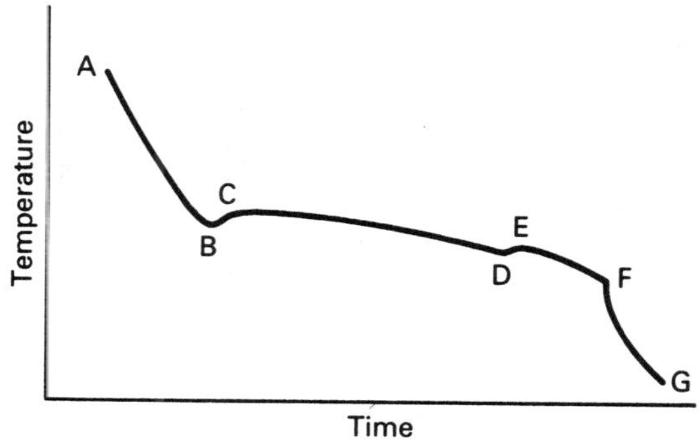

Figure 2.29 Typical freezing curve

the unfrozen material. However, some changes do occur very slowly during the freezing and frozen storage of foods. Because of differences in chemical composition and structure, different foods may exhibit different behaviour during freezing. In *Figure 2.29*, a typical pattern of change of temperature at the slowest cooling point (thermal centre) in a sample of food during freezing is presented. In the first stage, A-B, the food cools down to below its freezing point. The water may remain in the liquid state below the freezing point, i.e. it supercools. Once ice crystals start to form, the temperature rises to the freezing point, B-C. In the third stage, C-D, the water continues to freeze. The temperature falls slowly, due to an increase in the soluble solids content of the unfrozen material. At point D one or more of the solutes reaches the supersaturated state and crystallizes out and the temperature rises a little, D-E. During stage E-F the water and solutes continue to crystallize and the temperature continues to fall. At point F the crystallization is complete and during stage F-G the temperature of the solid/liquid mixture falls further and approaches that of the freezing medium. The amount of unfrozen water which remains at this stage varies from food to food but is usually in the range 5–25% of the total moisture content. The rate of freezing can affect the quality of the thawed product. A high rate of freezing (quick freezing) is desirable in the case of most foods. However, there are some foods which benefit from relatively slow freezing, e.g. strawberries, carrots, beans and some egg products. Very fast freezing, e.g. by immersion in liquid nitrogen, may lead to the break-up of the food due to the setting up of stresses within its structure. Freezing is the first stage in freeze drying and the final stage in dehydrofreezing (Fellows, 1988; Brennan et al, 1990). *See also* Dehydrofreezing, Freeze drying, Freezing methods.

Freezing methods

Freezing of food may be brought about by contact with a cooled solid surface, gas, liquid or a two-phase coolant (liquid/gas).

Freezing by contact with a cooled solid surface
The plate freezer employs this principle. Hollow refrigerated metal plates are mounted parallel to each other and the product to be frozen is placed between them. The space between the plates may be altered by a hydraulic or pneumatic system so that, after loading, the plates may be brought into close contact with the product. Horizontal plate freezers are used to freeze food in rectangular packages. They may be operated on a batch or semi-continuous principle. Vertical plate freezers are used to freeze unpackaged deformable materials such as meat and fish. These materials are formed into blocks by closing the plates on to limit stops prior to freezing. When freezing is complete, the plates are opened and heated to effect release of the frozen product and to defrost the plates. Ice cream is partly frozen in a scraped, or wiped, surface heat exchanger, with a refrigerated jacket. The partly frozen product is then packaged and freezing completed in a plate or blast freezer. Another type of continuous, cooled surface freezer consists of a moving, stainless steel band on the upper surface of which the product is placed. The underside of the band is sprayed with refrigerated brine. This type of freezer is suitable for thin products such as meat patties.

Freezing by contact with a cooled gas (blast freezing)
Refrigerated air at a temperature in the range −20 to −40°C is commonly used as a freezing medium. Foods of irregular shape may be frozen in blast freezers and individually quick frozen (IQF) products produced. Blast freezers may operate on a batch or continuous principle. A batch freezer consists of a well-insulated cabinet, equipped with a refrigerated air cooling system and fans. The product is loaded on to trays which are mounted on trolleys. The trolleys are placed in the cabinet where the refrigerated air flows mainly between the trays across the surface of the product. The equipment is similar in principle to a cabinet drier. Continuous blast freezers are of the tunnel, conveyor or fluidized-bed type. In tunnel freezers, the food is carried on trays, mounted on trolleys, through a well-insulated tunnel. Refrigerated air flows, mainly between the trays, across the surface of the product. The air may flow concurrently or countercurrently parallel to the direction of movement of the food. Alternatively, it may flow in a direction perpendicular to the product, i.e. a crossflow tunnel. The latter design has been more widely used in recent years. In a conveyor freezer, the food is loaded on to a perforated conveyor belt which carries it through a tunnel. Refrigerated air flows up or down through the product on the belt. The principle of the fluidized-bed freezer is similar to the fluidized-bed drier. Refrigerated air is blown upwards through a bed of food particles at a velocity sufficient to fluidize them. The bed may be supported on a moving perforated belt or vibrating deck to impact a forward motion to the particles. Bed depths of c. 15 cm are used. Peas, sliced beans, Brussels sprouts, potato chips (French fries) and berry fruits are frozen by this method. When blast freezing unpackaged food, some moisture will evaporate from the food and deposit as ice on to the refrigerated surfaces. It is necessary to incorporate a defrosting facility into such plants.

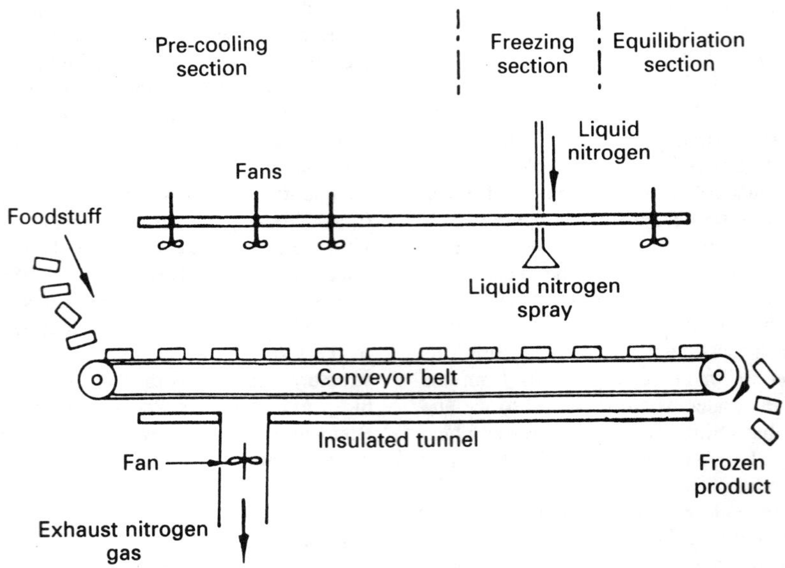

Figure 2.30 One type of two-phase freezer: liquid nitrogen is sprayed on to the food where it evaporates, freezing the food. The cold nitrogen gas moves upstream precooling the food prior to it contacting the liquid nitrogen spray (Brennan et al, 1990)

Freezing by contact with a cooled liquid (immersion freezing)
Rectangular blocks or irregularly shaped pieces of food may be frozen by immersing them in refrigerated liquids such as brines or syrups. IQF products can be produced in this way. In the case of unpackaged foods, problems of contamination arise. The product may be contaminated with the freezing liquid and so the latter must be edible and safe. On the other hand, the freezing liquid may be contaminated with fluids and soluble matter from the food. This makes long-term recirculation of the liquid difficult and so may increase the cost of the process. If suitable packaging is used this problem does not arise. Residual liquid can usually be washed off the surface of the packaged product.

Two-phase freezing (cryogenic freezing)
A freezing system in which the freezing medium is a subliming solid (solid carbon dioxide) or boiling liquid (usually liquid nitrogen). Pressurized liquid carbon dioxide may be sprayed through nozzles to atmospheric pressure, where it forms a mixture of cold CO_2 gas and solid CO_2 'snow'. This can be made to contact the food, thus freezing it. This system may be used as the main freezing source or in combination with blast freezing. Liquid nitrogen boils at –196°C at atmospheric pressure. It can be used for very rapid freezing. Using it as an immersion fluid is not widely practised. It is more economic to use the cold gas produced when the liquid gas boils to precool the product and to complete the freezing with a spray of the liquid gas. The principle of one type of two-phase freezer, using liquid nitrogen gas, is shown in *Figure 2.30*. The supply of liquid nitrogen to the spray nozzles is controlled to control the rate of freezing. Any liquid gas not vaporized after spraying can be collected in the bottom of the tunnel and recycled to the nozzles. The cost of this method of freezing is very dependent on the cost of the liquid gas itself (Fellows, 1988; Brennan et al, 1990). *See also* Cabinet, Tunnel, Conveyor and Fluidized-bed driers.

Froth flotation *see* Wet cleaning of raw materials.

Fuel efficiency
This is a measure of the efficiency of utilization of energy, or fuel, in a drier. Fuel efficiency, E_f, differs from thermal efficiency in that it takes into account the performance of the fuel burner as well as the drying chamber. E_f may be defined as:

$$E_f = \frac{\text{Energy used to evaporate water}}{\text{Energy content of fuel used}} \times 100$$

The term 'economy' is also used in a similar context but this is usually expressed as mass of fuel required to evaporate a unit mass of water, under specified conditions. *See also* Thermal efficiency.

G

Gas-ballasting (air-ballasting) of vacuum pumps
Oil-sealed vacuum pumps are used in series with refrigerated condensers to create and maintain the low pressure in cabinet and tunnel freeze driers. Although most of the water vapour produced by sublimation condenses as ice on to the condenser, some will reach the vacuum pump(s). This may condense in the pumps and result in failure of the oil seal and a reduction in the performance of the pumps. By providing a controlled bleed of filtered air into the pumps, the water vapour pressure may be reduced sufficiently to prevent the vapour condensing in the pumps. This measure is known as gas-ballasting. *See also* Cabinet freeze drier, Vacuum pump.

Gas constant
The proportionality constant in the law for ideal gas behaviour:

$$V = \frac{MRT}{p}$$

where V is volume, p absolute pressure, M mass and T absolute temperature.

Gauge pressure
Pressure above atmospheric pressure. Gauge pressure = absolute pressure – atmospheric pressure. In British units gauge pressure is often identified by the letter g, e.g. psig = pounds per square inch gauge, as distinct from psia = pounds per square inch absolute.

Gear pump *see* Pumping of foodstuffs.

Glass transition
Many dehydrated foods contain amorphous components in the form of a glass. This is a non-equilibrium state with a higher energy relative to the corresponding equilibrium state. If the temperature of a material in this state rises above a certain critical value then it transforms into a rubber. This phenomenon is known as glass transition and the temperature at which it occurs is the glass transition temperature. This transition results in an increase in mobility in the rubbery state and this in turn can lead to changes in physical and chemical properties of the material. Carbohydrates, including sugars, starch and hemicelluloses, can exist in the amorphous glassy state in dried foods. Low molecular weight sugars in the glassy state are usually extremely hygroscopic and have low glass transition temperatures. This can lead to problems in processing and handling of such materials. The addition of high molecular weight compounds to them can lead to an improvement in their processing and handling characteristics. Proteins, including gelatine, elastin, gluten, glutenin and lysozyme, are also found in the amorphous glassy state in dried foods. In the dry state they have relatively high glass transition temperatures. Water acts as a plasticizer of amorphous food components and as a result an increase in moisture content, even by small amounts, usually causes a decrease in glass transition temperature in amorphous materials. At temperatures below the glass transition temperature, most dry amorphous materials are quite stable. Crystallization of sugars or starch will not occur. Diffusivities in such materials are relatively low, which accounts for low drying rates towards the end of a drying cycle and retention of some volatiles. Transition to a rubbery state by raising the temperature above the glass transition temperature results in a decrease in viscosity of such materials. This can cause collapse of the structure of the materials and an increase in their stickiness. Structural collapse can occur during freeze drying of some foods, particularly those with high sugar content, if the temperature is high enough to cause a significant reduction in viscosity. This can lead to some melting of the material and a product of poor quality (King, 1973; Lorentzen, 1974). Stickiness in dry material occurs immediately after the completion of the transition from glass to rubber. This can cause handling problems in the case of dry food powders. It can also cause difficulties during the drying operation itself.

Brennan et al (1971) studied the effect of wall temperature on the deposition of powder on the wall of a laboratory spray drier when drying concentrated orange juice. They found that when the wall temperature was kept below the *sticky point temperature* of the powder, deposition was reduced. Presumably, the *sticky point temperature* was equivalent to, or slightly above, the glass transition temperature of the powder. The addition of liquid glucose to the concentrate also reduced wall deposition and improved the handling characteristics of the powder. Thus the phenomenon of glass transition is important in the context of food dehydration. During the drying of foods with a high sugar content, the glass transition temperature of the material should not be exceeded, to avoid structural collapse or deposition due to stickiness. Dry powders should be maintained in the amorphous glassy state by handling and storing at low temperature and humidity (Slade and Levine, 1991; Roos, 1992). *See also* Freeze drying, Spray drying.

Grading of raw materials
This preparatory operation involves selecting raw materials which are suitable for subsequent processing, in the context of this work for dehydration, on the basis of their overall quality. Thus, it usually involves the simultaneous evaluation of a number of properties. These may include: *size and shape* – e.g. of fruits, vegetables and cereals; *colour* – e.g. of fruits, as an index of maturity in bananas, apples and citrus fruits; *freedom from blemish* – e.g. bruises on fruit, blood spots in eggs, insect holes in beans and cereals; *freedom from contaminants* – e.g. insect parts in cereals, toxic metals in shellfish, spray residues on fruit; *texture* – as an index of maturity and freshness, e.g. peas, apples, melons; *taste and aroma* – as an index of maturity and freshness, e.g. pears, milk, meat. Thus, many of the techniques used in sorting foods for processing may also be used as part of a grading scheme. These include size, shape and colour sorters. In addition, samples may be taken, on a statistical basis, for chemical or physical analysis or sensory evaluation. Very often, grading is best carried out by trained human operators who assess a number of characteristics simultaneously. Very often the food units are carried on belts or rotating tables past the graders who visually assess their quality and remove unacceptable items manually or activate rejection mechanisms. Permanent colour standards may be made available to the graders. These may consist of colour charts, e.g. for judging the ripeness of bananas, coloured cards, e.g. for apples, or models, e.g. spheres of plastic dipped in lacquer for cherries. Meat carcases may be examined by inspectors for size, shape, fat distribution and evidence of disease. Eggs may be candled, i.e. examined in front of a strong light in a darkened booth for evidence of blood spots, rot or other undesirable conditions. The methods of sorting on the basis of size, shape and colour, mentioned above, may be used instead of, or in combination with, manual grading. Samples of raw materials may be taken to a quality control laboratory for chemical, physical, sensory or microbiological testing. Examples of chemical tests include the measurement of pH or acidity of milk, alcohol insoluble solids of peas and beans and fat content of meat. Physical testing usually involves the measurement of texture, e.g. using the pea tenderometer or fruit pressure tester. Testers or taste panels may be on standby to assess the organoleptic characteristics of the samples. Microbiological tests include total counts of viable microorganisms, detection and counting of particular microorganisms such as coliform or salmonella bacteria, yeasts or moulds. Many rapid test methods have been developed so that the results of laboratory testing can be available quickly to minimize hold-ups in the processing line (Brennan et al, 1990). *See also* Cleaning of raw materials, Sorting of raw materials.

Gray (Gy)
An SI unit of adsorbed dose of ionising radiation equivalent to $J\,kg^{-1}$.

Greenhouse effect *see* Solar drying.

H

Heat exchanger
A device in which heat is transferred from a hot fluid to a colder fluid through a solid, usually metal, wall. The design of the heat exchanger will depend on the nature of the fluids. Heat exchangers used to heat air entering a drier are usually of the finned tube or coil type. In the case of steam-heating of air, the saturated steam is introduced inside the tubes while the air is forced over the fins fitted to the outside of the tubes, within an insulated metal casing, by means of fans. The purpose of the fins is to increase the area through which heat is transferred as compared with a plain tube. The steam condenses inside the tubes, giving up its latent heat, which is transferred to the air outside the tubes by virtue of the temperature difference between the steam and the air. A number of finned tubes would be located inside the insulated metal casing to provide the necessary area for heat transfer. It is important that the condensate formed inside the tubes be removed as quickly as possible. This is achieved by means of steam traps. The efficiency of such a steam heater may be expressed as follows:

$$E_h = \frac{C_H G\ (\theta_o - \theta_i)}{M_c(H_s - H_c)}$$

where E_h is the efficiency of the heater, C_H the humid heat of the air, θ_o and θ_i the outlet and inlet temperature of the air respectively, G the air flow rate, M_c the flow rate of condensate, H_s and H_c the enthalpy of the steam and condensate respectively. The efficiency of such heaters may be as high as to 99%. Steam heaters should be well maintained as they are prone to corrosion, distortion and leakage.

Similar heaters may be used for indirect heating with fuel oil. The combustion gases flow over or through the tubes and heat is exchanged between them and the air. Access must be provided to permit periodic cleaning of the tubes. Heater efficiency may be expressed as follows:

$$E_h = \frac{C_H G\ (\theta_o - \theta_i)}{FQ_r}$$

where F is the rate of combustion of the fuel and Q_f its calorific value. Such heaters have efficiencies in the range 70–85%.

Indirect gas heaters are of similar design and perform within the same efficiency range as indirect fuel oil heaters.

In addition to heating the inlet air to drying chambers, heat exchangers are also used to recover heat from exhaust air leaving drying chambers. Finned tube exchangers are commonly used for this purpose. The thermal wheel, which is a rotating heat exchanger, is also used for heat recovery from hot-air driers (Van Arsdel et al, 1973; Green, 1984; Strumillo and Lopez-Cacicedo, 1987; Masters, 1985). *See also* Air heating, Energy for drying, Thermal wheel.

Heat pump
A device which recovers heat from a low temperature source and uses it to heat air or water for the purposes of evaporation or dehydration. An example is the refrigeration cycle evaporator. In this equipment a refrigerant gas condenses in the calandria tubes to supply the necessary heat to evaporate the water. The liquid gas then evaporates in the condenser, thus cooling and condensing the water vapour produced in the calandria. A heat pump may be incorporated into a solar drying system to improve its overall thermal efficiency (see *Figure 2.64*, page 99) (Imrie, 1987; Brennan et al, 1990). *See also* Solar drying.

HEPA filter *see* High efficiency particulate air filter.

Hertz (Hz)
An SI unit of frequency equivalent to cycles s^{-1}.

High efficiency particulate air (HEPA) filter
A filter which removes microorganisms from air or other gases to produce a sterile fluid. The air is pushed through pads of unwoven fibrous material by means of a fan. Such pads are made of glass fibres of the order of 5–20 μm in diameter or of other suitable fibrous material. Acetylated polyvinyl alcohol, coated with a heat-resistant resin, is frequently used. Such pads must withstand repeated sterilization. The mechanism whereby microorganisms are removed from the gas is not simple filtration. The pore sizes in these pads are larger than most microorganisms. Diffusion, settling and electrostatics may also be involved. The efficiency of such filters is high. Since organisms may be adsorbed on to dust particles, prefiltering the air or gas through a bag filter, or some other air filter, prior to HEPA filtration can greatly reduce the work to be done by the sterilizing filter. HEPA filters are used to sterilize the drying air, and to atomize air if a two-fluid nozzle is used, in aseptic spray-drying systems. They are also used in laminar flow chambers. In these chambers, a gentle flow of sterile air is directed either horizontally or vertically across a working area, in a laminar flow pattern, to avoid turbulence or eddying. The air is sterilized through HEPA filters. The working area is swept clean of particles, including microorganisms, and so maintained in a sterile condition. The sterile powder from an aseptic spray drier is usually discharged into a laminar flow chamber where it is packaged under aseptic conditions. *See also* Aseptic spray-drying system.

Homogenization
This term is used to describe the operation in which the size of the droplets in the internal phase of an emulsion are reduced. A pressure homogenizer is the equipment most commonly used for this purpose. In such a homogenizer, a high-pressure pump forces the crude emulsion, usually formed by premixing, between a valve and its seat. The gap through which the liquid passes is of the order 15–300 μm, and the pumping pressure $15–40 \times 10^3$ kN m^{-2}. The droplets of the internal phase are distorted out of their stable spherical shape and broken up into smaller droplets. For some applications, better performance is attained by homogenizing in two stages, the first stage at a high pressure and the second at a lower pressure. In the second stage, clumps of small droplets, which may have formed during the high pressure homogenization, are broken up. The applications for homogenization include the homogenization of milk to improve its stability and prevent the formation of a cream layer on the surface, the manufacture of ice-cream mix, synthetic creams, coffee/tea whiteners, salad dressings and other oil-in-water emulsion systems (Brennan et al, 1990).

Hot water peeling
Peel may be loosened on some fruits and vegetables by immersing them in hot water and removing the peel by mechanical means. For example, dipping tomatoes in boiling water for 30 seconds loosens the skin, which is then removed by passing the tomatoes over rollers which pinch the skin and remove it from the tomato. The skins of beetroot are loosened by immersing them in near-boiling water for 30 minutes and then removed with a rubber-tipped scrubber (Woodroof and Luh, 1975; Arthey and Dennis, 1991). *See also* Abrasive peeling, Flame peeling, Lye peeling, Steam peeling.

Humid heat
The amount of heat (energy) necessary to raise the temperature of a unit mass of dry air plus the water vapour it contains, by a unit amount.

$$C_H = C_A + C_v H$$

where C_H is the humid heat of an air-water vapour mixture, C_A the specific heat of dry air, C_v the specific heat of water vapour and H the absolute humidity of the mixture. Common units are kJ K^{-1} (kg dry air)$^{-1}$.

Humidification
The process of increasing the water vapour content, or humidity, of air.

Humidity
The amount of water vapour in an air-water vapour mixture. It may be expressed in different ways. *See* Absolute humidity, Percentage absolute humidity, Relative humidity.

Humidity chart *see* Psychrometric chart.

Humid volume
The volume of a mass of an air-water vapour mixture, containing a unit mass of dry air, at a pressure of one atmosphere and a specified temperature.

Assuming ideal gas behaviour and expressed in SI units:

$$V_H = \frac{22.4}{273}(\theta + 273)\left(\frac{1}{M_A} + \frac{H}{M_w}\right)$$

where V_H is the humid volume, θ the temperature of the mixture, in °C, M_A the molecular weight of air, M_w the molecular weight of water and H the absolute humidity of the mixture. When $H = \theta$, V_H = specific volume of dry air; when $H = H_s$, V_H = specific volume of saturated air (Green, 1984; McCabe et al, 1985; Brennan et al, 1990).

Hysteresis
Lagging of effect when cause varies in amount, e.g. magnetic induction behind magnetizing force. *See* Hysteresis in sorption isotherms.

Hysteresis in sorption isotherms
This is the phenomenon whereby adsorption and desorption isotherms of the same material are of different shapes. Often a closed hysteresis loop is apparent (see

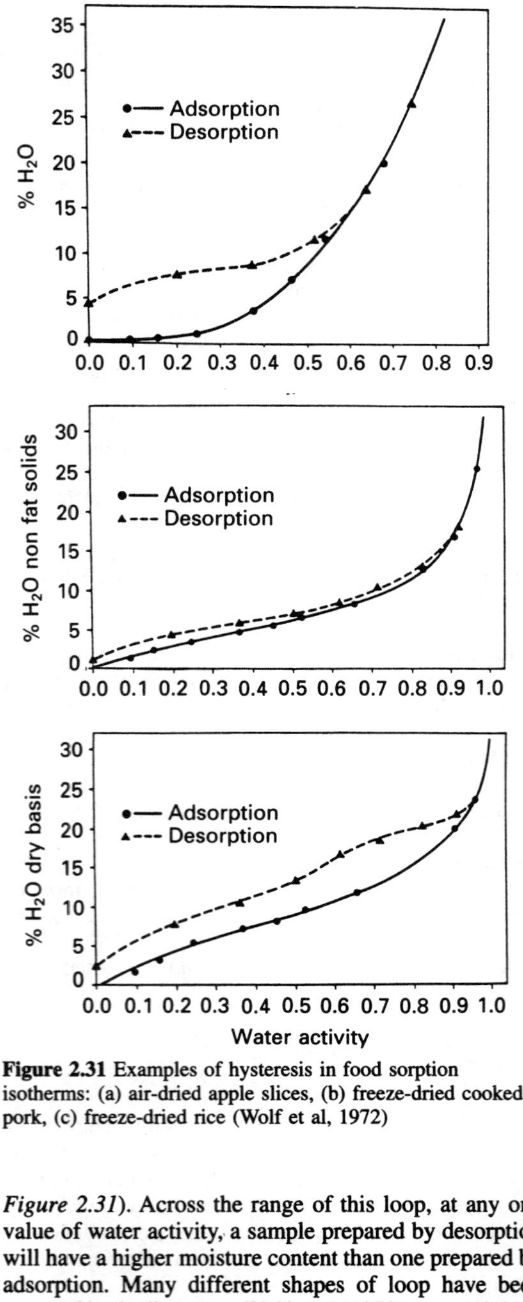

Figure 2.31 Examples of hysteresis in food sorption isotherms: (a) air-dried apple slices, (b) freeze-dried cooked pork, (c) freeze-dried rice (Wolf et al, 1972)

Figure 2.31). Across the range of this loop, at any one value of water activity, a sample prepared by desorption will have a higher moisture content than one prepared by adsorption. Many different shapes of loop have been reported in the literature. Wolf et al (1972) reported three general types as shown in *Figure 2.31*. In high sugar, high pectin foods, hysteresis occurs mainly in the monomolecular region. Although the total hysteresis is large none occurs above a water activity of 0.65. In high protein foods, a moderate hysteresis starts about a water activity of 0.85 and extends to 0. In starchy foods, hysteresis is relatively large, with a maximum at a water activity of about 0.7. Many attempts have been made to explain hysteresis but no one explanation has earned universal acceptance. One theory is known as the *incomplete wetting theory*. Due to the presence of impurities, such as dissolved gases, the contact angle of the receding film in the capillaries during desorption is smaller than that of the advancing film during adsorption. Thus, capillary condensation along the adsorption branch of the isotherm will be at higher vapour pressure. The *ink bottle neck theory* assumes that the structures within the solid consist of relatively large pores with narrow entrances and exists, represented by an ink bottle (*Figure 2.32*). During adsorption, condensation first takes place in the large diameter pore. During desorption the exit channel from the pore is blocked by a meniscus which will evaporate only when the pressure has fallen to that corresponding to the diameter of the channel. Then the whole pore empties immediately. For a given amount of water adsorbed, the pressure is greater during adsorption than during desorption. Thus, at any given value of water activity, within the range of the hysteresis loop, there will be less water present if approached by adsorption as compared with desorption. Changes of phase, e.g. from solid to liquid, or state, e.g. from amorphous to crystalline, can also affect the shape of the hysteresis loop. Changes in temperature also have an effect. In most cases, an increase in temperature will decrease the total hysteresis and reduce the span of the loop. *See also* Sorption isotherm, Water activity.

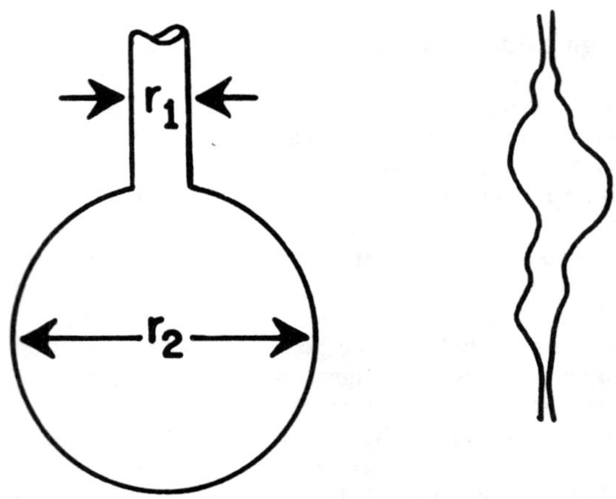

Figure 2.32 A representation of the ink bottle explanation of hysteresis: (a) the bottle model, (b) a pore within the food

I

Immersion freezing
A freezing method in which the food is immersed in a cooled liquid, usually a brine or syrup. Freezing is brought about by heat transfer from the food to the cooled liquid. *See also* Freezing methods.

Impingement drying
A drying method in which the heated air is blown against the product surface by means of nozzles. It is most suited to removing surface and unbound moisture and is mainly used for drying very thin sheets of material. If the drying rate is controlled by moisture movement within the material, impingement drying is generally not appropriate. Air velocity leaving the nozzles may range from 5–100 m s^{-1}. High rates of heat and mass transfer are attainable at the material surface. Part impingement and part throughflow drying may be used for shallow layers of particulates, on conveyor driers. Impingement drying, combined with other heating methods, e.g. microwaves, may find application for drying thin layers of food material (Mujumdar, 1987; Sturgeon, 1987; Brennan, 1992).

Indirect rotary drier *see* Rotary drier.

Individual quick blanching (IQB) *see* under Blanching.

Infrared drier
The principles of infrared heating are discussed on pages 12 and 13. Its application to food dehydration is limited to removing small amounts of moisture from low-moisture granular or powder materials. These are conveyed beneath banks of heating elements on belts or vibrating decks. Short-wave lamps are used as sources of heat for heat-sensitive foods while long-wave bar heaters are used for less sensitive materials. Products dried in this way include breadcrumbs, flour, starches, custard powders, cake mixtures, tea, spices and nuts (Williams-Gardner, 1976; Fellows, 1988; Brennan, 1989; Brennan et al, 1990).

In-package desiccant
To maintain a low water vapour content in packages of dried foods which are susceptible to deterioration as a result of small changes in moisture content, a desiccant may be included in the package. This is normally done in relatively large catering packs such as metal cans, or composite containers (wound body made of laminate containing paperboard and foil with metal or plastic ends). The desiccant may be calcium oxide, calcium chloride, silica gel or alumina. It is contained within a pouch made of material highly permeable to water vapour. The pouch may be fixed by adhesive to the underside of the lid of the container. Water vapour in the headspace of the container is absorbed by the desiccant and, over an extended storage period, more moisture may be released from the food and absorbed.

Instantizing of dried powders *see* under Reconstitutability of dried powders.

Insulation (lagging)
It is common practice to insulate drying chambers and associated ducting carrying heated air or other hot thermal fluids. Pipes carrying steam to heat exchangers and condensate back to the boiler are also lagged. The objective of lagging is to reduce the amount of heat lost by radiation and so improve the overall efficiency of utilization of energy. Insulating materials should have very low thermal conductivities. For example, the thermal conductivity of fibreglass and mineral wool in binder is of the order of 0.035 and 0.043 W m^{-1} K^{-1}, respectively, while that of stainless steel and copper is about 17.0 and 337.0, respectively, in the same units. Lagging materials should withstand the high temperatures involved without softening, melting or fusing. They must not give rise to a health hazard. It is for this reason that the use of asbestos has been discontinued. Fibreglass and mineral wool (rockwool) in binder are widely used as insulating materials at high temperatures. Insulating large areas such as spray-drying chambers is quite expensive. The thicker the layer of insulating

material, the lower will be the heat losses but the cost will increase. A careful balance has to be struck between the cost of wasted energy and the cost of insulation. Green (1984) gives some guidance on selecting the optimum thickness of insulating material. In food plants, the insulation must have an outer cladding. In high risk areas, stainless steel is used as cladding. In other areas, aluminium or painted mild steel may be used. This cladding must be well sealed to ensure that the insulating material does not become contaminated with food material. This is particularly important in wet areas of the plant. In such areas, the insulation should be removable for inspection and should be supported in such a way as to provide an air gap between it and the chamber wall so that it does not become contaminated with any product which might enter through cracks or pinholes in that wall. In some modern spray-drying plants, the chamber is not insulated but the drier is located inside a special building. In some spray-drying applications, where the product is sticky when hot, insulation of the chamber may be reduced or removed altogether, to lower the wall temperature and prevent a product being deposited on it (Green, 1984; Masters, 1991). *See also* Energy used in drying, Spray drying.

Integrated fluidized-bed (IFB) spray drier

This type of spray drier features two-stage drying in a single chamber. In the IFB chamber (*Figure 2.33*), rotary atomization is usually used and the air/product flow is concurrent. There is a stationary fluidized bed in the form of a ring at the bottom of the chamber surrounding the outlet air duct. The diameter of the fluidized bed is approximately half of the overall diameter of the chamber. Air is introduced into the top of the chamber at temperatures in the range 180–250°C, and also into the bottom, fluidizing the powder in the annular bed. This fluidizing air enters at temperatures in the range 100–150°C. Exhaust air temperature is usually in the range 60–80°C. Further air is introduced tangentially into the chamber to sweep the vertical wall. Because of the strong rotary movement of the air, powder separation is good and the fraction going to the cyclone is small. The feed is partially dried in the main body of the chamber, typically to 10–12% moisture (wet-weight basis), and partially in the fluidized bed. The depth of the fluidized bed is approximately 0.5 m and the residence time of the powder in the bed is about 30 min. Drying may be completed in these two stages or alternatively the product may be discharged on to vibrating fluidized beds for final drying and cooling. The advantages of this type of spray drier, compared to a simple, single-stage design include: energy savings of 20–40%, smaller space requirements, better control over product characteristics such as bulk density and solubility, and suitability for products which are difficult to dry (Masters and Pisecky, 1983; Pisecky, 1987; Brennan, 1989; Fergusson, 1989; Masters, 1991). *See also* Energy used in drying, Spray drying.

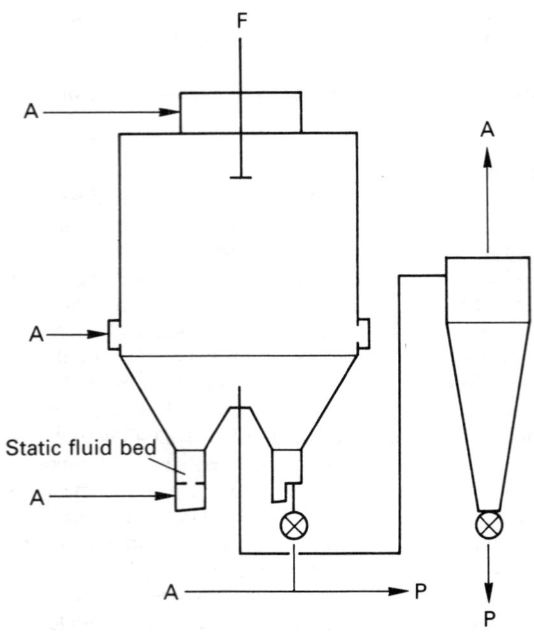

Figure 2.33 Spray drier with integrated static fluidized bed in drying chamber: A = air flow, F = feed, P = dried product (Brennan, 1989)

Intermediate moisture food (IMF)

This term describes a group of shelf-stable foods with a moist texture which do not normally need to be reconstituted before eating. The extended shelf life is attained by reducing the water activity, a_w, level to the point where microbial and enzymic activity and undesirable chemical changes such as oxidation and non-enzymic browning are inhibited. The low a_w is brought about by the incorporation of additives which reduce a_w. Such additives, known as humectants, include sugar, salt, propylene glycol, sorbitol, corn syrup invert sugar, and ethanol. IMF products have moisture contents in the range 10–40% (wwb) and a_w values in the range 0.6–0.9. Many traditional foods could be described as IMF. These include dried prunes, dates and figs; candied (crystallized) fruits, jams, jellies and some pie fillings; certain types of ham and sausage and fruit cake. Sugar and salt are the main humectants in these products. Specially formulated IMF products, with a_w values at the top end of the range, have antimicrobial agents, antioxidants and browning inhibitors incorporated, in addition to the humectants. Antimicrobial agents used include potassium sorbate, sorbic acid, propionic acid, sodium and calcium propionate, parabenz, benzoic acid and pimaricin butanediol. Antioxidants used may be butylated-hydroxy-anisole (BHA), butylated-hydroxy-toluene (BHT), propyl gallate, ascorbic acid and metal chelators such as citric acid and ethylene diamine tetraacetate (EDTA). Sulphur dioxide and sulphites are used as browning inhibitors. Many of the moist pet foods on the market could be described as IMF. A typical formulation for such foods would be: meat and/or meat by-product 30–70%, sodium caseinate 7–20%, sugar 15–30%, propylene glycol 2–10%, starch 0.5–10.0%, nutritional supplement, flavour and colour. IMF products for human consumption have mainly been used in space programmes and rations for the military, explorers, etc. These take the form of meat cubes, dried fruit bars, with or without cereal, and cake bites. Examples of some IMF products described in the literature are food sticks which contain: peanut butter, vegetable oil, soya protein, sugar, starch, corn syrup, glycerol, emulsifiers, stabilizers, salt, vitamins and minerals (a_w 0.60) and meal replacement bars which are comprised of milk protein, soya protein, fat, flour, starch, sucrose and other sugars, sorbitol, salt, propylene glycol, emulsifiers, antioxidants, antimicrobial agents, vitamins and minerals (a_w 0.60). IMF products may be produced by moist infusion in which solid food pieces are soaked in and/or cooked in a solution until the desired a_w is reached; by dry infusion in which solid food pieces are first dehydrated by an appropriate method, including freeze drying, and then soaked in a solution of the additives or by blending the ingredients, cooking the mix and extruding it (Burrows and Barker, 1976; Karel, 1976; Flink, 1977; Troller, 1978). *See also* Water activity, Water activity and food quality.

Internal energy
The energy possessed by a body or substance arising from its internal state, i.e. the degree of agitation of the molecules. As the temperature of the body increases so does the internal energy. Under ideal conditions, a change in internal energy may be represented thus:

$$\Delta U = mc_s \Delta T$$

where ΔU is the change in internal energy, m is the mass of the substance, c_s the specific heat of the substance and ΔT the temperature change. Often internal energy is expressed as a value above an arbitrarily chosen reference temperature. For water and steam, internal energy is taken to be zero at the triple point of water 0.01°C (often quoted as 0°C). Specific internal energy is the energy associated with unit mass of the substance. Common units are $kJ\,kg^{-1}$. Values of specific internal energy for water and steam are published in thermodynamic tables (steam tables). Internal energy is also known as thermal energy (Haywood, 1968).

International system of units *see* SI units.

J

Jet-spray drier
This design of spray drier, introduced in the 1960s, featured a very high air inlet temperature, 400°C, and velocity, 390 m s^{-1}. Very high rates of heat transfer were attainable. The drying chamber was relatively small compared to more conventional chambers (*Figure 2.34*). To ensure adequate and uniform drying in such a drier, the feed had to be atomized into a very fine spray with droplet diameters of the order of 10 µm. The powder produced from this fine spray could not be efficiently collected in a cyclone, was difficult to handle and had poor reconstitution characteristics. The jet-spray drier found little application in the food industry (Hayashi, 1989). *See also* Spray drying.

Joule (J)
An SI unit of energy. It is the work done when the point of application of a force of 1 newton is displaced through a distance of 1 metre in the direction of the force. Thus 1 J is equivalent to 1 Nm.

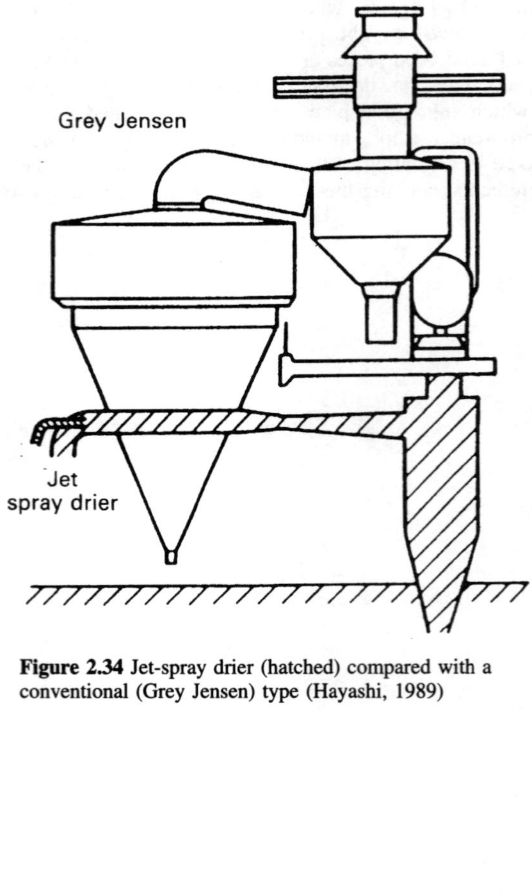

Figure 2.34 Jet-spray drier (hatched) compared with a conventional (Grey Jensen) type (Hayashi, 1989)

K

Kelvin (K)
The basic SI unit of thermodynamic temperature. It is the fraction $1/273.16$ of the thermodynamic temperature of the triple point of water.

Kelvin-Thompson equation
Represents the relationship between water activity and factors affecting the water held in capillaries within a wet solid:

$$a_w = \exp \frac{\gamma^{-2}}{r} \cos \theta \; C_1$$

where γ is the surface tension of the water, θ the contact angle of the water in the pores, r the radius of the capillary and C_1 a constant. C_1 is related to molar volume, V_o, thus:

$$C_1 = \frac{V_o}{RT}$$

where R is the gas constant and T absolute temperature. *See* Sorption isotherm, Water activity.

Kiln drier
An early type of drier consisting of a two-storey building with a furnace or burner located on the ground floor. The heated air and products of combustion rise by natural convection through the slatted floor of the second storey on which the food material is spread in an even layer, usually 0.1–0.2 m thick. After passing through this layer the humidified air is exhausted through a flue in the upper storey. Regular raking or turning of the food is necessary to promote uniform drying. In such a basic unit there is little control over the drying conditions and drying rates are relatively low. The inclusion of an exhaust fan can increase the rate of drying. Indirect heating of the drying air by means of a heat exchanger will eliminate problems of contamination of the food with the products of combustion. The main products dried in kiln driers are hops in the UK, and apple slices in the USA (Van Arsdel et al, 1973).

Kilogram (kg)
The basic SI unit of mass. It is the mass of the international prototype which is in the custody of the Bureau International des Poids et Mesures (BIPM) at Sèvres near Paris, France.

Kinematic viscosity
Dynamic viscosity/density. Common units are $m^2 \, s^{-1}$ or St.

Kinetic energy
The energy a body possesses through its motion. May be termed velocity energy or velocity head.

For linear motion, kinetic energy (KE) may be calculated from the relationship:

$$KE = \frac{1}{2} mv^2$$

where m is mass and v velocity of the body.

For rotational motion, kinetic energy (KE) may be calculated from the relationship:

$$KE = \frac{1}{2} Iw^2$$

where I is inertia and w is angular velocity of the body.

L

Lagging *see* Insulation.

Laminar (streamline) flow of fluid
When a fluid flows through a pipe or over a solid surface a pattern of flow will develop which will depend on the velocity of flow and the physical properties, notably viscosity, of the fluid and the geometry and surface characteristics of the pipe or surface. At low flow rates, through a smooth surface pipe, the fluid flow is in the form of parallel streams which do not interfere with each other. There is no bulk movement in a direction at right angles to the direction of flow of the fluid. Such flow is known as laminar or streamline flow. The higher the viscosity of the fluid, the more likely it is to exhibit laminar flow. If the flow rate is increased, a point is reached at which eddies are formed, causing mixing across the pipe section. This is known as turbulent flow. A dimensionless quantity, known as the Reynolds number (R_e), is used to characterize the pattern of flow of a fluid in a pipe. The Reynolds number may be calculated as follows:

$$R_e = \frac{Dv\rho}{\mu}$$

where D = diameter of the pipe (m), v = average velocity of the fluid ($m\,s^{-1}$), ρ = density of the fluid ($kg\,m^{-3}$), μ = viscosity of fluid ($N\,s\,m^{-2}$). A Reynolds number of 2100 or less describes laminar flow while a Reynolds number of 4000 or more indicates turbulent flow. Values of R_e between 2000 and 4000 represent transitional flow, which may be predominantly laminar or turbulent. Turbulent flow across heated surfaces will promote higher rates of heat transfer than laminar flow and so is desirable in heat exchangers. The formation of deposits on heated surfaces (fouling), and the settling of solids in pipelines, is less likely if liquid flow is turbulent. On the other hand, friction losses in pipes are lower under laminar flow conditions and so less energy is required for pumping (Fellows, 1988; Toledo, 1991).

Latent heat
The amount of heat (energy) required to convert a unit mass of a solid to a liquid at constant temperature is known as the *latent heat of fusion*. The amount of heat (energy) required to convert a unit mass of a liquid to a vapour is known as the *latent heat of vaporization*. SI unit $J\,kg^{-1}$.

Latent heat of fusion *see* Latent heat.

Latent heat of vaporization *see* Latent heat.

Layer drying
A technique used when drying cereal grains in bin driers. A layer of grain is placed in the bin and dried with heated air. When this layer is dry another layer is placed on top of it and dried. This procedure is repeated until the bin is full (Van Arsdel et al, 1973b). *See also* Cereal grains.

Lecithinization
The process of adding lecithin, a surface active agent, to spray-dried, whole milk powder to improve its wettability.

Litre (l)
A metric unit of volume equivalent to 1 dm^3.

Lobe pump *see* under Pumping of foodstuffs.

Louvred (rotary) drier *see* Rotary drier.

Low-NO_x burner (low-NOX burner)
A type of gas burner designed to minimize contamination of the air, heated directly by natural gas, with undesirable products of combustion. *See* Air heating.

Lye peeling
This chemical method of peeling involves immersing fruits or vegetables in hot sodium hydroxide solution for

a specified time. Chemical erosion of the skin and underlying tissue loosens the peel. The fruits or vegetables are conveyed through a bath of sodium hydroxide by a paddle wheel or some other mechanical device. High pressure water jets usually remove the skin and wash away the lye. The concentration and temperature of the lye solution and the dwell-time of the food in the lye influence the effectiveness of this peeling method. Lye concentration ranges from 5–20%, temperature from 60–95°C and dwell-time from 30 to 7 min. There will be an optimum set of conditions for each food. Some hand-finishing may be necessary but peeling losses are generally less than with abrasive peeling (Luh and Woodroof, 1975; Woodroof and Luh, 1975; Arthey and Dennis, 1991). *See also* Abrasive peeling, Flame peeling, Hot water peeling, Steam peeling.

Lyophilization *see* Freeze drying.

M

Magnetic separator

A magnetic device for separating ferrous metals from foods as part of the cleaning operation prior to drying. *See* Dry cleaning of raw materials.

Maillard reactions

A group of chemical changes involving aldehydic or ketonic compounds and amino compounds which react to form coloured polymeric products in dried fruits and vegetables. This colour change is generally undesirable and is known as non-enzymic browning. It may be reduced by sulphuring the food before dehydration. *See* Sulphuring, Water activity and food quality.

Materials of construction

In general, materials of construction for those parts of food-processing equipment which come in contact with the product should be inert with respect to the food and to cleaning and sanitizing materials likely to be used on it. Inertness in this context means corrosion resistance and freedom from catalytic activity. There should be sufficient mechanical strength so as not to easily deform, chip, flake or delaminate. Surfaces should be smooth, non-porous and free from cracks, crevices and pitting, which could trap food residues and support the growth of microorganisms. The surface finish should make it possible for soil to be easily and completely removed during cleaning. This surface finish should not change as a result of cleaning or sterilizing operations and should not develop porosity. Surfaces not normally directly in contact with food or cleaning materials should be mechanically stable and have a sufficiently smooth finish to facilitate thorough cleaning.

Stainless steel is the most commonly used food contact surface material in food-processing equipment. It is used for tanks, pipelines and pumps which handle liquid foods. It also finds uses in drying equipment. The belts in conveyor driers usually consist of woven stainless steel mesh or hinged or interlocking perforated stainless steel plates. Fluidized-bed driers are usually constructed of stainless steel with the bed support made of woven stainless cloth or perforated plate. Spray-drying chambers, cyclones and associated ducting are normally made of stainless steel. Shelving and belts in vacuum driers are usually made of this material and in some cases the vacuum cabinets themselves. Drum driers made of stainless steel are available. However, most drums consist of chromium-plated cast iron or mild steel.

Various other construction materials are used for those parts of drying equipment which do not normally come in contact with the food or cleaning materials. Cabinets of vacuum driers and freeze driers are usually made of thick mild steel reinforced by steel rings attached to the outside of the shell. Cabinets and tunnels for hot-air driers are made of a variety of materials including insulated steel panelling, bricks and concrete blocks (Jowitt, 1980). *See also* Insulation, Stainless steel.

Membrane concentration

Removing water and some solutes by means of a semi-permeable membrane is one way of concentrating liquid foods prior to dehydration. The advantages of this method of concentration compared to vacuum evaporation include: (i) low temperatures prevail during the process which leads to better retention of nutritional and organoleptic quality, (ii) no change of phase occurs so that energy requirements are less, sometimes as little as one tenth of those required in vacuum evaporation, (iii) the large volumes of cooling water used in vacuum evaporation are not required. On the other hand, the limitations include: (i) a limited degree of concentration; usually the concentrate will contain less than 30% total solids, (ii) the capital cost of membrane equipment is relatively high, and (iii) the fouling of membranes necessitates frequent cleaning with resultant interruption of the process. When a semi-permeable membrane separates solutions of different concentrations, solvent molecules pass through the membrane from the region of lower concentration to the region of higher concentration, under the influence of osmotic pressure. The lower the molecular weight of the solute the greater the osmotic pressure developed across the membrane. If pressure is applied across the membrane, in opposition to the osmotic pressure, then the solvent may be made to flow from the region of higher concentration to the region of lower concentration, thus further increasing the solute concentration on the high concentration side. For large molecules, e.g. solutes with molecular weights greater than 500, the osmotic pressure developed is so small that a low applied counter-pressure is sufficient to bring about concentration. This is known as ultrafiltration (UF). Pressures in the region of 0.1–1.0 MN m^{-2} are sufficient to cause solvent and small molecular weight solutes to flow in the direction of the applied pressure drop, thus increasing the concentration of the larger molecules upstream of the membrane. Membranes used for UF have a relatively 'loose' structure and a high porosity. With low molecular weight solutes, which develop high osmotic pressures, applied counter-pressures in the range 5–10 MN m^{-2} are necessary to effect concentration. This is known as reverse osmosis (RO). Membranes with a relatively 'tight' structure are used for RO.

UF and RO membranes are of two types. Asymmetric membranes are cast from a single material and consist of a very thin layer (0.1–1.5 µm) of dense polymer, supported on a thicker layer of porous polymer, the overall thickness being 0.1–0.2 mm. The membrane may be further supported by a rigid, porous backing structure. Composite membranes also consist of a very thin active film but placed on top of a highly porous sub-layer of a different material. Membranes must be strong enough to withstand the pressure developed during UF and RO and be capable of being cleaned and sanitized. Flexible membranes are cleaned with acid and alkaline cleaning solutions. UF membranes are made of polysulphones, polyamides (nylons), polyvinyl chloride (PVC), polystyrene, polycarbonates, polyethers and rigid cellulose esters. They are thicker and more porous than RO membranes. In recent years, solid UF membranes consisting of zirconium oxide supported on carbon tubes have been developed. These can be sterilized with steam. RO membranes are made of cellulose acetate, mixed cellulose esters, polyamides, polyurethanes and polyacrylonitrile. They are relatively thin and have a tight structure. Tubular membranes line perforated stainless steel tubes. Numbers of these tubes are fitted inside an outer stainless steel cylinder. The feed flows inside the membrane tubes and the low molecular weight permeate passes radially through the tubes. The higher molecular weight retentate, or concentrate, leaves through an outlet header plate. Flat-sheet membrane systems consist of assemblies of parallel layers of membrane material supported on plastic plates. By means of seal rings and entry and exit holes at the top and bottom of the plates, the liquid flows through the spaces between successive pairs of membranes. The permeate passes through the membranes and exits from behind them through narrow drain channels (*Figure 2.35*). In the spiral cartridge design, alternating layers of membrane material and plastic support material are wound around a perforated central tube. The liquid flows longitudinally between the layers of membrane. Permeate passes through the membrane and into the perforated central tube from which it is removed. The retentate flows out of the other end of the cartridge (*Figure 2.36*). UF is used to concentrate milk to about 30% total solids content and to separate and concentrate proteins from whey to produce whey protein concentrate (WPC). This may be further concentrated by vacuum evaporation prior to spray drying. It has also been used to concentrate sucrose solutions and tomato paste, and in the treatment of effluents to make them less polluting. RO has been used to concentrate whey, fruit juices, egg white, coffee extract and citric acid and to clarify wine and beer and

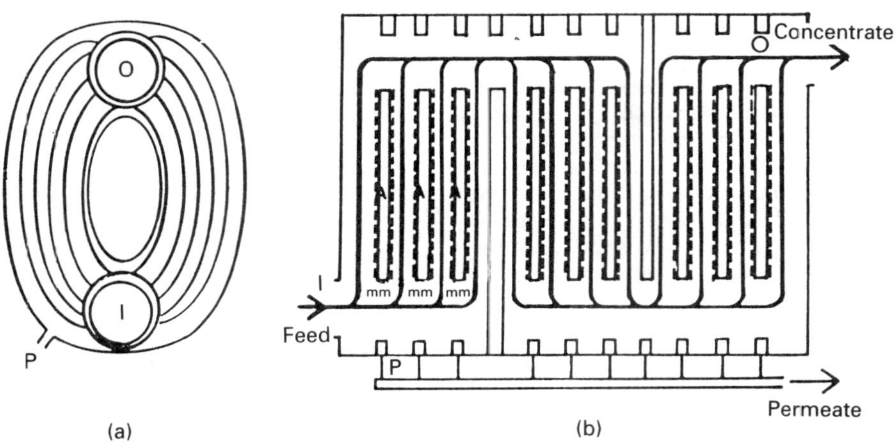

Figure 2.35 Flat sheet ultrafiltration module: (a) membrane support plate: O inlet port, P permeate outlet, (b) membrane assembly: A support plates, m membranes (Brennan et al, 1990)

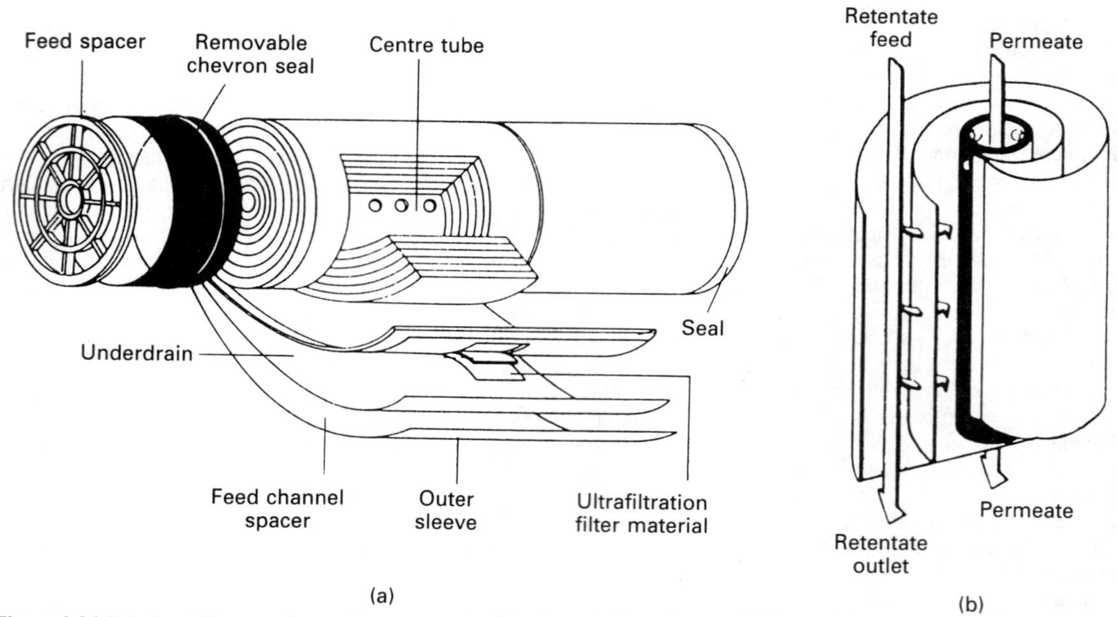

Figure 2.36 Spiral cartridge membrane: (a) components, (b) schematic flow diagram (Fellows, 1988)

purify water (Fellows, 1988; Brennan et al, 1990). *See also* Freeze concentration, Evaporation (concentration).

Metal detector
The use of magnets to detect and separate metal contaminants in food is discussed under Dry cleaning of raw materials. After leaving such separators, the food material may pass under electronic metal detectors which can detect ferrous and non-ferrous metal fragments. These detectors generate a powerful electromagnetic field through which the food passes. The metal fragments distort this field, the disturbance is amplified and the signal used to activate an alarm or an automatic rejection device. Such detectors can also detect other contaminants such as clinker, some stones, carbonized grease and metal-impregnated grease. X-rays may also be used to detect solid contaminants, including metal fragments, in foods. The food is conveyed beneath an X-ray machine and the rays pass through it and are converted into visible light by a phosphor strip. This light is magnified and transmitted by fibre optics to an image intensifier and video camera and thence to a VDU. When solid contaminants are detected, an alarm or automatic rejection device is activated. Such detectors may be located downstream of a packaging machine as a final check on the product (Fellows, 1988; Brennan et al, 1990). *See also* Dry cleaning of raw materials.

Metre (m)
The basic SI unit of length. It is the length of the path travelled by light in vacuum during a time interval of 1/299 792 458 of a second.

Microencapsulation *see* Encapsulation.

Micron or micrometre (μm)
A metric unit of length equivalent to 10^{-6} m.

Microwave drying
Microwaves are located in the electromagnetic spectrum between far infrared and the radio frequency region. The term usually applies to waves in the frequency range 300 MHz to 300 GHz. Frequency allocations for industrial, scientific and medical uses are agreed internationally. For food applications, two frequencies are used, 915 MHz and 2450 MHz. Microwave energy is generated by radio frequency power tubes, using high voltage direct current. As the waves pass through the food they cause the molecules to oscillate about their axis and heat is generated by intermolecular friction. The penetration of microwaves into materials such as foods depends on the frequency and the characteristics of the material. Penetration may be quantified by half-power depth, i.e. the thickness of the material that reduces the wave energy to 50% of the incident level. The half-power depth for water is about 12 mm at 2450 MHz and 75 mm at 100 MHz. For foods it is less than for water. The mechanism of drying using microwave energy differs from that which occurs in heated air or in contact with a hot surface. Because of the rapid internal heat generation moisture evaporates quickly within the food. This creates a total pressure gradient which promotes rapid movement of liquid water and water vapour towards the surface of the food. This results in shorter drying times and lower product temperatures as com-

pared with conventional methods of drying. There is less movement of solutes within the food. The efficiency of utilization of energy may be higher than in hot-air or direct contact drying. The use of microwave energy for drying foods on an industrial scale is limited to removing relatively small amounts of water. It has been used to finish the drying of potato chips (French fries) and crisps (potato chips), coming from the fryer if the potatoes have a high sugar content to avoid darkening. There are also reports of it being used to complete the drying of fruits and vegetables, partially dried in heated air. Pasta products have been dried by means of microwaves. They have also been applied in vacuum driers. Fruit juice concentrates, tea extract and malt-based drinks have been dried in this way. Final drying of cereals such as malt flakes and other grains has been effected under vacuum using microwave energy. The application of microwaves as the heating method in freeze drying has been the subject of much research. This application is discussed under Freeze drying (Schiffmann, 1987; Brennan, 1989, Brennan et al, 1990; Salunkhe et al, 1991). *See also* Cabinet freeze drier, Freeze drying.

Mixed flow spray drier *see* Spray-drying chamber.

mmHg (torr)
A metric unit of pressure equivalent to 133.0 N m^{-2}.

Moisture content
The amount of moisture present in a wet material may be expressed on a dry weight basis or wet weight basis. *See* definitions of these terms. *See also* Bound moisture, Equilibrium moisture content, Free moisture, Monomolecular moisture content, Unbound moisture.

Monolayer moisture content *see* Monomolecular moisture content.

Monomolecular moisture content
That moisture which is strongly bound to specific sites on the solids in a wet material. This occurs at low water activity, usually below 0.35. This moisture is unavailable as a solvent to support microbiological activity and/or chemical reactions. Many foods are most stable when their moisture content corresponds to the monomolecular moisture content. Water held in this way exhibits a large specific heat of interaction and energy in excess of the latent heat of evaporation is required to remove it from the solid. It is also known as monolayer

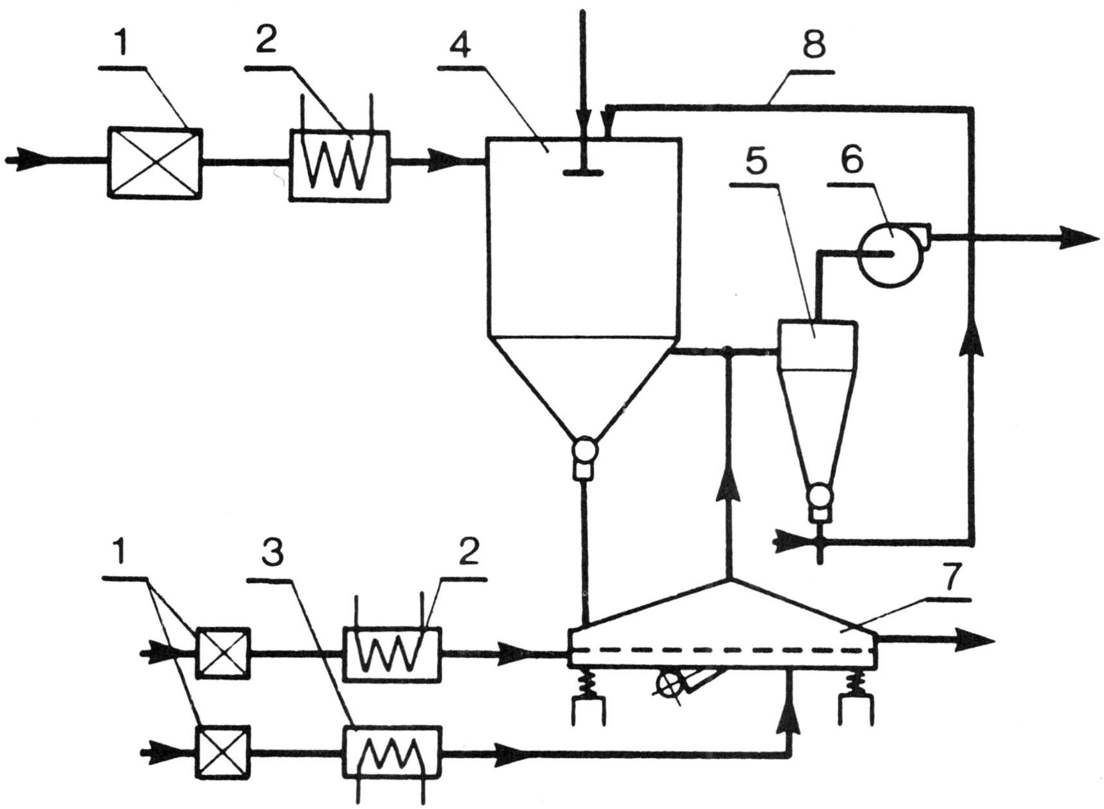

Figure 2.37. A two-stage spray-drying process for milk: (1) air filter, (2) heater, (3) cooler, (4) spray drier, (5) cyclone, (6) exhaust fan, (7) fluidized-bed drier, (8) return line for fine powder (Filkova and Mujumdar, 1987)

moisture content. *See also* Adsorption, Bound moisture, Sorption, Sorption isotherm, Water activity.

Mono pump *see* Pumping of foodstuffs.

Multicabinet freeze drier
This consists of two or more cabinet freeze driers which are operated with staggered, overlapping drying cycles. In a typical unit, four cabinets are connected to two vacuum manifolds. One manifold is used for rapid pumpdown in the early stages of a freeze-drying cycle. The other is used to maintain the low pressure during the rest of the cycle. The cabinets are loaded sequentially at appropriate time intervals, typically 2–3 hours, and the vacuum system used to maintain the optimum pressure in each cabinet throughout the drying cycle. This arrangement can cope with high throughputs more economically in comparison to operating four cabinets independently (Lorentzen, 1975; Brennan et al, 1990). *See also* Cabinet freeze drier, Freeze drying, Tunnel freeze drier, Vacuum-spray freeze drier.

Multiple-effect evaporation *see* Evaporation (concentration), Evaporator.

Multistage drying
Drying in two or more stages is frequently practised when drying foods in heated air. Such multistage drying can enable better control over product characteristics, may lead to savings in energy and more economic use of expensive equipment, as compared with single-stage drying. Two-stage tunnel drying, concurrent followed by countercurrent, can enable low moisture contents to be reached without excessive heat damage to the product. Multistage conveyor drying enables efficient use to be made of the equipment as the partially dried material is redistributed on to successive belts in deep layers. The mixing which occurs when the material is transferred from one belt to the next facilitates uniform drying. Material may be partially dried on a conveyor and then finished off economically in a bin drier. Multistage fluidized-bed drying can enable better control to be exercised over the drying conditions and lead to improved quality products, as compared with one-stage drying. Multistage pneumatic drying of small particles can also lead to improvements in quality. Two- or three-stage spray drying is widely practised. In a two-stage system, the powder is discharged from the main chamber at a higher-than-usual moisture content, 7%, on to a vibrated fluidized-bed drier where drying is completed (*Figure 2.37*). Higher air inlet temperatures to the main spray-drier chamber may be used without heat damaging the product, as compared with single-stage spray drying. This can lead to 15–20% savings in energy due to the increased thermal efficiency of the chamber. Some agglomeration of the powder particles in the fluidized-bed drier can improve the reconstitution characteristics of the final product. Two-stage spray drying in one unit is possible in the integrated fluidized-bed drier (IFB) and the Filtermat drier. Three-stage drying may be carried out by using an external fluidized-bed drier after an integrated fluidized-bed spray drier (Filkova and Mujumdar, 1987; Pisecky, 1987; Anon, 1992; Masters, 1991). *See also* Conveyor drier, Energy used in drying, Fluidized-bed drier, Integrated fluidized-bed spray drier, Pneumatic drier, Spray drying, Tunnel drier.

N

Newton (N)
An SI unit of force. It is that force which, when applied to a body having a mass of 1 kilogram, gives it an acceleration of 1 metre per second. Thus 1 N is equivalent to $1\,kg\,m\,s^{-2}$.

Newtonian flow
The flow pattern of fluids in which there is a linear relationship between shear stress and shear rate. *See* Shear stress, Dynamic viscosity.

Newtonian fluids *see* Newtonian flow.

Non-enzymic browning
An undesirable colour change which occurs in some dehydrated fruit and vegetable products, as a result of a group of chemical changes involving aldehydic or ketonic compounds and amino compounds. These reactions lead to the formation of coloured polymeric products. Sulphuring of the food prior to dehydration can inhibit browning. *See* Sulphuring, Water activity and food quality.

O

Osmotic drying

This method of drying involves immersing pieces of fresh food, mainly fruits and vegetables, in a solution with a higher osmotic pressure, and hence lower water activity, than the food. Water will pass from the food pieces into the solution, under the influence of osmotic pressure. For fruits, sugar solutions, with or without the addition of small quantities of salt, are used. For vegetables, solutions of salt are mainly used. In this process, the cell walls of the food act as semi-permeable membranes. However, these membranes are not completely selective and there may also be movement of solute, in both directions. Up to a 50% reduction in the fresh weight of fruits or vegetables may be brought about by osmosis. The process is more accurately described as concentration rather than drying, although the latter term is commonly used. In osmotic drying, the rate of diffusion is rapid initially but slows up significantly after 60–120 minutes. The operating conditions which most influence this process are solute concentration and temperature. As can be seen from *Figure 2.38*, an increase in either of these factors will increase both the initial rate of water loss and the maximum water loss attainable. Reports in the literature indicate that, not only water loss, but also solids gain by the fruit from the solution contribute to the reduction of water activity. Different solutes in the osmotic solutions, e.g. sucrose, fructose, glucose, cornstarch (cornflour), give rise to different drying patterns. The addition of small quantities, 0.5–2.0%, of salt to sugar solutions increases the driving force of the drying process. Osmotically dried foods are seldom shelf-stable without further treatment. Some such products can be stabilized by a mild heat treatment. They can be frozen to produce a dehydrofrozen product. They may be further dried in heated air, under vacuum or by freeze drying. They may be formulated into intermediate moisture products (Lerici et al, 1985; Brennan, 1989; Salunkhe et al, 1991). *See also* Dehydrofreezing, Intermediate moisture food, Osmovac process.

Osmovac process

A term sometimes used to describe two-stage drying of fruits. In the first stage, some moisture is removed by osmotic drying. The fresh weight of the fruit may be reduced by 50% in this stage. Drying is then completed by another technique, usually vacuum drying in a vacuum-shelf drier. When applied to some fruits, notably apples, pears and peaches, this method can result in dried products with better colour and flavour compared with hot-air dried fruit. Interesting textures may also be developed in the fruit (Woodruff and Luh, 1975; Jagtiani, 1988). *See also* Osmotic drying, Vacuum-shelf drier.

Oxygen scavenging agent

For dried foods highly susceptible to oxidation, a low oxygen content (down to 1 ppm) may be maintained in the package by including an oxygen scavenging agent in it. This is normally only done in relatively large catering packages such as metal cans or composite containers (wound body of a laminate containing foil and paperboard with metal or plastic ends). The product will have been vacuum packaged, or more likely, gas packaged in nitrogen. The scavenging agent is contained within a pouch made of material highly permeable to gas. This pouch may be fixed with adhesive to the underside of the lid of the package. The scavenging agent may consist of finely powdered iron. This oxidizes using up the oxygen in the headspace. Enzyme systems are also used as scavenging agents, e.g. a glucose/glucose oxidase system. Water vapour may be produced by the agent and so an in-package desiccant may also be included in the package. *See also* In-package desiccant.

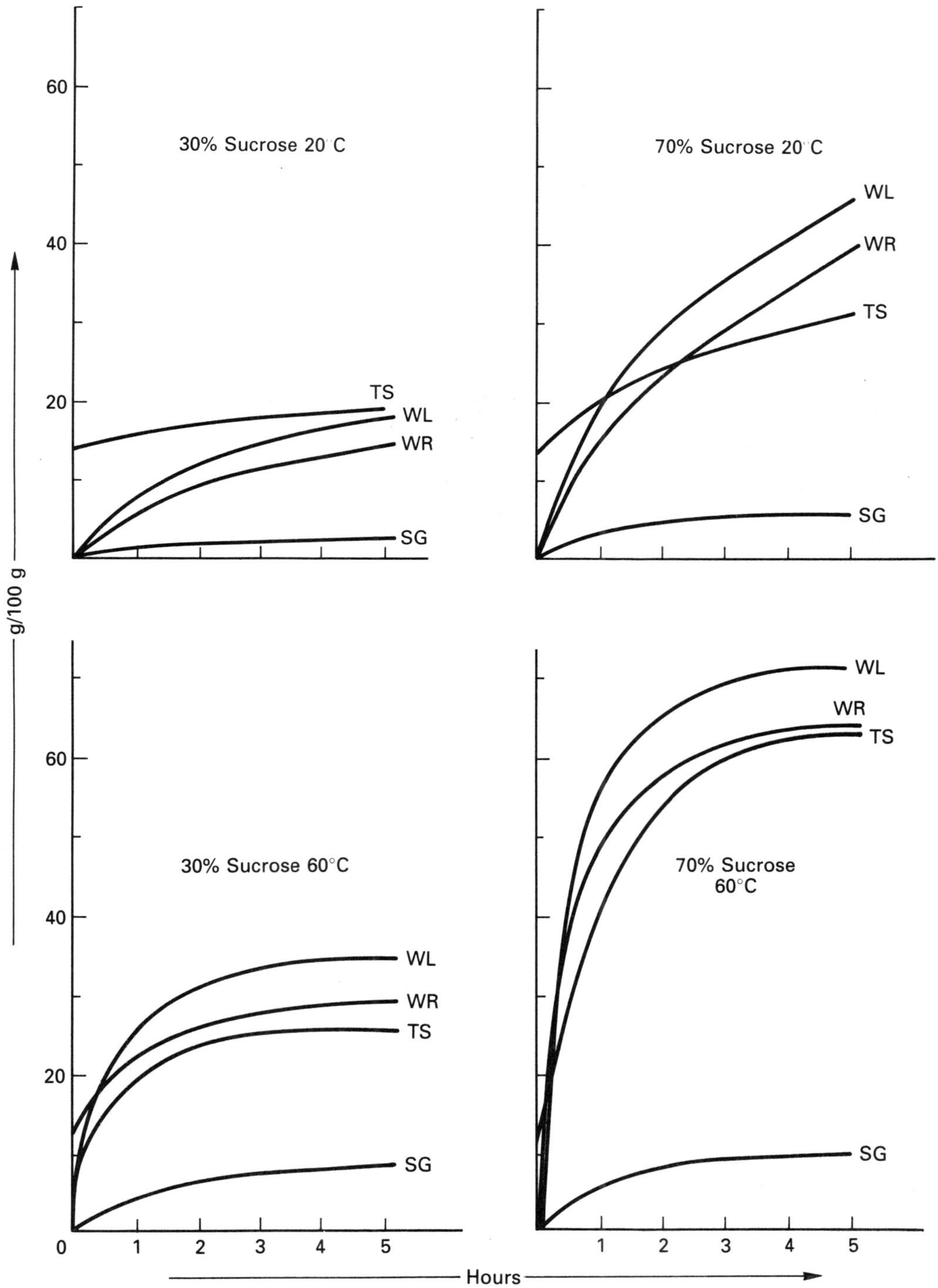

Figure 2.38 Effects of solute concentration and temperature on the osmotic drying pattern of apple (Brennan, 1989)

P

Packaging of dried foods

The main deteriorative changes which can occur in dried foods during handling, transport and storage are (i) pick-up of moisture which can lead to caking of dry powders if it occurs at low levels, or chemical and microbiological spoilage if it occurs at high levels; (ii) oxidation which can occur at a low moisture content in the presence of oxygen; (iii) non-enzymic browning which occurs at moisture contents above the monomolecular value; (iv) loss of volatile flavour compounds resulting in a reduction in the odour and taste of the product; (v) absorption of foreign odours from other food sources (oranges, onions, fish) and non-food (paint, petrol, detergents) or from packaging materials, adhesives or printing inks; (vi) colour changes due to exposure to light; (vii) contamination with microorganisms; (viii) infestation by insects; (ix) physical damage resulting in the fracturing of dehydrated food pieces, the breaking up of agglomerates and the creation of fines. Appropriate packaging can help to limit these undesirable changes in dried foods. The package and its contents must be chemically compatible but this is not normally a problem with dehydrated foods. Leaching of undesirable compounds from the packaging material into the food is not likely to occur nor are chemical interactions. In the case of dried foods with a high fat content some migration might occur. Supplies of packaging materials should be consulted as to the suitability of their products for packaging dried fatty foods. Other factors have to be considered when selecting a packaging system for a dried food product. These include: (x) is the package sufficiently greaseproof?; (xi) is it tamper evident?; (xii) can it be handled, transported and stored easily, including display on the supermarket shelf?; (xiii) does the labelling comply with relevant regulations?; (xiv) is the package attractive and likely to entice the customer to purchase the product?

The following are desirable characteristics of a packaging material/container for dried foods:

The packaging should be a good barrier to water and water vapour so as to minimize moisture pick-up by the product and so limit caking of powders and non-enzymic browning. Metal, glass and thick-walled plastic containers, suitably sealed, offer the best protection. Metal cans usually have lever lids. The closures on glass and plastic jars are usually screw caps. However, a membrane usually made of a foil laminate, may be sealed by heat or adhesive over the mouth of the container underneath the cap. This provides a barrier to water vapour and gases during handling, transport and storage outside the home. It is removed by the consumer when the container is first opened. The screw cap provides the protection from that point on, until the container is empty. When a non-rigid package is required, laminates of low permeability films (*see* below) and foil offer good protection. Paper or paperboard may be incorporated into the laminate where a degree of rigidity is desirable, e.g. pouches for soup mixes. Laminates of two or more low-permeability films provide adequate protection for many applications. Single films vary in the degree of protection they provide. Orientated polypropylene (OPP) and polyester (OPET), polyvinylidene chloride (PVDC) and high-density polyethylene (HDPE) are very good barriers to water vapour. Unorientated polypropylene (PP) and polyester (PET), unplasticized polyvinyl chloride (UPVC) and low-density polyethylene (LDPE) are quite good barriers. Plasticized polyvinyl chloride (PVC) and polystyrene (PS) have relatively high permeabilities to water vapour. Regenerated cellulose coated with a PVC/PVDC copolymer (MXXT) is quite a good barrier but wax-coated versions (MS) provide less protection. Metallized films, such as PET are very good barriers. When using laminates and single films it is important that the seal is effective. The most appropriate sealing method for any particular film should be used and sealing conditions (temperature, pressure and time) should be optimum.

To minimize oxidative changes during storage dehydrated products may be vacuum or gas packed. Gas packing using nitrogen is widely practised. This can be done in both rigid and non-rigid packages. When such techniques are used the container must be a good barrier to gases. Rigid containers again offer the best protection. Laminates containing foil are good gas barriers. Film only laminates offer quite good protection provided low permeability materials are used. PVDC, used as a copolymer with PVC, or as a coating on PP or PET, is a very good barrier to gases. PVDC coated regenerated cellulose is also a good barrier. So also are polyamides (nylon) and metallized PP. A very wide range of laminates, comprised of different combinations of low permeability films, is available. Again, a good heat seal is essential when using such materials.

To minimize loss of desirable flavours or pick-up of undesirable taints from external sources, packages should be good barriers to the volatile compounds that are responsible for such flavours. Metal, glass and thick-walled plastic containers give the best protection, with laminates containing foil the next best. The effectiveness of film laminates or single films for this purpose depends on the choice of film. Volatile flavour compounds are generally non-polar and will readily pass through non-polar films such as PE. Polar films such as regenerated celluloses are good barriers to such compounds. Thus, the inclusion of a layer of regenerated cellulose in a laminate can improve its resistance to the passage of volatiles. PVC/PVDC coated regenerated cellulose on its own is a good volatile barrier. So also are OPP, OPET and metallized PP. Some packaging materials may cause tainting of foods. Paper, paperboard and fibreboard may cause problems in this respect, imparting a 'cardboardy' taint, particularly to fatty foods. Storing the packaging material before use in well-ventilated stores can reduce this problem. Some inks and adhesives give off odorous vapours as they dry and these can cause tainting of food. These should be avoided where possible. Printing on packaging should be thoroughly dry before the package is filled with product. Photosensitive products can be protected by using opaque packaging materials such as rigid metal, laminates containing foil, metallized film or paperboard, as appropriate. Pigmented films can also provide such protection.

Unless protected, dried foods can become contaminated with microorganisms during handling and storage. Such contamination is unlikely to cause any significant deterioration in the product as long as its water activity is maintained at a level which inhibits the growth and development of microorganisms. However, if the moisture level should rise because of poor packaging or when the food is reconstituted with water to prepare it for consumption, and if it is held at ambient temperature in that condition for extended periods of time, then the contaminating microorganisms will become active. This could lead to a health hazard to the consumer and/or spoilage of the product. Therefore, it is desirable that contamination should be kept to a minimum. Most of the packaging systems discussed above, when effectively sealed, should prevent contamination with microorganisms. Wood-based materials such as paper, paperboard and regenerated cellulose may be attacked by moulds which could lead to contamination of the packaged product. This can be prevented by protecting the product in a polymer film, inside the paper/paperboard package.

Insect infestation causes considerable losses during storage and transportation of foods. Packaging can help to reduce such losses. However, control of insect infestation is largely a question of good housekeeping. Dry, clean storage conditions, good ventilation, adequate turn-around of warehouse stocks and the controlled use of fumigants and insecticides all reduce the likelihood of infestation. Insects can infest packaged foods in two ways. Some types (invaders) gain access through openings, as small as 0.1 mm, in the package. Others (penetrators) can gnaw their way through some packaging materials. Good package design, avoiding pinholes in corners and near sealed areas, overwrapping with insect-resistant film and other similar measures can limit infestation by invaders. Rigid metal, glass and thick-walled plastic containers, suitably sealed, resist penetration by insects. Laminates containing foil are also good barriers. Packaging films vary in the resistance they offer to penetrators. OPP and OPET offer high resistance. PE, PVDC and regenerated cellulose are relatively easily penetrated. Paper, paperboard and fibreboard are easily penetrated. Insecticide may be applied to some packaging materials, e.g. to the outer surface of multi-wall paper sacks. However, such applications are strictly limited by regulations.

Physical damage is most likely to occur during handling and transportation of packaged material. Packages may be subjected to sudden shocks, impacts, vibration and compression which could lead to damage to the product. In this context, we are concerned with the consumer package and outer or secondary packaging. In the case of consumer packages rigid metal, glass, strong plastic or paperboard can protect the contents against crushing. In-package movement can be restricted by tight filling or retaining a pressure below atmospheric within the package. The most common type of outer packaging for consumer packages of dried food is the fibreboard case. This should be sufficiently strong and rigid to prevent it deforming or collapsing when stacked in warehouses. The use of cushioning materials such as corrugated board, paper or foamed plastic can reduce the likelihood of damage due to impacts and vibration.

When packaging dried fatty foods, some degree of greaseproofing is usually necessary. Very often, good quality greaseproof or vegetable parchment paper provides adequate protection. If not, films that are resistant to fats and oils may be used. Regenerated cellulose, PVC/PVDC, OPP and OPET are good barriers to fats and oils.

Tamper-evident packaging is now widely used to counter the growing incidence of deliberate contamination of foods with hazardous substances. Most attention has been paid to liquid foods in glass or plastic bottles.

However, dried foods could also be the subject of sabotage. The following are some of the methods that can be used to make dried foods in glass or plastic jars, with screw-cap closures, more secure: the jars, including the closure, can be shrink-wrapped in a thermoplastic film; a band of thermoplastic material may be positioned around the cap and then heat-shrunk, to seal the cap to the top of the jar (the band has to be torn to open the jar); a membrane may be sealed, by heat or adhesive, to the top of the jar, under the cap; there are many designs of breakable cap made of metal or plastic. In one system, the cap has a perforated security band which is tucked under the rim of the jar. To unscrew the cap the perforations have to be broken, leaving evidence that the jar has been opened. Pouches made of films and laminates are tamper-evident except for small pinholes, e.g. made with a syringe.

Ease of handling and storing applies to consumer packages and outer packaging. The design of consumer packages should be such that they can be handled without trouble on high-speed conveyors, filling and closing lines. They should be easy to stack for display in the supermarket without falling over or off the shelf. The design of outer packaging should be such that the units can be conveyed, palletized and stacked in warehouses without difficulty or damage. The labelling should comply with all legal requirements and should convey to the consumer all the information the manufacturer wishes to impart, in a clear, easily read format. The design of the package, in particular the decoration on the outside, should appeal to the consumer, attract attention in the supermarket and persuade him or her to purchase the product.

In common with all other forms of packaging, the environmental implications of food packaging should be considered. Is the packaging material recyclable or biodegradable? Is a lot of energy used during its manufacture? Are undesirable by-products created which would pollute the atmosphere?

Paine and Paine (1983) classified dried foods into four groups on the basis of their packaging requirements. *Group 1* consists of dried foods with very low moisture content, 1–3% (wwb) and low equilibrium relative humidities, below 20%. Examples are freeze-dried coffee and tea extract, vacuum-dried fruit juices and chocolate or malt-based powders. Such foods very often have a porous structure. They are hygroscopic and readily absorb water when exposed to even a moderately humid atmosphere. They are also very susceptible to oxidation and may be vacuum or gas packaged. Thus the package must be a very good water vapour and gas barrier and impermeable to volatile flavour compounds. The packaging operation may need to be carried out in an air-conditioned, low humidity atmosphere. Catering packs may consist of large metal cans with lever lids or composite containers consisting of a spirally wound cylindrical body, comprised of a paperboard/foil/film laminate with metal ends. In-package desiccants and/or oxygen scavenging agents may be included in such packs. Consumer packages may be metal cans with lever lids, glass or thick-walled plastic jars with screw caps or pouches of laminate containing foil. *Group 2* consists of dried foods with a moisture content in the range 2–8% (wwb) and equilibrium relative humidities in the range 10–30%. Examples of such foods are spray-dried milk and coffee powders, dehydrated vegetables and soup mixes, many potato and maize-based snack foods and roasted nuts. A high level of protection is required for products in this group and materials impermeable to water vapour, gases and volatiles are used. Many dehydrated vegetables contain SO_2 which should be retained within the package. Catering packs similar to those used for group 1 products are appropriate. Consumer packages are mainly pouches made from low permeability laminates or single films, although cans and jars are also employed. Some of the more expensive snack foods may be gas packaged. Anti-caking agents may be added to powders which are to be dispensed from the package or from vending machines, if that is permitted by law. *Group 3* is made up of products with moisture contents in the range 6–30% (wwb) and high equilibrium relative humidities, 30–70%. Most cereals, nuts and dried fruits fall into this category. Such products are relatively stable and generally do not require sophisticated packaging. Multi-wall paper sacks and polyethylene sacks are used for large quantities (20–50 kg). Consumer packages may consist of pouches made from laminates or single films, paper sachets or bags and paperboard cartons. Insect infestation can be a problem with such foods. Some dried fruits contain added SO_2 to inhibit non-enzymic browning. A barrier to this gas is required in such cases. *Group 4* consists of high moisture fruits, 25–40% (wwb), with very high equilibrium relative humidities, greater than 80%. Some types of prunes fall into this category. They are susceptible to spoilage by yeasts and moulds and to drying out. They may be hot-filled at 80–85°C. Film pouches are used for this product. The material must withstand this high temperature and be a good barrier to water vapour (Brennan, 1971; Paine and Paine, 1983; Bakker, 1986; Guise, 1989; Brennan et al, 1990; Paine, 1990; Robertson, 1993). *See also* Water activity and food quality.

Particle density *see* Absolute density.

Pascal (Pa)
An SI unit of pressure equivalent to $N\,m^{-2}$.

Peeling methods
Most vegetables and many fruits have to be peeled prior to dehydration. Peel may be removed mechanically by abrasion or with knives, chemically using hot sodium hydroxide or thermally using a flame, hot water or steam. *See* Abrasive peeling, Flame peeling, Hot water peeling, Steam peeling.

Percentage absolute humidity
The ratio of the absolute humidity of moist air at a given temperature to its saturation humidity at the same temperature.

Assuming ideal gas behaviour

$$H_p = 100 \frac{p_v (P - p_w)}{p_w (P - p_v)}$$

where H_p is the percentage absolute humidity, p_v the partial pressure of water vapour in the system, P the total pressure of the system and p_w the vapour pressure of water at the same temperature.

Percentage relative humidity *see* Relative humidity.

Peristaltic pump *see* Pumping of foodstuffs.

Pipe blancher *see* Blanching.

Piston pump *see* Pumping of foodstuffs.

Plate freezing
A freezing method in which the food is brought into contact with a cooled metal surface. Freezing is brought about by heat transfer from the food to the cooled plate. *See also* Freezing methods.

Pneumatic conveying
This method of moving solid particles in an air stream is widely used for handling dehydrated powders and granules. One particular application is for conveying spray-dried powders from discharge points in spray-drying systems to storage silos or packaging areas. Pneumatic conveying is also used to recycle fine powder collected by cyclones back into the wet zone of the spray-drying chamber. Theoretically, particles will be conveyed in an air stream flowing at a velocity in excess of the entrainment velocity for those particles. This velocity may be calculated, for spherical particles, by the equation on page 53. However, in practice velocities higher than this have to be used for pneumatic conveying systems. In such systems, particles must be accelerated not just suspended in the air. The velocity must be high enough to overcome frictional forces between the particles and between the particles and the surface of the ducting. Different velocities may be required for vertical runs as compared with horizontal runs. In many cases, the critical velocity for pneumatic conveying may exceed the entrainment velocity by a factor of at least four. Velocities in the range 15–25 m s^{-1} are used for most food applications.

There are three arrangements for pneumatic conveying. In one arrangement, a blower is located near the air inlet and the air is 'pushed' through the system which is under a positive pressure. This arrangement is used when delivery is being made from one inlet to several outlets. In another arrangement, a fan is located near the air outlet and the air is 'pulled' through the system which operates under a negative pressure. This method is used when a delivery is being made from several inlets to one outlet. The third arrangement is a 'push-pull' system. The blower is located at some mid-point in the system. This is used when delivery is being made from a number of inlets to a number of outlets. An enclosed arrangement is also used in which the air is recirculated. This reduces product contamination by the air and limits dehydration. This type of system can be difficult to control and a cooler may need to be incorporated into the system to extract heat generated by the blower. A negative pressure system is usually used to convey spray-dried powder away from the chamber. A positive pressure system is usually used for recycling of fines. The higher the solid-air ratio in a conveying system the more effective it is. Too high a ratio may result in settling out of the solid and blockage of the system. Ratios of 80 kg m^{-3} are used for fine particles such as flour, while 30 kg m^{-3} is the limit for wheat grains. Single-stage centrifugal fans operate at a pressure of about 120 kN m^{-2}. Multistage centrifugal fans can reach pressures of about 140 kN m^{-2}. Roots blowers (*Figure 2.81* (page 124)) can attain pressures of up to 200 kN m^{-2}. Ducting may be of circular or rectangular cross-section. It should have a smooth internal surface and sharp bends should be avoided. The feed particles may be introduced into the air stream by means of venturi feeders or rotary valves. The particles are recovered from the air stream by means of cyclones or bag filters. Powders with high moisture contents or at high temperatures may be difficult to convey. Low humidity air should be used to avoid agglomeration. High potential electrostatic charges may be developed in such conveyors. These may lead to dust explosions, cause shock to operatives and agglomeration of particles. All sections of ducting should be earthed. Conditions in pneumatic conveying systems may be such as to cause fire or explosion hazards. Precautions similar to those outlined under safety aspects of drying, should be taken (Kraus, 1980; Brennan et al, 1990; Masters, 1991). *See also* Bag filter, Cyclone, Fan, Safety aspects of drying, Spray drying.

Pneumatic (flash) drier
A hot-air drier in which particulate feed material is introduced into a fast-moving stream of heated air. The particles become entrained in the air stream and are conveyed in it through ducting of sufficient length to give the required drying time. The dried particles are removed from the air stream by a dry cyclone or filter and the air exhausted to the atmosphere or partially recycled as appropriate. The principle of a simple vertical pneumatic drier is shown in *Figure 2.39*. The air velocity must exceed the entrainment velocity for the particles. This velocity may be calculated, for spherical particles, by the equation on page 53. Air velocities used in pneumatic drying are usually in the range 10–40 m s^{-1}. Ducting may be circular or rectangular in cross-section and should have a smooth internal surface.

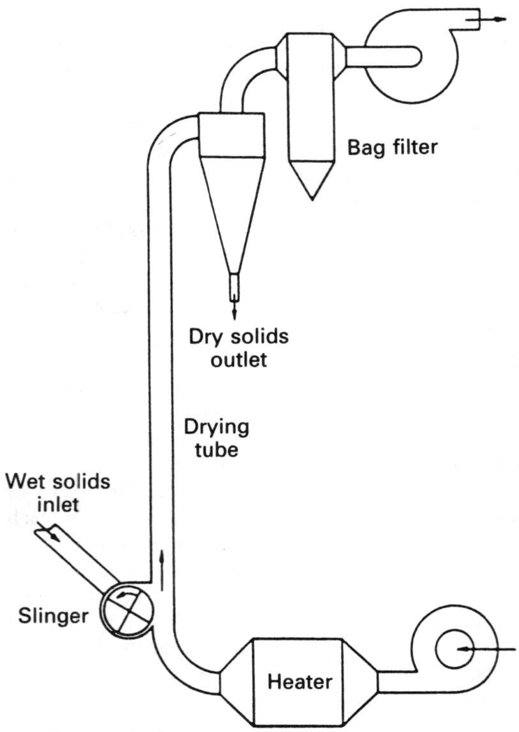

Figure 2.39 Principle of vertical pneumatic drier, after Nonhebel and Moss (1971)

High rates of heat and mass transfer are attainable in pneumatic driers. If it is mainly surface moisture that is to be removed very short drying times, 0.5–3.5 s, are adequate. This rapid evaporation of surface moisture has led to the term 'flash drying' being applied to this method of drying. Where significant internal moisture movement occurs, longer drying times are required, extending up to several minutes. The length of the ducting may range from 5 to 50 m. Where short drying times are required, simple vertical ducting, as shown in *Figure 2.39*, may suffice. For longer drying times, a number of vertical columns may be arranged in a serpentine format. Vertical ducting consisting of two concentric tubes is another option. The particles are conveyed upwards through the central tube and downwards through the annular channel between the tubes. One or more expansion chambers may be incorporated into vertical ducting. Thus, the gas velocity is alternately reduced and increased as it rises up the ducting. Heavier particles will remain in the expansion chambers until they are sufficiently light to be entrained in the outgoing air. Turbulent air-flow patterns are maintained in these chambers, which are usually biconical in shape. Vertical pneumatic driers, with or without expansion chambers, are also known as airlift driers. Such driers have been used for second-stage drying of powders coming from a spray drier. However, for this application they have been largely replaced by vibrated fluidized-bed driers. Multistage or multipass pneumatic systems are also used (*Figure 2.40*). In the pneumatic ring drier (*Figure 2.40*), the ducting is in the form of a closed loop. The food

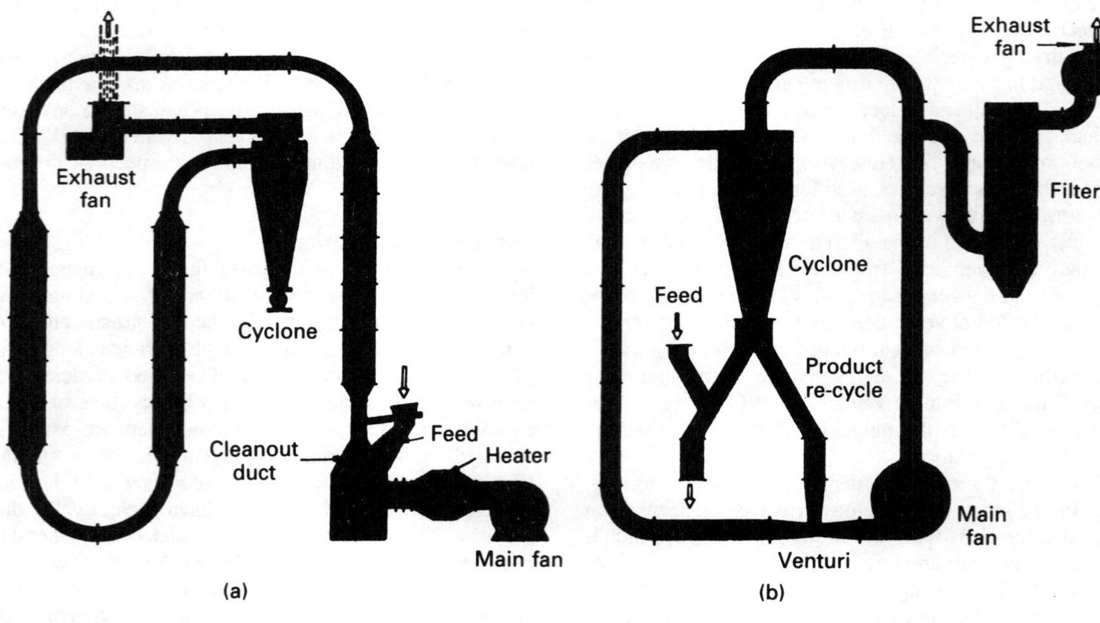

Figure 2.40 Pneumatic driers: (A) multipass (two-stage) design, (B) air recycle (ring) design (Anon, 1992)

particles travel a number of times around the loop, spending up to several minutes in the drying zone. Humid air is being exhausted continuously from the drier and replaced with fresh air, entering through the heater. Particles with a satisfactorily low moisture content are removed from the drier through a cyclone. A common method of introducing feed particles into pneumatic driers is through a venturi. Screw and vibrating conveyors, ribbon mixers and rotary air-locks are also used for this purpose. Materials to be dried in pneumatic driers should be in the form of small particles of uniform size and shape; they must flow freely in the wet condition; they should not be very heat sensitive and should not be abrasive nor tend to break up into finer particles. Mills may be incorporated into such driers to convert larger particles to a size suitable for pneumatic conveying. Back-mixing of a proportion of dry material into the feed may convert it into a form which can be pneumatically conveyed. For applications which may give rise to a fire and/or explosion hazard a self-inertizing pneumatic drying system may be used (*Figure 2.41*). This is similar in principle to the self-inertizing spray-drying system (Nonhebel and Moss, 1971; Williams-Gardner, 1976; Kisakurek, 1987; Anon, 1992). *See also* Fluidized-bed drier, Self-inertizing spray drying system.

Pneumatic nozzle *see* Two-fluid nozzle.

Pneumatic ring drier *see* Pneumatic drier.

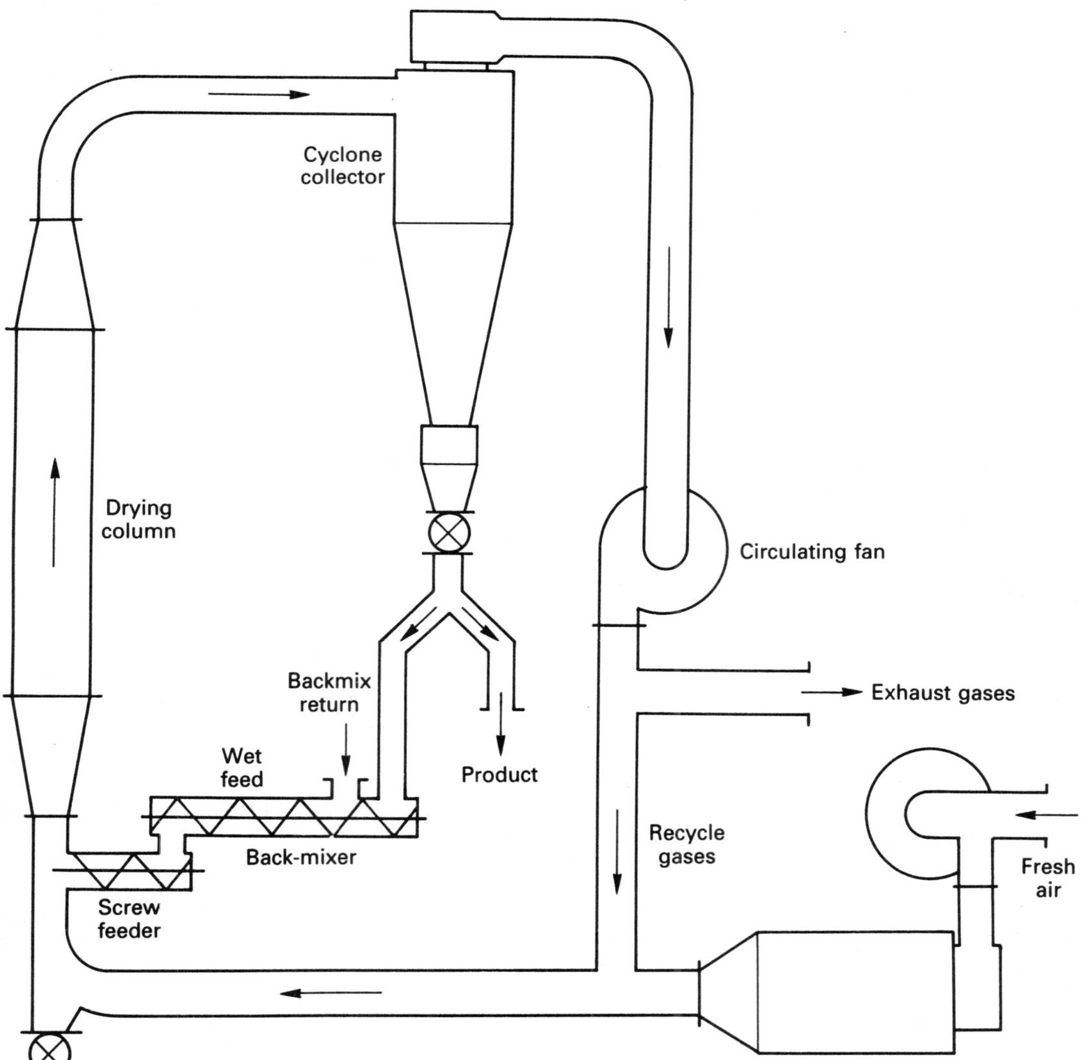

Figure 2.41 Self-inertizing pneumatic drier with backmix facility (Anon, 1992)

Poise (P)
A metric unit of dynamic viscosity equivalent to $0.1\,N\,s\,m^{-2}$.

Poisseuille (Pl)
An SI unit of dynamic viscosity equivalent to $1.0\,N\,s\,m^{-2}$.

Positive displacement pump *see* Pumping of foodstuffs.

Potential energy
The energy possessed by a body or substance due to its position, e.g. the head of a column of liquid above a reference level or datum. Potential energy may be converted into kinetic energy.

Power
The rate of doing work. Common units are W or $J\,s^{-1}$.

Pressure
Force per unit area. Common units are $kN\,m^{-2}$ or MPa or bar.

Pressure homogenizer *see* Homogenization.

Pressure jet *see* Pressure nozzle.

Pressure nozzle
This is one type of device used to convert a liquid or slurry into a spray for the purpose of spray drying. In such a nozzle, the feed is pumped at a relatively high pressure, in the range $5–50\,MN\,m^{-2}$, through a small orifice, 0.4–4 mm in diameter. A grooved core insert, or swirl chamber, positioned before the orifice, imparts a spinning motion to the liquid and a hollow cone of spray is produced (*Figure 2.42*). The energy in the pressurized liquid is converted into surface energy. Various empirical relationships have been suggested to predict the droplet size produced from such a nozzle. One of the simplest, which gives a rough prediction, was given by Filkova and Mujumdar (1987) as follows:

$$D = \frac{9575}{\Delta P^{1/3}}$$

where D is the average drop diameter (µm) and ΔP is the pressure drop across the nozzle (Pa). Pressure nozzles are capable of producing uniform droplets within a particle diameter range of 10–800 µm. The dried particles often have hollow centres. Single nozzles have a relatively small capacity, but a number of them may be used in a drying chamber. They are particularly suitable for use in tall, relatively narrow, drying chambers. Orifices may become blocked or abraded by insoluble solid particles in the feed. High pressures are required to atomize viscous feeds (Van Arsdel et al, 1973; Coulson et al, 1978; Green, 1984; Brennan et al, 1990; Masters, 1991). *See also* Centrifugal atomizer, Spray drying, Spray-drying chamber, Two-fluid nozzle.

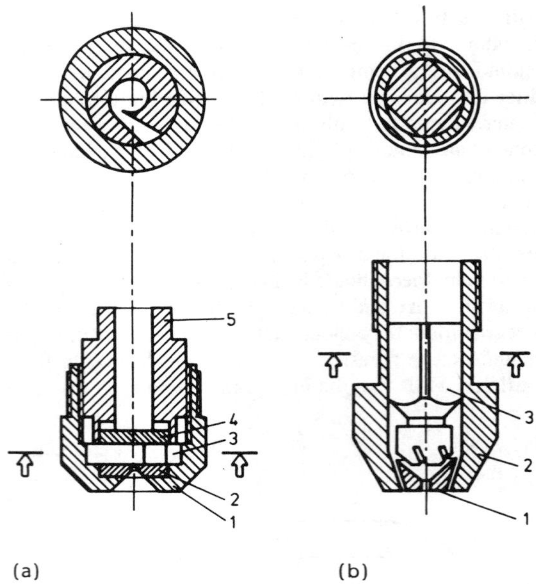

Figure 2.42 Pressure nozzles: (a) With swirl chamber: (1) nozzle body (2) orifice insert, (3) swirl chamber, (4) end plate, (5) screw pin. (b) With grooved insert core: (1) orifice insert, (2) nozzle body, (3) grooved core insert (Pisecky, 1987)

Psychrometrics
The study of gas-vapour mixtures. In this text, psychrometrics is restricted to the study of air-water vapour mixtures.

Psychrometric chart
A chart on which the properties of air-water vapour mixtures are presented. Also known as humidity chart. One form of psychrometric chart is shown in *Figure 2.43*. Many other forms are available. The use of the chart may be explained with the aid of *Figure 2.44*. Point 'b' represents a sample of moist air at a dry-bulb temperature θ_1 and absolute humidity H_1. The percentage of absolute humidity of this air, H_{p1}, is obtained from the curve of constant percentage absolute humidity passing through point 'b'. If 'b' does not lie exactly on one such curve then interpolation between curves is necessary. A horizontal line through 'b' represents what happens if that air is heated or cooled at constant absolute humidity. Thus point 'c', on the 100% percentage absolute humidity curve, is the dewpoint of the air and the dewpoint temperature is indicated by 'd'. Line a-b is an adiabatic cooling line, i.e. a locus of points representing air samples with a constant adiabatic saturation temperature. Point 'a' on the 100% percentage absolute humidity curve represents the original air sample at 'b' if it cooled adiabatically until it was saturated. Thus θ_{s1} is the adiabatic saturation temperature and H_{s1} the saturation humidity of that air. The humid heat of the original sample of air represented by

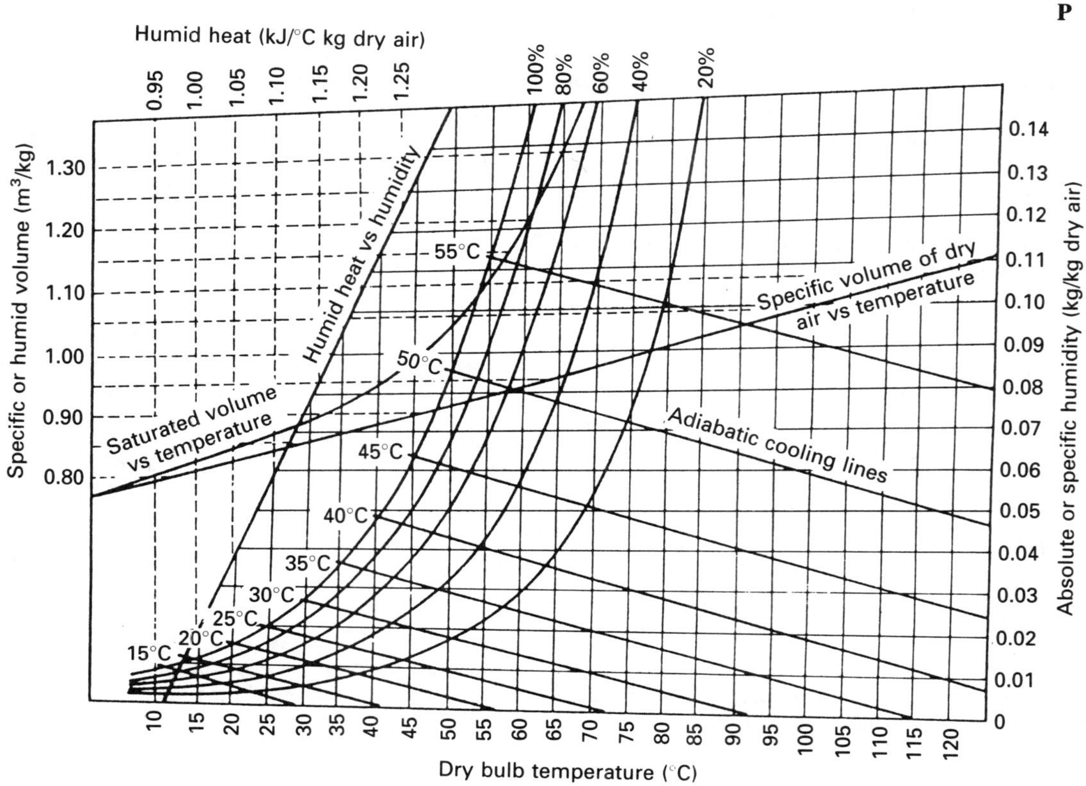

Figure 2.43 Psychrometric chart (Brennan et al, 1990)

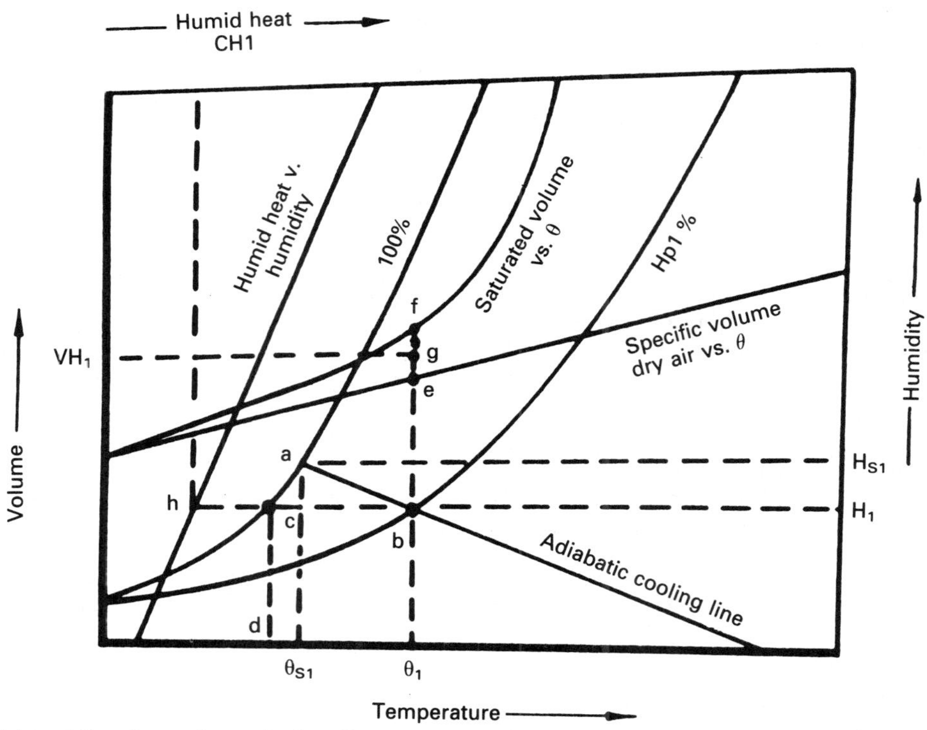

Figure 2.44 Interpretation of a psychrometric chart (Brennan et al, 1990)

point 'b' is obtained with the use of the humid heat versus absolute humidity line. Point 'h' represents the humid heat of the sample and its value is C_{h1}. The humid volume of the original air sample is obtained by interpolating between the straight line plot of specific volume of dry air versus temperature and the curved plot of saturated volume versus temperature. Point 'g' represents the humid volume of the original air sample and is located so that the distance ratio is:

$$\frac{\overline{eg}}{\overline{ef}} = \frac{H_{p1}}{100}$$

Other useful properties of air/water-vapour mixtures may be included in a psychrometric chart such as enthalpy, latent heat of evaporation and vapour pressure. Similar charts are available for other gas/vapour mixtures (Van Arsdel et al, 1973a; Hall, 1978; Green, 1984; Brennan et al, 1990).

Puff drying

This term is used to describe either (a) explosion puffing or (b) vacuum puff drying. These techniques are discussed elsewhere in this text under these titles.

Pumping of foodstuffs

Pumps are used in the food industry for filling and emptying, transporting and dosing duties. Low viscosity liquids, slurries, highly viscous liquids and pastes may all be pumped. In recent years, there has been considerable interest in pumping relatively large food particles without breaking them down or damaging them. Many factors have to be considered when selecting a pump for a particular duty including:

Capacity

The volumetric flow rate required will often determine the size of pump.

Delivery conditions

The pressure that needs to be developed by the pump, e.g. to what height does it have to raise the product or, if it is supplying a pressure nozzle in a spray drier, what is the optimum pressure for that purpose? Is the pump required to deliver the product at a fixed flow rate against a varying head (pressure)? If so, a positive displacement pump is required.

Suction conditions

How is the feed supplied to the pump, e.g. is the supply tank on a level with or above the pump or is the pump required to raise the feed from a tank below the level of the pump? In the latter case a self-priming pump is required.

Important properties of the product to be considered are:

Viscosity

Generally speaking the more viscous the product the slower the pump should run. Large-diameter suction piping is needed for viscous materials. If the product is shear sensitive, i.e. if it thickens or becomes less viscous when agitated, a positive displacement pump is usually required.

Nature of solids in a slurry or suspension

If the solid particles are abrasive, relatively low pump speeds may need to be used, clearances within the pump should be large enough to accommodate the solid particles and the construction material of the system should be suitably hard-wearing. If relatively large particles, susceptible to mechanical damage, are being pumped, low pumping speeds need to be used and the action of the pump should be such as to prevent the particles being subjected to high shear rates.

Acidity of the product

When pumping acid foods, the materials of construction of the system must be acid resistant. Type 316 stainless steel is used for all metal parts coming in contact with the food. Rubber and synthetic rubber resilient materials, of recognized food quality, may also be used.

Temperature

When pumping hot liquids, there is a danger that bubbles of vapour may form on the suction side of the pump and subsequently collapse within the pump, causing vibration and noise and perhaps damage to the pump. This is known as cavitation. Account must also be taken of expansion and contraction of parts as a result of variations in temperature.

Hygienic considerations

Pumps and associated pipework should be designed to avoid build-up of soil, and should be capable of being cleaned-in-place (CIP) or easily dismantled for cleaning.

The following are examples of types of pump used for foods:

Centrifugal (rotodynamic) pump

An impeller consisting of a series of curved vanes rotates within a casing. The feed enters the pump casing axially, is picked up by the vanes, accelerated and thrown from the tips of impeller into a chamber of gradually increasing cross-section. It is finally discharged through

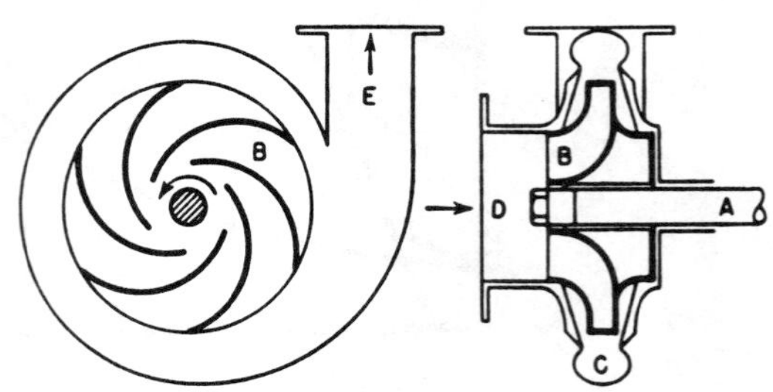

Figure 2.45 Simple centrifugal pump: (a) shaft, (b) impeller, (c) stationary casing, (d) liquid inlet, (e) liquid outlet (Green, 1984)

a tangential outlet (*Figure 2.45*). This is not a positive displacement pump. The pressure developed reduces as flow rate increases. It is not self-priming. This type of pump is best suited to handling relatively low viscosity liquids, less than 500 cp (0.5 N s m^{-2}) or slurries containing relatively small particles. It does not develop very high pressures. It is possible to reach higher pressures by having two or more in series. However, multistage systems are not widely used for food applications. Fairly high shear rates develop in this type of pump and so it is not suitable for handling shear sensitive materials. The centrifugal pump is robust, easy to clean and is relatively cheap to purchase and maintain.

Piston pump
This is a reciprocating positive displacement pump. A piston moves inside a close-fitting cylinder which is equipped with inlet and outlet valves. On the suction stroke, liquid is drawn into the cylinder and on the delivery stroke it is discharged through the outlet valve under pressure. In single-acting piston pumps, the liquid is admitted only to the portion of the cylinder in front of the piston. In double-acting pumps, the liquid enters both sides of the piston. Single-acting piston pumps are more commonly used. The output from such a pump is pulsating. The flow from a double-acting pump is more uniform. Multi-cylinder pumps are also available. These consist of three, five or seven cylinders each containing a piston. The pistons operate in sequence and are driven via a crankshaft. The output from a multi-cylinder pump varies up or down by 20%. This variation can be further reduced by using high-speed, short-stroke pistons. With piston pumps, the volume delivered is accurate and so they can be used as metering pumps. They can handle high viscosity feeds and operate at high pressure, up to 100 MN m^{-2}. Multi-cylinder piston pumps are used to supply feed to pressure-nozzle atomizers in spray driers.

Diaphragm pump
This is another type of reciprocating positive displacement pump. The pump is in two sections separated by a resilient diaphragm. In one section, a piston operates in a cylinder containing a non-corrosive fluid. The feed enters and leaves the other section via inlet and outlet valves. When the fluid is compressed by the piston, the flexible diaphragm is extended into the second section, compressing the feed and forcing it out through the outlet valve. When the piston is withdrawn, the diaphragm contracts and feed enters through the inlet valve. The flow from the pump is pulsating. The diaphragm is made of a non-corrosive material, suitable for contact with food. It is replaceable. This type of pump is suitable for handling viscous liquids and slurries and corrosive materials at pressures up to 200 kN m^{-2}. The liquid is subjected to very little shear and so it is useful for pumping agglomerates without breaking them down.

Gear pump
This is a rotary positive displacement pump. Two gear wheels rotate inside a casing. One wheel is driven while the other rotates in mesh with it (*Figure 2.46*). The liquid is carried round in the space between consecutive teeth in the casing and is ejected as the teeth come into mesh. There are no valves and there is only a small clearance between the gear wheels and the inner surface of the casing. The flow from this pump is uniform and it can operate at pressures up to 35 MN m^{-2}. It is not suitable for abrasive slurries which could promote wear of the teeth with resultant loss in performance of the pump.

Lobe pump
This is another type of rotary positive displacement pump similar in design to the Roots vacuum pump (*Figure 2.81*, page 124). As the two shaped rotors turn within the casing they touch at their tips and also touch the inner surface of the casing. Liquid is carried in the spaces formed by the lobes and discharged at increased pressure. The action of this pump is gentle and it is used for handling viscous liquids and liquids containing soft, solid particles and agglomerates.

Screw pump
A helical screw rotates within a close-fitting barrel. The liquid is carried along in the channel between the screw and the wall of the barrel and discharged at an increased pressure through an opening at the other end of the barrel. This type of rotary positive displacement pump is most suited to handling very viscous liquids.

Mono pump
A specially shaped helical metal rotor revolves eccentrically within a double-helix, resilient rubber stator

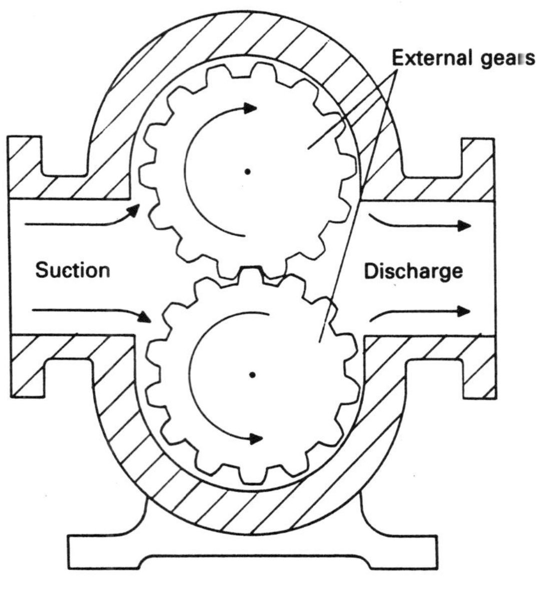

Figure 2.46 Gear pump (Brennan et al, 1990)

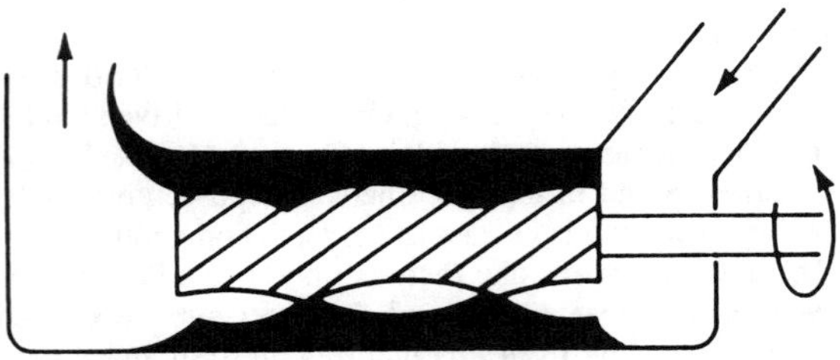

Figure 2.47 Mono pump (Lewis, 1987)

(*Figure 2.47*). A cavity is created as the rotor revolves and the material is carried along in this cavity and discharged at increased pressure through an opening at the end of the stator. The outflow is uniform and it can operate at high pressures. It can handle corrosive and abrasive materials.

Peristaltic pump (flow inducer)
The liquid is contained in a flexible tube which is fitted to a curved track mounted concentrically with a rotor equipped with three or four rollers (*Figure 2.48*). As the rotor turns the rollers flatten the tube against the track at the points of contact. A chamber full of liquid is formed between each pair of rollers. As the front roller of a pair moves away from the tube the liquid is moved along the tube towards the discharge point by the second roller. Pressures of up to $1.0\,\mathrm{MN\,m^{-2}}$ may be developed. At high rotational speeds, the outflow is reasonably uniform but at low speeds it pulsates. The liquid being pumped only contacts the inner surface of the tube which should be made of a material compatible with it. Thus it is a very hygienic pump for foods. It has a very gentle action and can handle large pieces of food. In one commercial application (Anon, 1985) the tubing is made up of an inner core of food quality perbunan and an outer covering of hard rubber reinforced with braided nylon. While the pump is running, the tube passes continuously through a sealed housing which is partially filled with a glycerin/glycol mixture to lubricate the outer surface of the hose. Any leak into or out of the hose causes a change in level of the mixture. In the event of a leak, a level detector stops the pump and sounds an alarm. The tubing is cleaned by passing suitably sized absorbent sponge balls through it at intervals. In this application pieces of chicken and cheese, orange segments and walnuts were transported by this pump (Anon, 1985; Lambert, 1989; Brennan et al, 1990; Coulson et al, 1990).

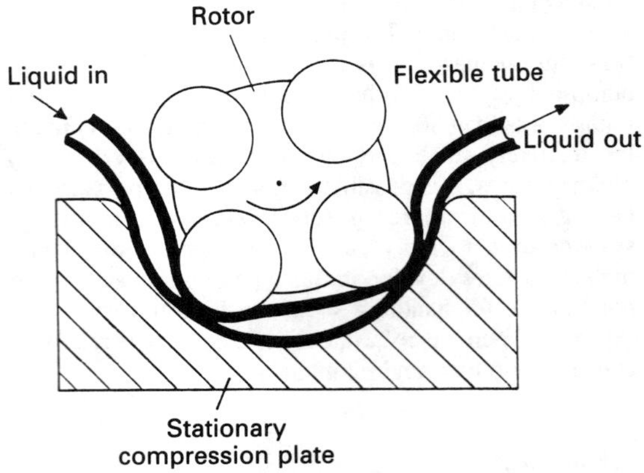

Figure 2.48 Peristaltic pump (Brennan et al, 1990)

R

R *see* Gas constant.

Raoult's law
Relates the partial vapour pressure of component 'a' in a solution, p_a, to the vapour pressure of pure 'a', P_a, and the mol fraction of 'a' in the solution, x_a, at a fixed temperature. Under ideal conditions this relationship may be written:

$$p_a = P_a x_a$$

See also Sorption isotherm.

Reconstitutability of dried food powders
This term describes the rate and extent to which dried powders pick up and absorb water and revert to a condition resembling the undried material, when placed in contact with a surplus of water, or other aqueous liquid. Properties of dry powders which influence their reconstitutability include the following:

Wettability
The ability of the powder particles to absorb water on their surfaces and so start the reconstitution process. Fine particles present a large surface area to the liquid. The whole of this area may not become wetted. Instead, the particles clump together, sharing a wetted surface. These clumps are difficult to break up. Increasing the size of the dried particles or agglomerating them can improve wettability. When agglomerates are put in contact with liquid, the interstitial air is quickly replaced with the liquid and rapid wetting takes place. Free fat at the surface of particles will also impair wettability. The application of a surface active agent to the surface of the particles may improve wettability. The addition of lecithin to whole milk powder is an example of this.

Sinkability
The ability of the powder particles to sink quickly into the liquid. The larger the size and the higher the density of the particles the more readily they will sink. Hollow particles, containing a lot of occluded gas, will exhibit poor sinkability. Centrifugal atomizers with provision for injecting steam reduce the air content of particles and increase their densities (*Figure 2.12*).

Dispersibility
The ease with which individual powder particles become distributed across the surface and throughout the bulk of the liquid. Powders with good wettability and sinkability exhibit good dispersibility.

Solubility
The rate and extent to which the components of the dried powder dissolve in the liquid. This depends mainly on the chemical composition and physical state of the powder and is greatly influenced by the drying conditions. Generally, exposure of foods to high temperatures reduces their solubility, but there are exceptions to this behaviour. Powders with a good combination of these qualities will reconstitute rapidly and are known as instant powders.

Improvements in the reconstitutability of dried powders may be achieved in different ways including:

(i) Adjustment of the drying conditions to produce relatively large, dense particles. These properties are influenced by the solids content and viscosity of the feed to the drier, the type of atomizer used, the atomizing conditions, the air inlet and outlet temperatures. By finding the best combination of these conditions, the reconstitution characteristics of the product can be optimized.

(ii) Special treatments before or after drying. Examples of such treatments are the inclusion of a crystallization stage(s) in the drying of whey and the use of lecithin to improve the wettability of whole milk powder.

(iii) Reducing the amount of very fine particles in the product. This can be achieved by recycling the fines from the cyclone of a spray drier back into the wet zone

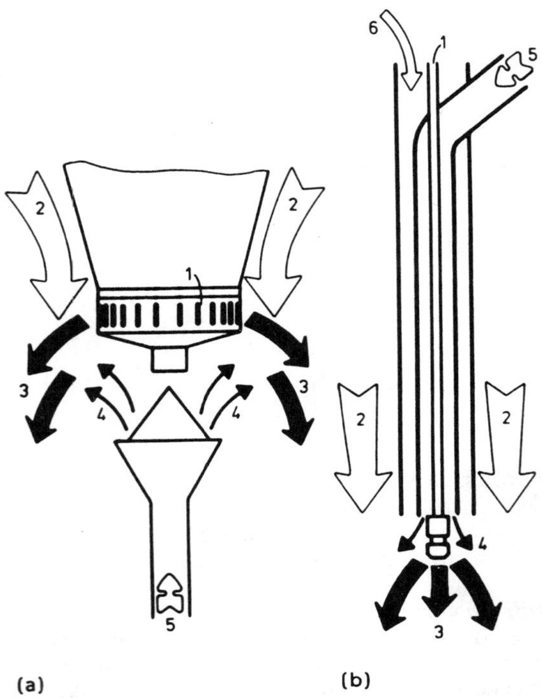

Figure 2.49 Introduction of fines to the wet zone of a spray-drying chamber for agglomeration (recycling of fines): (a) rotary wheel (atomizer), (b) pressure nozzle. (1) Feed supply, (2) drying air, (3) atomizing cloud, (4) fines cloud, (5) fines air transport, (6) cooling air (Pisecky, 1987)

of the drying chamber (*Figure 2.49*). The fine particles form agglomerates with the droplets in the spray.

(iv) The powder may be removed from the drying chamber at a moisture content of 6–8% (wwb) and drying completed in a vibrating fluidized-bed drier. Due to the sticky nature of the particles leaving the spray-drier chamber, they form agglomerates in the fluidized bed drier. These agglomerates may be cooled in a second vibrating fluidized bed and then size-graded through screens. Fine powder may be recycled as in (iii) above. This system is sometimes known as the straight-through instantizer.

(v) Spray-dried particles, produced in the conventional way, have their surfaces rewetted by treatment with steam, warm humid air, a fine mist of water or the original feed. The particles are agitated to promote interparticle collisions and agglomerates are formed. This may be done in agglomerator tubes, cyclones or fluidized beds. These agglomerates are dried back down to a table moisture content. After cooling, the agglomerates are size-graded through screens and fine particles recycled into the wet zone of the agglomerator. This process is known as rewetting. The procedure of improving the reconstitution characteristics of dried powders, mainly by methods (iv) and (v) above, is known as instantizing (Brennan et al, 1990; Masters, 1991). *See also* Agglomeration, Centrifugal atomizer, Spray drying, Vibrating fluidized-bed drier.

Recycling of fines
The small particles collected from the exhaust air leaving a spray-drying chamber, by means of a cyclone, are returned to the wet zone of the drying chamber (*Figure 2.49*). There they form agglomerates with the spray. In this way, the fine particles are removed from the product and its handling and reconstitution characteristics improved. *See* Reconstitutability of dried food powders.

Relative humidity
The ratio of the partial pressure of water vapour in moist air at a given temperature to the vapour pressure of pure water at that temperature, usually expressed as a percentage.

$$H_R = 100 \frac{p_v}{P_w}$$

where H_R is the relative humidity, p_v the partial pressure of water vapour in the system and P_w the vapour pressure of pure water at the same temperature. It is sometimes known as the percentage relative humidity. *See* Equilibrium relative humidity, Equilibrium moisture content, Water activity.

Reynolds number (R_e)
A dimensionless quantity which is used to characterize the flow pattern of a fluid in a pipe. One application is to determine whether the flow of a fluid is laminar or turbulent. *See* under Laminar flow of fluid.

Reverse osmosis *see* Membrane concentration.

Ring drier *see* Pneumatic drier.

Rod washer *see* Wet cleaning of raw materials.

Roller drier *see* Drum drier.

Roots pump (blower) *see* Vacuum pump.

Rotary drier
Some experts classify rotary driers into direct and indirect categories. In a *direct rotary drier* (*Figure 2.50*) wet material is rotated in a cylindrical shell and heated air is passed through as the material is agitated. The shell, which may be up to 3 m in diameter and up to 30 m long, is mounted on rollers at an angle to the horizontal. The slope of the shell may be from almost horizontal to 1 in 20. The shell rotates at 4–5 rpm. Flights are fitted to the inner surface of the shell. As the shell rotates, these flights lift the material up and then it cascades down through the stream of heated air. The wet feed is

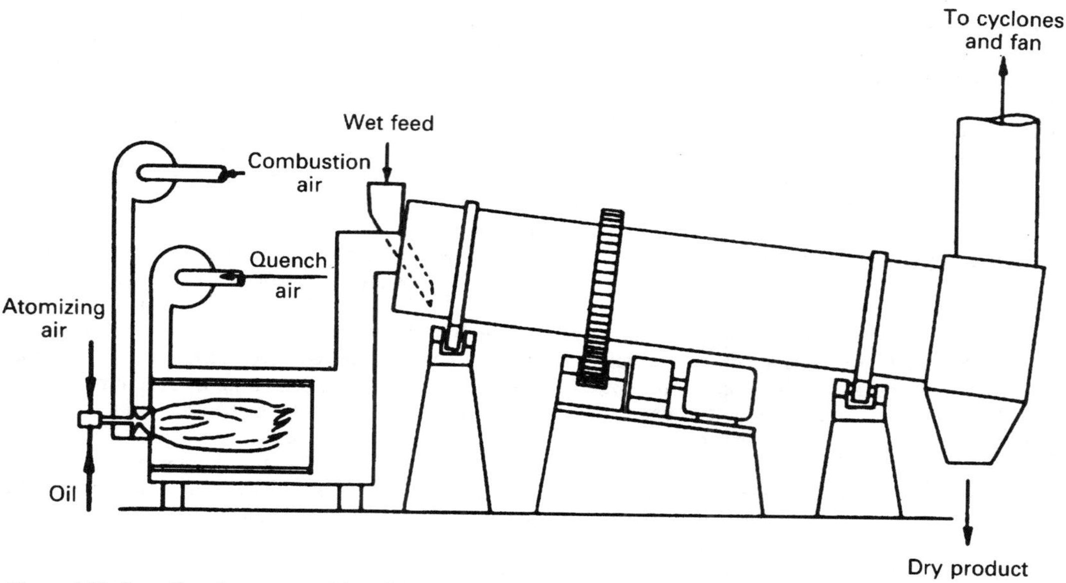

Figure 2.50 Cascading direct rotary drier with countercurrent flow (Nonhebel and Moss, 1971)

introduced continuously at the raised end of the cylinder while the dried product is discharged over a weir at its lower end. The air may be heated directly or indirectly and flow concurrently or countercurrently to the movement of the material. Air velocities in the range 1.5–2.5 $m\,s^{-1}$ are used and the residence time of the material in the cylinder may be from 5 to 60 minutes. This type of rotary drier is also known as the *cascading direct rotary drier*. The applications for this type of drier are limited to particulate material with reasonably good free-flowing characteristics. The particles must also be resistant to breakdown due to impact and abrasion forces. Where fines are likely to be formed during drying, the exhaust air is passed through cyclones or filters to recover these fine particles. Some damp, sticky materials may be converted into a form suitable for rotary drying by back-mixing a proportion of the dried material into the feed. Quite high rates of heat and mass transfer and uniform drying are attainable in such driers because of the continual mixing of the particles and the intimate contact between them and the heated air.

In the *louvred (rotary) drier* (*Figure 2.51*), the heated air enters the shell through specially shaped louvre plates and passes through a relatively deep bed of particles. As the shell rotates, the bed is subjected to a gentle rolling action which continuously mixes the particles, exposing new surfaces to the drying air and promoting uniform drying. The movement of particles in the bed is less likely to cause their breakdown to form fines, as compared with the cascading action in the rotary driers described above. This type of drier is most suited to drying coarse granular material. In the *indirect rotary drier* conduction is the main mechanism of heat transfer. The rotating shell may be jacketed and/or fitted with tubes. The products of combustion from an oil or gas burner, or steam, or hot water is circulated through the jacket and/or tubes. The solid particles are heated by direct contact with the inner surface of the shell or the outer surfaces of the tubes. They tumble over these surfaces as the shell rotates. Sufficient air is drawn through the shell to carry the water vapour formed out of the drier. This air flow may be induced by natural convection or a fan, at a relatively low velocity, typically 0.3 $m\,s^{-1}$. The air may be heated or unheated. Such driers are suited to drying materials that cannot be exposed to the products of combustion from a direct air heater, which give rise to dust formation and which exhibit a

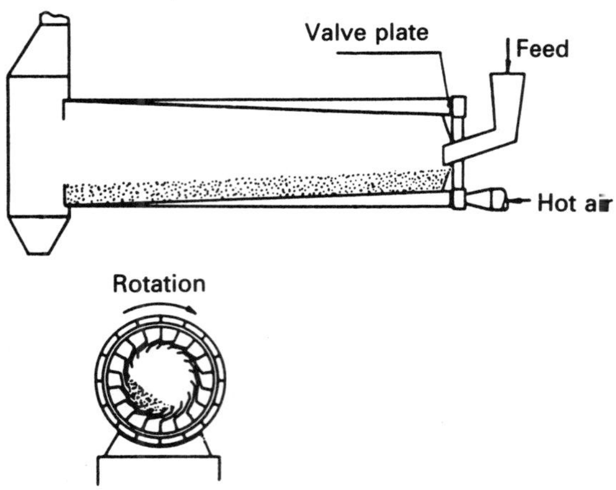

Figure 2.51 Principle of louvred rotary drier (Nonhebel and Moss, 1971)

long falling-rate drying period. Particles can be held at a constant temperature for long periods in this type of drier. Rotary driers are available which feature both heated air flow and heated jackets or tubes, so-called *direct-indirect rotary driers* (Nonhebel and Mass, 1971; Williams-Gardner, 1976; Kelly, 1987; Anon, 1992).

Rotary feeder *see* Rotary valve.

Rotary piston pump *see* under Vacuum pump.

Rotary valve
A rotor carrying six or eight blades turns within a cylindrical casing. As the rotor turns the product, usually a dry powder, it is carried in the space between successive blades from an inlet port at the top to a discharge port at the bottom of the casing. The clearance between the tips of the blades and the inner surface of the casing should be as small as possible to minimize leakage of air back into the valve. If this air is humid it may cause caking of the powder. The size of the clearance will depend on the particle size of the product. A relatively large clearance is required if the product consists of coarse granules. A smaller clearance is used with fine powders. Rotary valves are used to discharge spray-dried powder from drying chambers and cyclones. For this application the shaft bearings should be dust-tight and the shaft sealed. The drive system is isolated from the valve interior. If the powder is hygroscopic and/or sticky the valve may not operate effectively. In some valve designs, the clearance between the blades and casing is adjustable. The rotor is tapered and can be moved horizontally to change the clearance. Rotary valves are also used for discharging powders from storage bins and silos. Sophisticated designs, with very small clearances, are used for discharging dry product from vacuum chambers, e.g. from a vacuum band drier. Rotary valves (or feeders) may be used to introduce dry solids at a controlled rate into pneumatic driers or pneumatic conveying systems (Brennan et al, 1990; Masters, 1991). *See also* Pneumatic drier, Pneumatic conveying, Spray drying, Vacuum band drier.

Rotary vane pump *see* Vacuum pump.

S

Safety aspects of drying
The main hazards associated with drying are fire and explosion. A dust explosion may occur when fine solid particles are dispersed in air and ignited. This hazard is most often associated with spray driers. However, other types of driers in which fine particles are handled, or produced as a result of impact and abrasion between larger particles, are also susceptible to such hazards. Pneumatic, fluidized-bed and rotary driers are the most susceptible but dust explosions have also occurred in tray and conveyor driers and even vacuum-shelf and band driers. For a dust explosion to occur, three main conditions must be met. (1) There must be a sufficient concentration of fine particles in the air, (2) there must be sufficient oxygen in the atmosphere to support combustion and (3) there must be an ignition source strong enough to start combustion. The smaller the particle size of the solids and the higher the temperature, the more likely it is that an explosion will occur.

One of the most common causes of ignition is spontaneous combustion in layers of powder allowed to build up in drying chambers, or other parts of a drying system, where high temperatures prevail. Food powders contain constituents which are oxidizible, such as fats, proteins and sugars. Such oxidation reactions are exothermic and heat is generated. The rate of heat generation increases very rapidly with increase in temperature. Such heat is lost to the atmosphere from thin layers of powder. However, such heat loss may not occur quickly enough in thicker layers of powder and so the temperature in such layers increases rapidly to values which may cause ignition. Ignition temperatures for layers of food powders, reported in the literature, range from 130–300°C. The critical thickness for ignition of a powder layer is dependent on temperature. Critical thicknesses reported in the literature for dairy powders range from 7–17 mm at 200°C and from 100–320 mm at 100°C.

To reduce the chances of spontaneous combustion occurring, drying equipment should be designed and operated to avoid the build-up of powder layers. Feeds with high fat or sugar contents are more likely to form deposits. With such difficult feeds, devices such as hammers, chains or air brooms may be used. Cooling the drying chamber wall is another step that may be taken. Regular and thorough cleaning of drying chambers and ancillary equipment should be practised. Good measurement and control of temperature throughout the drying system will help to avoid ignition of powder layers. Other sources of ignition are possible. Heat generated by friction between moving parts may be sufficient to cause ignition. Correct assembly, adequate lubrication, good maintenance and regular cleaning of parts should minimize this source of fire. Electric sparking due to failure of equipment is another possible cause which, again, can be minimized by good maintenance. Static generated in the drier may cause sparks. All parts of the drier should be earthed to eliminate this possibility. In the case of directly heated hot-air driers, incandescent particles may be carried in the air entering the chamber and lead to ignition. Filtering of the inlet air should prevent this happening.

Explosion venting is the most common technique used to reduce risk to personnel and damage to plant and buildings if an explosion does occur. Doors may be fitted which will open quickly, at a relatively low pressure, limiting the explosion. Relatively weak panels may be fitted into drying chambers, and other parts of the drying system. These are designed to burst open at an early stage in the explosion cycle and relieve the pressure. Data, in the form of nomographs, are available in the literature to assist in the design of vents (Schofield, 1984). Containment and suppression of explosions may also be practised. Containment involves designing the drier to withstand an explosion without rupturing. This can be very expensive and is not common in commercial-scale food dehydration plant. Suppression involves the detection of the explosion at an early stage and the injection of an inert material into the chamber to halt the spread of the flame and hence the build-up of pressure. A dry, inert powder or a halogenated compound may be used as suppressant. This technique is normally only suitable for small-scale driers and is not often used. Information on containment and suppression of explosions is available in the literature (Schofield and Abbott, 1988).

For drying highly flammable materials, such as some pre-gelatinized starches, self-inertizing driers or closed-cycle driers may be used with an inert gas, usually nitrogen, as the drying medium (Beever and Crowhurst, 1989; Narayan and Mujumdar, 1987). *See also* Closed-cycle spray-drying system, Self-inertizing spray-drying system, Semi-closed-cycle spray-drying system, Spray drying.

Saturated air
Moist air in which the water vapour pressure equals the vapour pressure of pure water at air temperature. Such air contains the maximum amount of water vapour which it can normally hold at the prevailing temperature and pressure.

Saturation humidity
The absolute humidity of saturated air, usually expressed on a dry-weight basis, e.g. kilograms of water vapour per kilogram of dry air.

Assuming ideal gas behaviour

$$H_s = \frac{M_w p_w}{M_A(P - p_w)}$$

where H_s is the saturation humidity, M_w is the molecular weight of water, M_A is the molecular weight of air, P is the total pressure of the system and p_w is the vapour pressure of pure water at the same temperature.

Saturation temperature *see* Dewpoint.

Screw pump *see* Pumping of foodstuffs.

Second (s)
The basic SI unit of time. It is the duration of 9,192,631,770 cycles of the radiation corresponding to the transition between the two hyperfine levels of the fundamental state of the cæsium-133 atom.

Self-inertizing spray-drying system
A semi-closed-cycle spray-drying layout in which direct air heating is used. The combustion is controlled so that the recycled air has a low oxygen content. This reduces the risk of fire or explosion in the system. *See also* Semi-closed-cycle spray-drying system.

Semi-closed cycle spray-drying system
In this spray-drying layout, part of the exhaust air from the drying chamber is recycled through the heater and drier. If the main purpose of recycling is to recover heat from the exhaust air, then a conventional spray drier may be used. After passing through a cyclone or filter, the exhaust air is divided into two streams. One of these streams is recycled back through the heater and into the drying chamber. The other stream is released to the atmosphere, through a wet scrubber if desired. Up to 50% of the exhaust air may be recycled. However, the more air that is recycled the lower the evaporative capacity of the drier. Such recycling is usually only feasible at high exhaust air temperatures, greater than 120°C, and so its application to food dehydration is limited.

If the objective of partial recycling is to contain toxic material or unpleasant odours, larger proportions of the exhaust air are recycled. If the drying air is indirectly heated, a very small amount of the air is released to the atmosphere from a condenser/scrubber, as used in closed-cycle spray drying. The amount of air vented is equivalent to the amount which may enter the chamber due to leakage as a result of the slightly negative pressure in the chamber. The air vented from the system may be burnt off, if necessary. This layout is not suitable for materials which represent a fire or explosion risk.

If the drying air is directly heated, a self-inertizing system is possible. Special burners are used, which can bring about complete combustion of the fuel, usually gas. By controlling the combustion, the oxygen content of the recycled air can be kept low. Such a system is suitable for handling materials which give rise to a fire and/or explosion hazard. Emission to the atmosphere is also low as only a small amount of air is released from a condenser/scrubber. The amount vented is equivalent in volume to the combustion products from the burners. Thus, the risk of fire or explosion is reduced and toxic or odorous emissions limited, without the need for a gas-tight system or the use of inert gas as the drying medium, as is the case in a closed-cycle system. This layout is only suitable for aqueous feeds. If organic solvents are involved, a closed-cycle system with indirect heating must be used.

A further development of the self-inertizing system is known as the environmental spray drier. In this, complete deactivation or deodorization of the vented air is achieved in a special design of direct air heater. Thus, emissions of powder, toxic or odorous material are controlled and the risk of fire or explosion is greatly reduced (Masters, 1991). *See also* Closed-cycle spray-drying system, Spray drying.

Shear rate *see* Shear stress.

Shear stress
If a shearing force F is applied over an area A of the surface of a fluid in contact with a flat stationary surface, then the upper layer of the fluid will move with velocity v, while the layers in contact with the surface will remain stationary. The upper layer drags the layer below, with a small reduction in velocity. This in turn drags the next layer down. Thus a velocity gradient is set up in the fluid. The shear stress that the fluid is subjected to is given by:

$$\tau = \frac{F}{A}$$

where τ is the shear stress. Common units are $N\,m^2$.

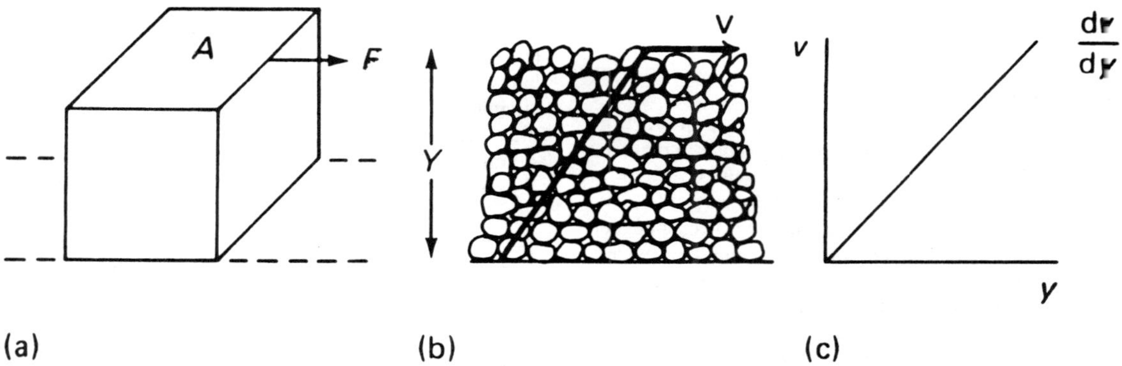

Figure 2.52 (a) Shear stress, (b) layers, (c) velocity gradient established in a fluid (Lewis, 1987)

The rate of shear is d*v*/d*y*, where *y* is the vertical distance from the flat surface (*Figure 2.52*). Common units are s^{-1} (Lewis, 1987).

Shrinkage *see* Structural changes during dehydration.

Single-drum drier *see* Drum drier.

Sinkability of dried powders *see* Reconstitutability of dried powders.

SI units
International system of units; Système International d'Unités.

Table 2.4 Seven basic SI units

Quantity	*Name*	*Unit*
Mass	kilogram	kg
Length	metre	m
Time	second	s
Temperature	degree Kelvin	K
Electric current	ampère	A
Luminous intensity	candela	cd
Amount of substance	mole	mol

Table 2.5 Derived SI units

Quantity	*Name*	*Symbol*	*Unit*
Force	newton	N	$kg\ m\ s^{-2}$
Pressure	pascal	Pa	$N\ m^{-2}$
Energy	joule	J	N m
Power	watt	W	$J\ s^{-1}$
Frequency	hertz	Hz	Cycles s^{-1}
Adsorbed dose of ionizing radiation	gray	Gy	$J\ kg^{-1}$
Area			m^2
Volume			m^3
Density			$kg\ m^{-3}$
Dynamic viscosity	poiseuille	Pl	$N\ s\ m^{-2}$
Kinematic viscosity			$m^2\ s^{-1}$
Specific heat			$J\ kg^{-1}\ K^{-1}$
Thermal conductivity			$W\ m^{-1}\ K^{-1}$
Specific energy			$J\ kg^{-1}$
Heat transfer coefficient			$W\ m^{-2}\ K^{-1}$
Velocity			$m\ s^{-1}$
Acceleration			$m\ s^{-2}$
Momentum			$kg\ m\ s^{-1}$

Table 2.6 Prefixes used with SI units

Prefix	*Symbol*	*Multiple*	*Prefix*	*Symbol*	*Multiple*
atto	a	10^{-18}	decca	da	10^{1}
femto	f	10^{-15}	hecto	h	10^{2}
pico	p	10^{-12}	kilo	k	10^{3}
nano	n	10^{-9}	mega	M	10^{6}
micro	u	10^{-6}	giga	G	10^{9}
milli	m	10^{-3}	tera	T	10^{12}
centi	c	10^{-2}	peta	P	10^{15}
deci	d	10^{-1}	exa	E	10^{18}

Table 2.7 Some non-SI metric units

Quantity	*Name*	*Symbol*	*Unit*
Mass	tonne	t	10^3 kg
Length	micron	µm	10^{-6} m
	ångstrom	Å	10^{-10} m
Temperature	degree	°	1°
	Celsius*	°C	1°K
Volume	litre	l	1 dm^3
Force	dyne	dyn	10^{-5} N
Pressure	bar	bar	$10^5\ N\ m^{-2}$
	standard atmosphere	atm	0.101 MN m^{-2}
	torr	torr	133.0 N m^{-2}
	mmHg	mmHg	133.0 N m^{-2}
Energy	erg	erg	10^{-7} J
	colorie	cal	4.187 J
Dynamic viscosity	poise	P	0.1 N s m^{-2}
Kinematic viscosity	stoke	St	$10^{-4}\ m^2\ s^{-1}$

* 0°C = 273K.

Solid density *see* Absolute density.

Solar drying

This is the oldest method of food preservation practised by man. For centuries, fruit and vegetables have been preserved by spreading them on the ground on leaves or mats and allowing them to dry slowly by exposure to the sun's rays. Strips of meat and fish have been hung on racks for a similar purpose. Foods dried in this way were exposed to the vagaries of the weather and open to contamination by insects, birds and animals. Long drying times were required and the possibility of spoilage through the activity of microorganisms and enzymes and undesirable chemical changes were high. Covering the food with a transparent material reduces these problems. Glass and transparent plastics will transmit most solar radiation incident on them. However, most radiation from heated surfaces within the enclosure will be of longer wavelength and will not pass outwards through the transparent material. Consequently, higher temperatures and shorter drying times are attainable within such enclosures as compared to those reached through direct exposure to the sun. This is known as the 'greenhouse effect'. Glass retains heat better than plastic. A simple design of solar drier is a plastic tent placed over the food material which is spread on a perforated shelf raised off the ground. Warm air moving by natural convection through the shelf and layer of food helps to speed up drying. Higher temperatures are attainable if the panels facing the sun are transparent and the others not, e.g. if they are made of pigmented plastic. A simple solar drier of sturdier construction is shown in *Figure 2.53*. Such a solar box has a relatively low capacity, the air flow rate, and hence the drying rate, is low. A solar drier of higher capacity is shown in *Figure 2.54*. In this air is heated by a flat-plate solar collector, percolates by natural convection up through a number of perforated shelves carrying layers of food and is exhausted from near the top of the chamber, through a chimney. The taller the chimney, the higher the flow rate of air. A fan can be incorporated in the drier to increase further the flow rate of air. However, a power supply is required in this case. Much research has been carried out on the design of flat-plate solar collectors to improve their efficiency. Other types of collectors are also used in solar drying. An example is an inflatable collector consisting of a black polythene tube inside a larger diameter tube of clear polythene. The black tube acts as the collector while the outer clear tube acts as an insulator. The tubes are inflated by a fan which also propels the heated air to the drying chamber. Solar heating for drying may be supplemented by other forms of heating such as a gas or oil flame. This promotes higher rates of drying. Drying can proceed in the absence of sunlight. Provision for the storage of heat from solar collectors is also possible. Heat may be stored in tanks of water (*Figure 2.55*) or in beds of pebbles or rocks. This heat can then be used to extend the drying period, when sunlight is not available. This arrangement also enables better control to be exercised over drying temperature, as compared to systems without a storage facility. As an alternative to water or rocks, heat-storing salt solutions or adsorbents may be used for storing collected solar energy. Thermal wheels, or other types of heat exchanger, can be used to recover heat from the

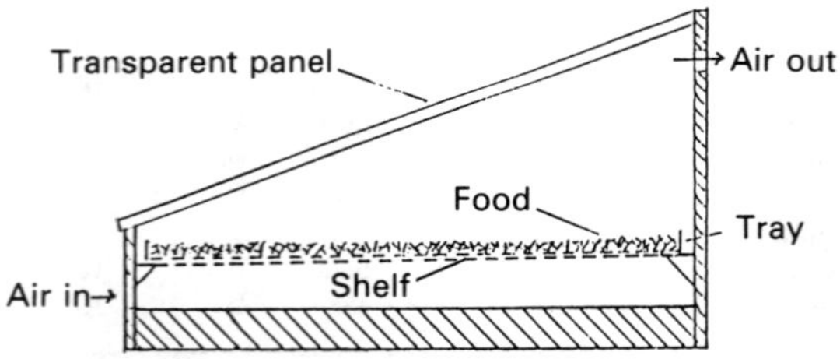

Figure 2.53 Simple solar drier (hot box) (Brennan, 1989)

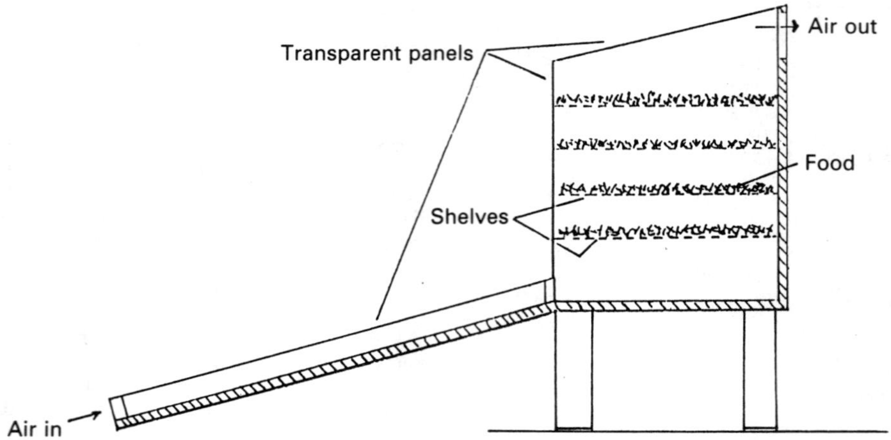

Figure 2.54 Solar drier fitted with shelves (Brennan, 1989)

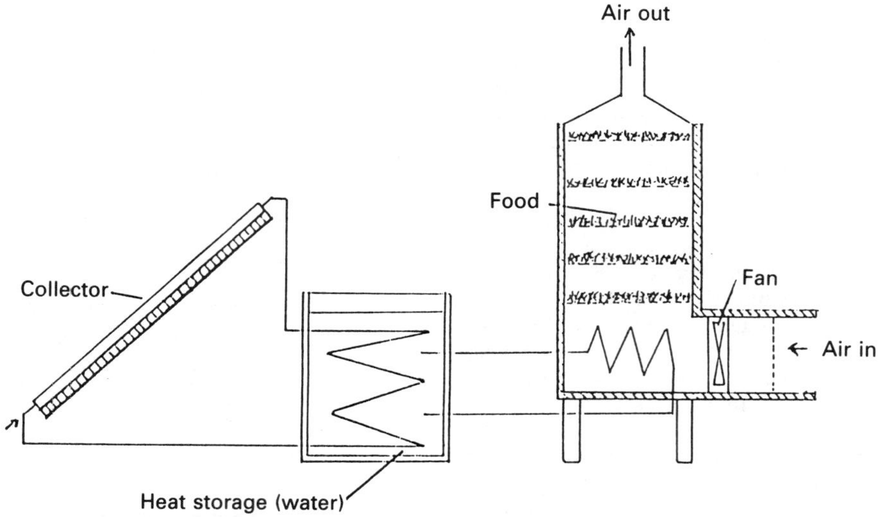

Figure 2.55 Solar drier with provision for heat storage (Brennan, 1989)

exhaust air leaving a solar drying system. Heat pumps may also be incorporated into such systems to improve their thermal efficiency. Such an arrangement is shown in *Figure 2.56*. 'A part of the enthalpy of the entering outside air is used 1 – interposing pump system 2 – for evaporating sprayed water in an evaporator 3. The water vapour goes over to the brine sprayed into tank 4. Pump 5 feeds the brine through a regenerator heat exchanger 6 into a high-pressure boiler 7. Water in the boiler is distilled with the help of solar energy obtained in a

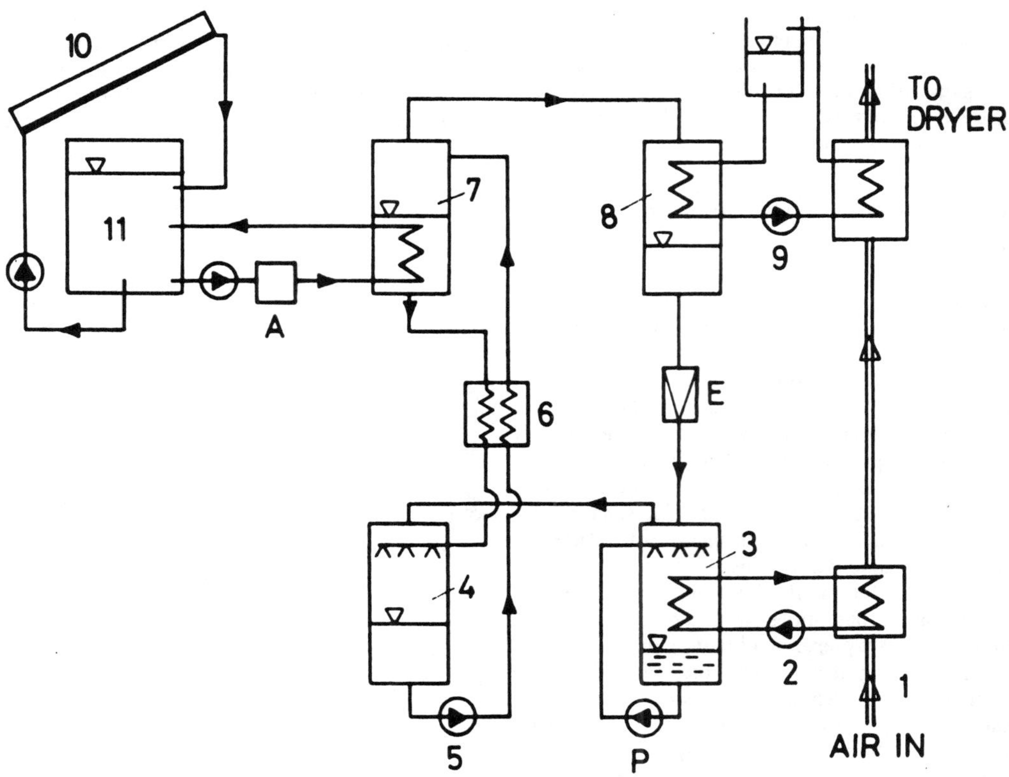

Figure 2.56 Solar drier equipped with adsorption heat pump (Imrie, 1987)

collector 10 and stored in water tank 11, and by using auxiliary energy A to the extent necessary the strong solution is led back into tank 4 through regenerator 6. The high-pressure water vapour condenses in condenser 8 and with the help of the pump heat exchanger system 9 warms the air of reduced moisture content, which is supplied to the dryer. The condensed high-pressure water flowing through an expansion valve E cools and arrives in evaporator 3. This system was originally designed for drying peanuts' (Imrie, 1987). To make good use of solar energy throughout the year, solar driers can be integrated into circuits supplying heat for other purposes, e.g. hot water. When drying is not taking place, solar energy from the collector can be used for water heating. Another use for solar energy in drying has been investigated. A desiccant was used to remove moisture from the air and solar energy to regenerate the desiccant. Activated alumina and anhydrous calcium sulphate were the most useful desiccants. Flat-plate solar collectors with tray inserts were used for regeneration (Bolin and Salunkhe, 1982; Imrie, 1987).

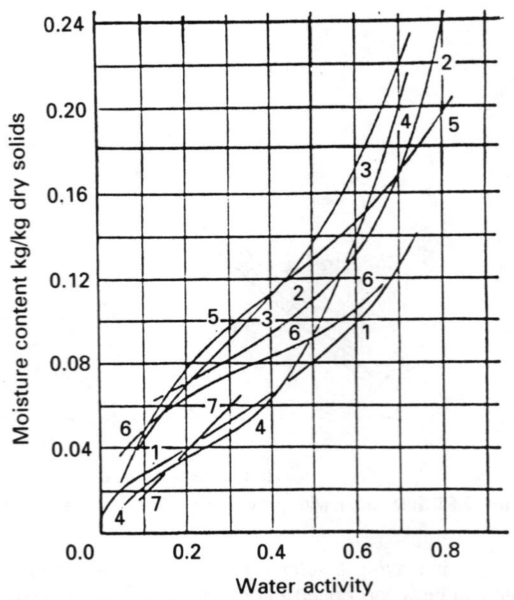

Figure 2.57 Typical water sorption isotherms of various foods: (1) egg solids at 10°C, (2) beef at 10°C, (3) cod fish at 30°C, (4) coffee at 10°C, (5) starch gel at 25°C, (6) potato at 28°C, (7) orange juice, adapted from Van Arsdel (1973a)

Solubility of dried powders *see* under Reconstitutability of dried powders.

Sorption
A general term used to describe the process whereby a solid gains or loses moisture without specifying the mechanisms involved. The terms adsorption and desorption are also used to describe a sorption process involving gain and loss of moisture, respectively. These terms also have more specific meanings which are given elsewhere in this dictionary.

Sorption hysteresis *see* Hysteresis in sorption isotherms.

Sorption isotherm
A plot of moisture content as a function of water activity, at a fixed temperature. See *Figure 2.57* for some typical food isotherms. Note the low values on the moisture content (dwb) scale. In many foods, as moisture content (dwb) approaches 1.0 kg (kg dry solids)$^{-1}$, the water vapour pressure approaches that of liquid water at the same temperature, i.e. the water activity approaches 1.0. Sorption isotherms are prepared by allowing samples of the food to equilibrate in atmospheres of different constant relative humidities, at constant temperature, and plotting the resultant equilibrium moisture contents against the relative humidities or water activities. An adsorption isotherm is prepared by starting off with a dry sample and exposing it to atmospheres of increasing relative humidity. To prepare a desorption isotherm, a wet sample is exposed to atmospheres of decreasing relative humidity. In the case of many foods if both adsorption and desorption isotherms are plotted, two different shapes of curve result, i.e. a hysteresis effect is observed (see *Figure 2.58). See* Hysteresis in sorption isotherms for further discussion of this phenomenon.

In many publications, isotherms are divided into two or three regions depending on the availability of water, as shown in *Figure 2.58*. In region A, at low levels of

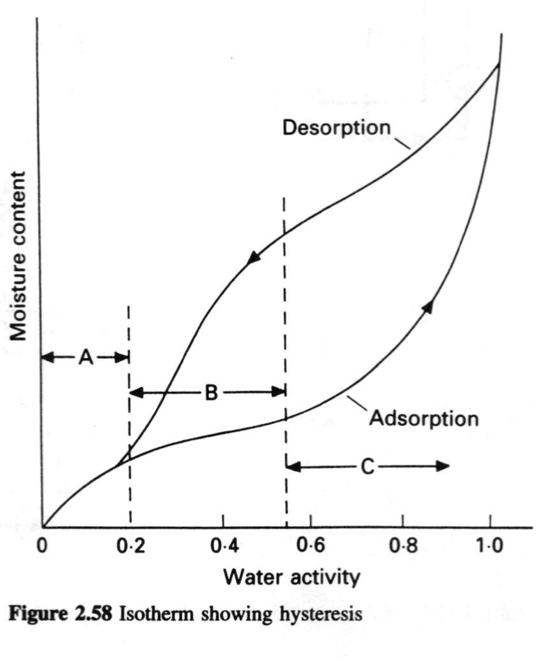

Figure 2.58 Isotherm showing hysteresis

water activity, water molecules are strongly bound to specific sites on the solid. *See* Adsorption for further discussion of this binding. Water bound in this way is difficult to remove and energy in excess of the latent heat of vaporization needs to be applied to overcome the bonds. It is usually assumed that this water is unavailable as a solvent to support chemical reactions and microbiological activity. The amount of water strongly bound in this way is known as the monomolecular moisture content. Foods are generally most stable when they contain this amount of moisture.

At moisture levels above the monomolecular moisture content, water may still be bound to the solid to some extent. Some authors state that two distinct regions exist above region A. In region B, further layers of water molecules are less strongly bound to the solid. In region C, solution and structural effects account for the lowering of the water vapour pressure of the wet material. The distinction of region B from C is dubious, as these effects occur over the whole isotherm.

The presence of solutes in solution will lower the water vapour pressure of the solution below that of pure water at the same temperature. Under ideal conditions, if Raoult's law applies, the water activity of a solution should equal the mol fraction of water in the solution as:

$$a_w = \frac{p_w}{P_w} = x_w$$

where a_w is water activity, p_w is the partial water vapour pressure exerted by the solution, P_w is the vapour pressure of pure water at the same temperature and x_w is mol fraction of water in the solution. In foods, there are usually substantial deviations from this ideal relationship. Many causes of such deviations have been put forward including: (a) not all the water present in the food acts as a solvent, (b) not all the solute present in food is in solution, (c) there may be interaction between different solutes. In Table 2.8 the a_w values of some food solutions are given. There is no clear relationship between concentration and a_w values. In the case of solid foods, even greater deviations from ideal behaviour are to be expected.

Table 2.8 Water activity of solutions of some food humectants

Solute	*Conc. % (w/w)*	*Approx. (a_w)*
Sucrose	58.4	0.90
	67.0[a]	0.86
Glucose	47.0[a]	0.82
Invert sugar	63.0[a]	0.82
Sucrose 37.6 + invert sugar 62.4	(75.0[a])	0.71
NaCl	9.3	0.94
	19.1	0.85
	27.0[a]	0.74
Lactose	14.5[a]	0.94
Maltose	48.8[a]	0.95

At values of a_w above 0.5, usually in the range 0.6–0.7, many foods exhibit a rapid increase in moisture content. This is usually assumed to mean that capillary condensation is taking place. In theory, the influence of such condensation on water activity can be represented by the Kelvin-Thompson equation. However, to make use of such a relationship, a knowledge of pore sizes in foods is necessary. Such information is not readily available. Most pore sizes in foods are relatively large, i.e. greater than 10 μm. However, there is some evidence that at relatively low values of water activity, i.e. less than 0.5, water is held in much smaller capillaries, i.e. less than 100 Å. It is likely that lowering of water activity occurs in such small capillaries. Another complication arises due to swelling of many foods as water is adsorbed. This will alter pore size and shape. For these and other reasons, the application of the Kelvin-Thompson equation has little practical significance.

Other factors affect the shape of sorption isotherms, including the state of the material, e.g. whether it is amorphous or crystalline and whether any change of phase occurs during the preparation of the isotherm. Temperature also affects the sorption behaviour of foods. The amount of adsorbed water at any given value of a_w decreases with increase in temperature (*Figure 2.59*). The relationship between water activity, a_w, and absolute temperature, *T*, may be expressed in the form:

$$\frac{d(\ln a_w)}{d(1/T)} = -\frac{Q_s}{R}$$

where Q_s is the heat of adsorption and *R* the gas constant. If data obtained from adsorption/desorption experiments at different temperatures are plotted as shown in *Figure 2.60* values of Q_s can be obtained from

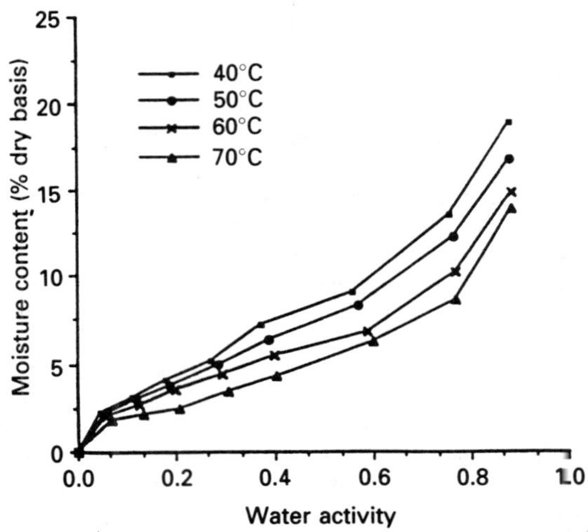

Figure 2.59 Influence of temperature on the adsorption isotherm of potato (Wang and Brennan, 1990)

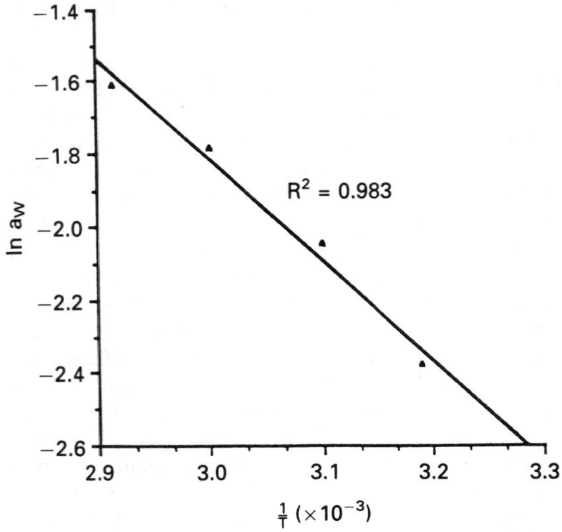

Figure 2.60 ln a_w vs $1/T$ plot for potato, adsorption at 40°C (Wang, 1992)

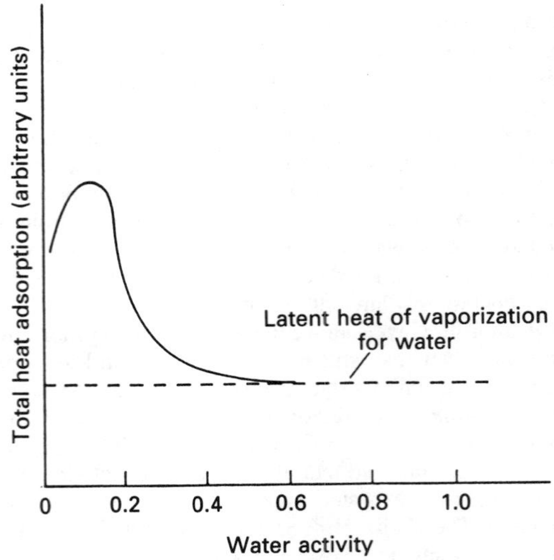

Figure 2.61 Dependence of total heat of adsorption on water activity (Brennan et al, 1990)

the slopes of the lines (Labuza, 1968 and 1977; Karel, 1974; Troller and Christian, 1978; Loncin and Merson, 1979; Chirife, 1983; Wolf et al, 1985).

Sorption isotherm equations

Many mathematical expressions to represent the sorption characteristics of foods have been published. Some are empirical in nature and are arrived at by fitting expressions to the isotherm curves. Others have been developed from a consideration of the chemistry of surfaces or the thermodynamics of adsorption. None of these expressions has received universal acceptance. Many have been found to apply only to a limited part of the water activity range. Since the suppression of water vapour pressure in foods may be due to a number of phenomena, i.e. adsorption, solution and structural influences, the sorption behaviour of foods arises from the integrated hygroscopic properties of individual components, and since changes in composition and structure occur while water is being absorbed, it is not surprising that no single mathematical model to represent the sorption behaviour of foods over the whole range of water activity has been found.

One of the earliest reported and most widely quoted equations is the BET isotherm equation (Brunæur, Emmett and Teller, 1938). This was developed from a consideration of the thermodynamics of adsorption. It is based on three main assumptions which are: (1) that the heat of sorption of the first layer is constant and equal to the sum of the heat of vaporization and a constant heat due to interaction, (2) that the heat of sorption for all layers above the monolayer is equal to the heat of vaporization, and (3) that the sorption occurs on specific sites. In *Figure 2.61* the usual behaviour of foods is represented. The BET equation may be expressed as follows:

$$\frac{a_w}{(1 - a_w)} = \frac{1}{W_m C} + \frac{a_w(C - 1)}{W_{mC}}$$

where a_w is water activity, W is moisture content (dwb), W_m is monomolecular moisture content (dwb) and C is a constant. C is related to the heat of adsorption on to homogeneous sites, Q_s, the gas constant, R, and absolute temperature, T, as follows:

$$C = k \exp (Q_s/RT)$$

where k is a constant usually assumed to have a value of unity. If $a_w/(1-a_w)$ is plotted versus a_w, values for W_m and C can be obtained from the intercept and slope of the line. Values of Q_s may be obtained from C. The BET equation has been found to apply in the water activity range 0.05–0.45.

Many reviews and discussions of sorption isotherm equations and their applications have been published (Chiriffe and Iglesias, 1978; van den Berg and Bruin, 1981; Iglesias and Chiriffe, 1982; Wang and Brennan, 1990).

Caurie (1981) proposed an equation to represent multilayer adsorption which was developed from the BET equation. This took the form:

$$\frac{1}{W} = \frac{1}{CW_m}\left(\frac{1 - a_w}{a_w}\right)^{2C/W_m}$$

This equation was tested on selected protein materials over the water activity range 0.05–0.95. Values of monomolecular moisture content calculated using this equation were higher than those obtained using the BET equation and values of the constant C were lower. The exponent $2C/W_m$ was related to the number of adsorbed layers. For a monolayer, this exponent has a value of 2 and for multilayer adsorption it had lower values. In a study of proteins and wheat it was found that most of the samples adsorbed at most 4 to 6 layers.

In recent years, the GAB (Guggenheim-Anderson-De Boer) equation has been widely used. This takes the form:

$$\frac{W}{W_m} = \frac{Cka_w}{(1 - ka_w)(1 - ka_w + Cka_w)}$$

where C is the Guggenheim constant and:

$$C = c' \exp[(Q_1 - Q_m)/RT]$$

where k is a factor correcting properties of the multilayer molecules with respect to the bulk liquid and:

$$k = k' \exp[(Q_1 - Q_q)/RT]$$

where Q_1 is the heat of condensation of pure water, Q_m is the total heat of sorption of the first layer of water molecules on primary sites and Q_q is the total heat of sorption of the multilayer water molecules. This equation was found to fit sorption data up to a water activity of 0.9 and was used and recommended by the European project group COST 90 on physical properties of food. Wang and Brennan (1990) found that the GAB equation fitted well to sorption data for potato (Bizot, 1983; Schar and Ruegg, 1985).

Sorting of raw materials

This involves the separation of raw foods into groups on the basis of physical properties including size, shape, weight and colour. It is an important stage in the preparation of foods for processing, including dehydration. Sorted foods are suited to mechanized operations such as peeling, pitting, coring and slicing. When dried whole, or when subdivided, uniformity in size and shape facilitates uniform processing. A product consisting of uniformly sized pieces is attractive to the consumer and easily served in uniform portions.

Sorting on the basis of size

Screens are commonly used for this purpose. They may consist of metal sheets with holes or slots, wire mesh or woven fabrics such as cotton or silk. The aperture size is chosen to suit the size of the food particles being sorted. Using a single screen, food particles may be sorted into two size fractions. By using a number of screens, in series or parallel, several size fractions may be sorted. Screens may be stationary and sloping, or flat-bed and vibrating. Drum screens are also used and may be arranged concentrically or consecutively, in series or parallel. Stationary screens are used for sorting vegetables, such as potatoes, carrots and turnips. Vibrating flat-bed screens and drum screens are used for raw materials such as peas, nuts and cereals. Size sorters featuring variable apertures are also available. One type consists of a pair of inclined driven rollers with a diverging gap between them. Smaller items fall between the rollers before the larger items. Different size fractions are collected in padded chutes or boxes located beneath the rollers. Instead of rollers, inclined driven ropes or cables may be used. Roller conveyors in which the gap between the rollers is arranged to increase from the inlet to the outlet end are also used. In belt sorters, the food is carried along a continuously diverging slot produced by driven inclined belts. A screw-sorter consists of two partially intermeshing helices. The gap between them increases stepwise, enabling size sorting to take place. Photoelectric sensing devices may be used to sort on the basis of length or width. The food items are conveyed at constant speed underneath a light source and the sensing device measures the time it takes each item to pass a fixed point. Video cameras, linked to computers, may be used in a similar manner for sorting on the basis of size. They may also detect surface defects and colour differences.

Sorting on the basis of shape

In a disc sorter, a number of discs are mounted on a rotating horizontal shaft. As they rotate, the discs are partially submerged in a bed of the food particles. Particles of the correct shape are collected in indentations on both sides of the discs. In a cylinder sorter, the internal surface of a rotating horizontal cylinder is machined with shaped indentations. As food particles are tumbled in the cylinder those of the correct shape are collected in the indentations, carried to the top of the cycle and discharged into a chute or on to a belt which conveys them out of the cylinder. Near-spherical food particles may be separated from less spherical particles by feeding on to a moving inclined belt. The less-spherical particles will be carried to the top of the belt and discharged, while the more spherical particles will roll down the belt.

Sorting by weight or density

Weight sorting is seldom used when preparing foods for dehydration. Some tropical fruits are sorted by weight. Aspirators may be used to sort small food particles on the basis of density (*Figure 2.20*).

Sorting on the basis of colour/appearance

Relatively small particulate foods may be sorted at high rates using photometric detectors. The particles are made to flow past such detectors. The signal generated by the light reflected from the particles depends on their colour and surface texture, i.e. matt or glossy. The signal is compares automatically with a pre-set standard signal. If it compares unfavourably with the standard a rejection mechanism is activated. The colour of the background and the type and intensity of the light used to illuminate the food particles are controlled for each product. Surface defects such as worm holes or bruises may also be detected by such devices. Cereals, nuts, peas and beans may be colour-sorted in this way. Larger items of food such as apples, potatoes and onions may be similarly colour-sorted but are carried past the detectors on roller conveyors, belts or on vacuumized pick-up wheels. Rejection of small particulates is effected by air jets. For larger units, diversion gates may be opened on conveyors or chutes. Where a pick-up wheel is used, rejected units are released on to a separate conveying system to the main food stream. In recent years video cameras, located above conveyors, have been used for colour-sorting. In one sophisticated system, the relative intensities of red, green and yellow light are sent to a computer which puts together a composite image of each food item. This is compared with pre-set specifications. If the comparison is unfavourable, a rejection system is activated. The transmission of light waves or X-rays through foods may be used to detect internal defects. The candling of eggs to detect blood spots or other defects is an example of this technique (Fellows, 1988; Brennan et al, 1990). *See also* Candling of eggs, Cleaning of raw materials, Grading of raw materials.

Specific gravity
The mass of a substance divided by the mass of an equal volume of water. It is a dimensionless quantity.

Specific humidity *see* Absolute humidity.

Spin flash drier
In this type of fluidized-bed drier the bed is mechanically agitated. It is designed to cope with very viscous liquids or solid feed materials that cannot be spray dried and are difficult to fluidize in conventional beds. It features a vertical, cylindrical drying chamber with an inverted conical bottom (*Figure 2.62*). The heated air enters tangentially through an annular inlet at the bottom of the chamber. This imparts a whirling motion to the particles in the bed. These particles are further agitated by an axially mounted rotor. The wet feed in the feed hopper may be agitated with a low-speed impeller to break up large lumps. It is delivered into the bed by a feed-screw. As they fall towards the bed the lumps of wet feed become coated with dry powder. These coated lumps are dispersed throughout the bed by the action of the mechanical agitator and the fluidizing air. As the particles dry, they tend to rise up the wall of the drying chamber. Some are back-mixed with the incoming feed. At the top of the chamber they must pass through a classification opening. Large particles fall back on to the fluidized bed and the finer ones are carried in the outgoing air stream to the bag filter where they are separated from the air. The air is discharged by an exhaust fan. The particular features of this type of drier that make it suitable for drying heat-sensitive materials are (1) that the dry powder leaves the bed when it becomes light enough and (2) the moist powder in the bed remains at a temperature below that of the exhaust air. A cooling jacket around the base of the chamber, above the inlet port, is an optional extra. This would assist in maintaining a low bed and wall temperature. Semi-closed and closed-cycle versions of this type of drier are available for handling non-aqueous solvents or drying gases other than air. These are similar in principle to semi-closed and closed-cycle spray-drying systems. The spin flash drier has so far found little application in the food industry but it has considerable potential. It offers the possibility of removing some of the moisture from wet foods by mechanical means, e.g. filtration or centrifugation, followed by drying of the cake in this type of drier. This should lead to significant savings in energy as compared with straight-through drying in more conventional equipment (Anon, 1992). *See also* Closed-cycle spray drying, Fluidized-bed drying, Semi-closed-cycle spray drying.

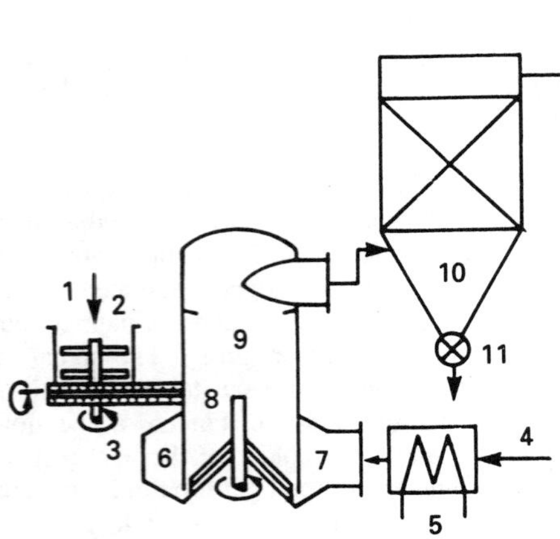

Figure 2.62 Spin flash drier: (1) feed vat, (2) agitator, (3) feed screw, (4) air in, (5) air heater, (6) hot-air inlet plenum, (7) annular inlet, (8) axially mounted rotor, (9) drying chamber, (10) bag collector, (11) rotary discharge valve, (12) exhaust fan (Anon, 1992)

Spiral cartridge reverse osmosis or ultrafiltration membrane *see* Membrane concentration.

Spouted-bed drier
A fluidized-bed drier in which a proportion of the drying air enters as a jet through the conical base of the chamber (*Figure 2.63*). This creates a spout of fast-moving particles up through the centre of the bed. The particles return to the base of the chamber in the form of a slow-moving annular bed surrounding the spout. Some of the drying air flares out into this annular bed and travels upwards, countercurrent to the flow of particles, at a relatively low velocity. High rates of heat and mass transfer are achieved in the spout, leading to high initial rates of drying. Evaporative cooling maintains the particle temperature at a moderate level. The final stages of drying, in which moisture movement within the particles determines the rate of drying, takes place mainly in the annular bed. Thus, this type of drier is suitable for drying heat-sensitive particulate solids. It is more suited to drying large particles, with diameters greater than 5 mm, than a conventional fluidized bed. A high-pressure drop needs to be developed to initiate

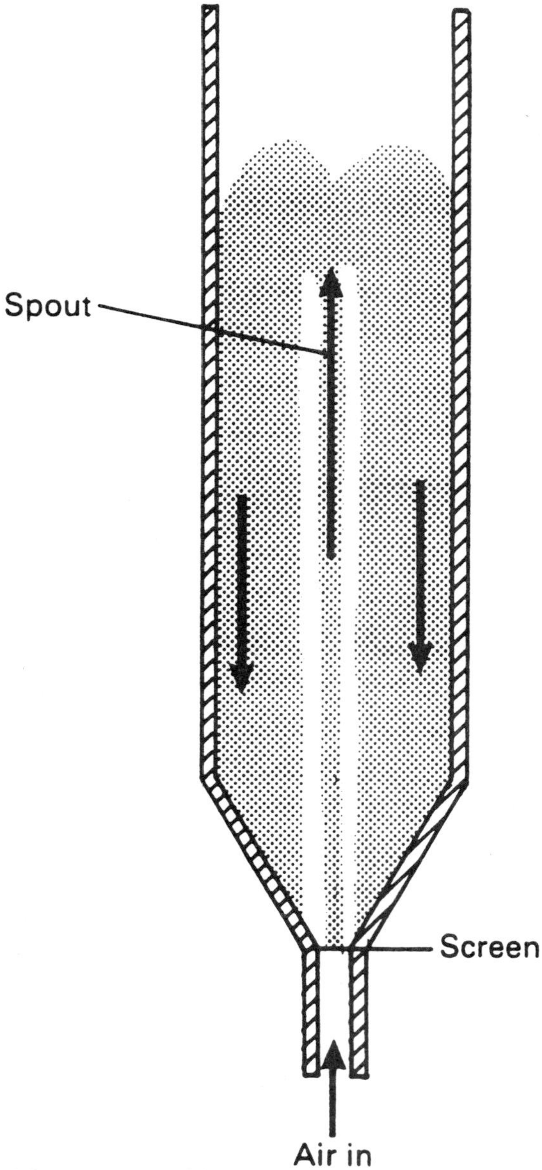

Figure 2.63 Spouted fluidized bed

spouting, but then the pressure drop reduces to a value less than that developed in a conventional fluidized bed. There is a minimum air velocity and maximum bed depth which will sustain the spouted bed. The values for these parameters will depend on the design of the chamber and the properties of the particles. The bed height can be increased if a draft-tube is introduced around the spout. Another design of spouted-bed drier features tangential air feeding and is known as a vortex-bed drier. This method of introducing the air reduces the high pressure loss caused by the central inlet in the conventional spouted bed. A screw conveyor may be positioned vertically in the centre of the drier to control the upward movement of the particles, independent of the air velocity (*Figure 2.64*). This arrangement is suited to drying relatively small particles.

The spouted-bed drier may also be used for drying liquids. The bed consists of inert particles. The liquid feed is sprayed on to these particles, coating them. Heat is supplied by conduction from the inert particles and convection from the heated air. The dry film formed is released from the particle surfaces by abrasion and impact between particles, and is separated from the exhaust air by means of a cyclone. Metal and glass spheres and various plastic particles have been used as inert material. Cattle blood and other liquid foods have

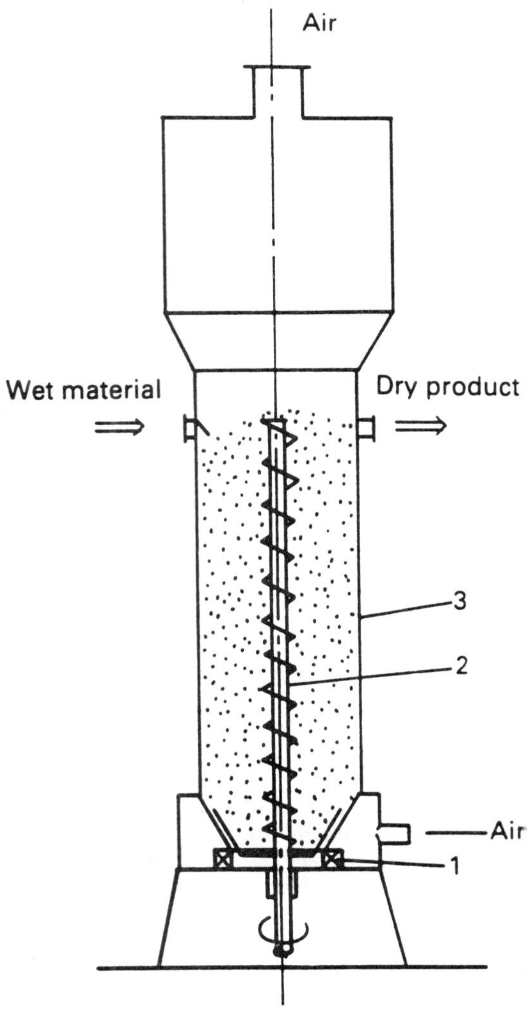

Figure 2.64 Spouted-bed drier with tangential air inlet and a central conveyor screw: (1) tangential air inlet, (2) conveyor screw, (3) drier body (Pallai et al, 1987)

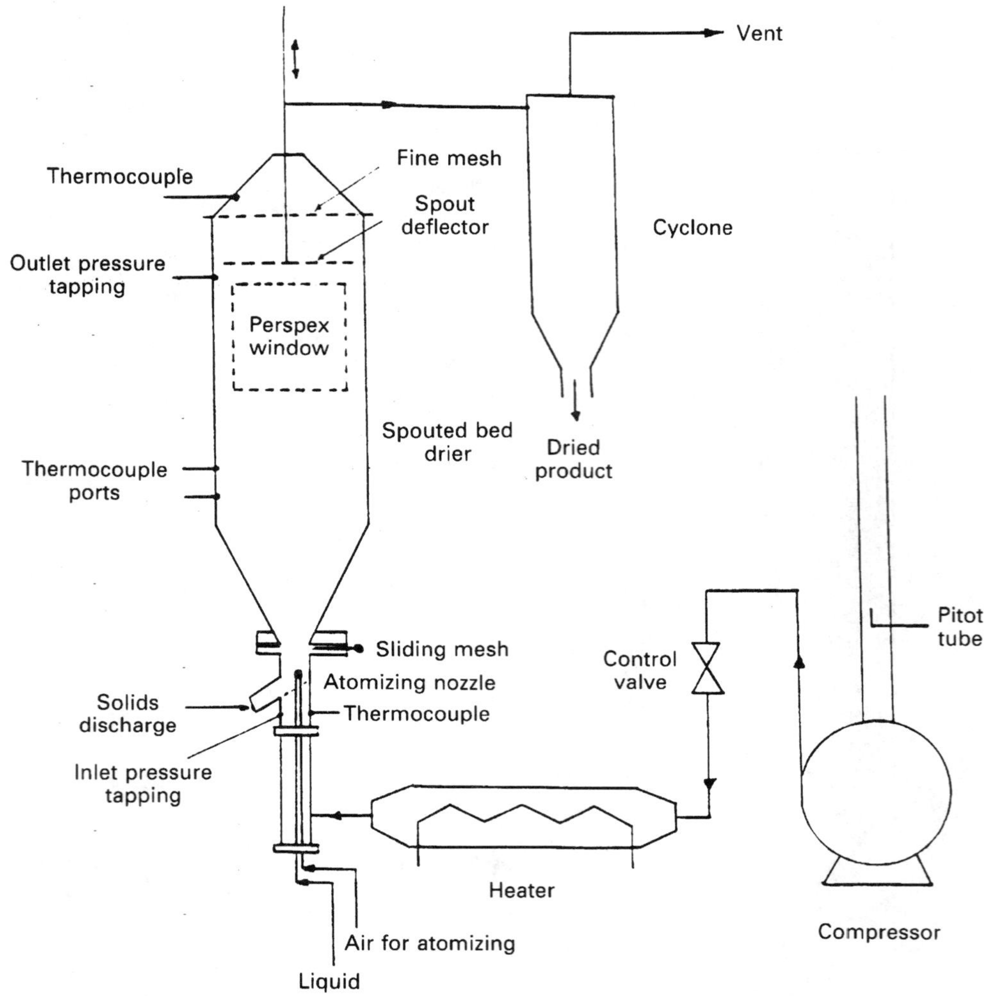

Figure 2.65 Experimental spouted-bed drier for liquid foods (Brennan, 1992)

been dried in this way. In a recent study (Osorio, 1990), the performance of an experimental drier of this kind (*Figure 2.65*) using polypropylene beads as inert material, was compared with that of a laboratory spray drier, using a model liquid food. The results indicated that the spouted-bed drying process was more severe than spray drying, that the spouted-bed equipment had some six times the capacity of the spray drier, although they were of comparable physical size and that the spouted bed was less thermally efficient than the spray drier (Fane et al, 1980; Strumillo et al, 1983; Hovmand, 1987; Pallai et al, 1987). *See also* Fluidized-bed drier.

Spray chilling
A technique used to encapsulate solid food particles in lipid material. It provides for controlled release of core material. It is applied to vitamins and minerals. *See* Encapsulation.

Spray drying
A method for drying food liquids and some slurries in which the feed is converted into a fine mist or spray and is mixed with heated air in a drying chamber. Rapid drying takes place and a dry powder is formed. The droplets have diameters in the range 10–300 μm and present a very large surface area to the drying air. The distance moisture has to migrate within the particles to the surface is relatively small. Drying times are short, of the order of 1–20 s. Evaporative cooling of the surface of the droplets maintains the surface temperature close to the wet-bulb temperature of the drying air until drying nears completion. If the particles are removed from the drying chamber quickly, once they are dry, heat damage to the product can be limited. The main components of a spray-drying plant are shown in *Figure 2.66*. Inlet fan A draws air in through filter B, then through heater C into drying chamber D. The feed material is pumped

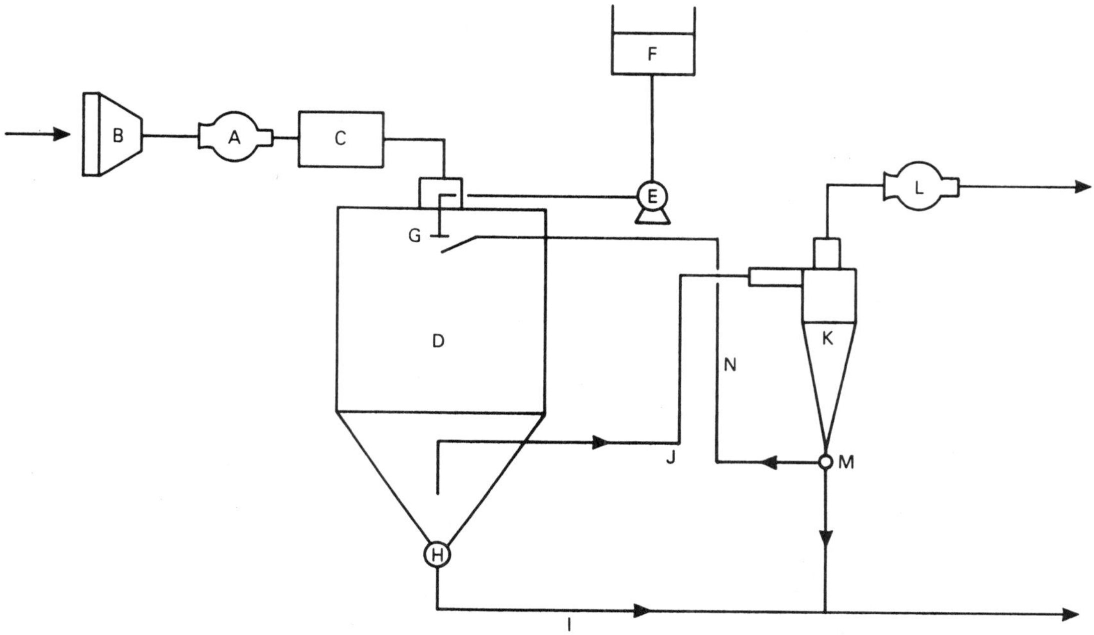

Figure 2.66 Single-stage spray-drying system with recycling of fines

from tank F by pump E to the spray-forming device G. The spray is formed and mixes with the heated air in the drying chamber D. The bulk of the powder produced is removed from the drying chamber through valve H and is conveyed away to storage in duct I. The air is exhausted from the chamber by outlet fan L via duct J through powder separator K. The fine powder recovered by the separator via valve M may be added to the main product stream or recycled to the wet zone of the drying chamber through duct N. The system described here is an open-cycle system. Semi-closed- and closed-cycle systems are used for certain applications (Van Arsdel et al, 1973; Coulson et al, 1978; Green, 1984; Masters, 1985; Filkova and Mujumdar, 1987; Pisecky, 1987; Brennan et al, 1990; Brennan, 1993). *See also* Air filters, Air heating, Air/powder separators, Closed-cycle spray-drying system, Fans, Heat exchangers, Jet-spray drier, Recycling of fines, Semi-closed-cycle spray-drying system, Spray-drying chambers, Spray-forming devices.

Spray-drying chamber

It is in the drying chamber of a spray drier that the feed contacts the heated air and drying takes place. The residence time of the droplets/particles in the chamber must be sufficient for them to reach a suitably low moisture content. However, once the particles are dry, they must be removed quickly from the chamber in order to avoid unnecessary heat damage. There are many designs of drying chamber in use. These may be classified into three groups, depending on the relative direction of flow of the feed and air through the chamber. The three classes of drying chamber, i.e. concurrent, countercurrent and mixed-flow, are represented in *Figure 2.67*. Concurrent chambers are most commonly used for food drying applications, as they are least likely to lead to heat damage to the product. They often consist of a cylindrical body with a conical base section. For drying relatively large droplets of a heat-sensitive feed and when the dried particles are sticky when hot and tend to adhere to the chamber wall, chambers with tall cylindrical sections are used (*Figure 2.67(a)*). The feed is introduced at the top of the chamber, often through a nozzle atomizer, and flows vertically downwards concurrently with the drying air. There is little radial movement of the droplets/particles and so little chance of them contacting the hot wall of the chamber. When they reach the conical section at the base of the chamber the particles may be cooled with air introduced into the chamber or into a jacket surrounding that part of the chamber. For foods which are less heat sensitive and/or when the particles are not tacky when hot, chambers with shorter cylindrical bodies may be used (*Figure 2.67(b)*). The air is introduced at the top of the chamber tangentially and develops a swirling flow pattern down through the chamber. The feed is also introduced at the top of the chamber, via a centrifugal atomizer. The jet-spray drier is an extreme example of this chamber design. The flat-bottom chamber (*Figure 2.67(c)*) has been available for quite some time and has been the subject of renewed interest in recent years. The powder is removed from the relatively flat bottom of the

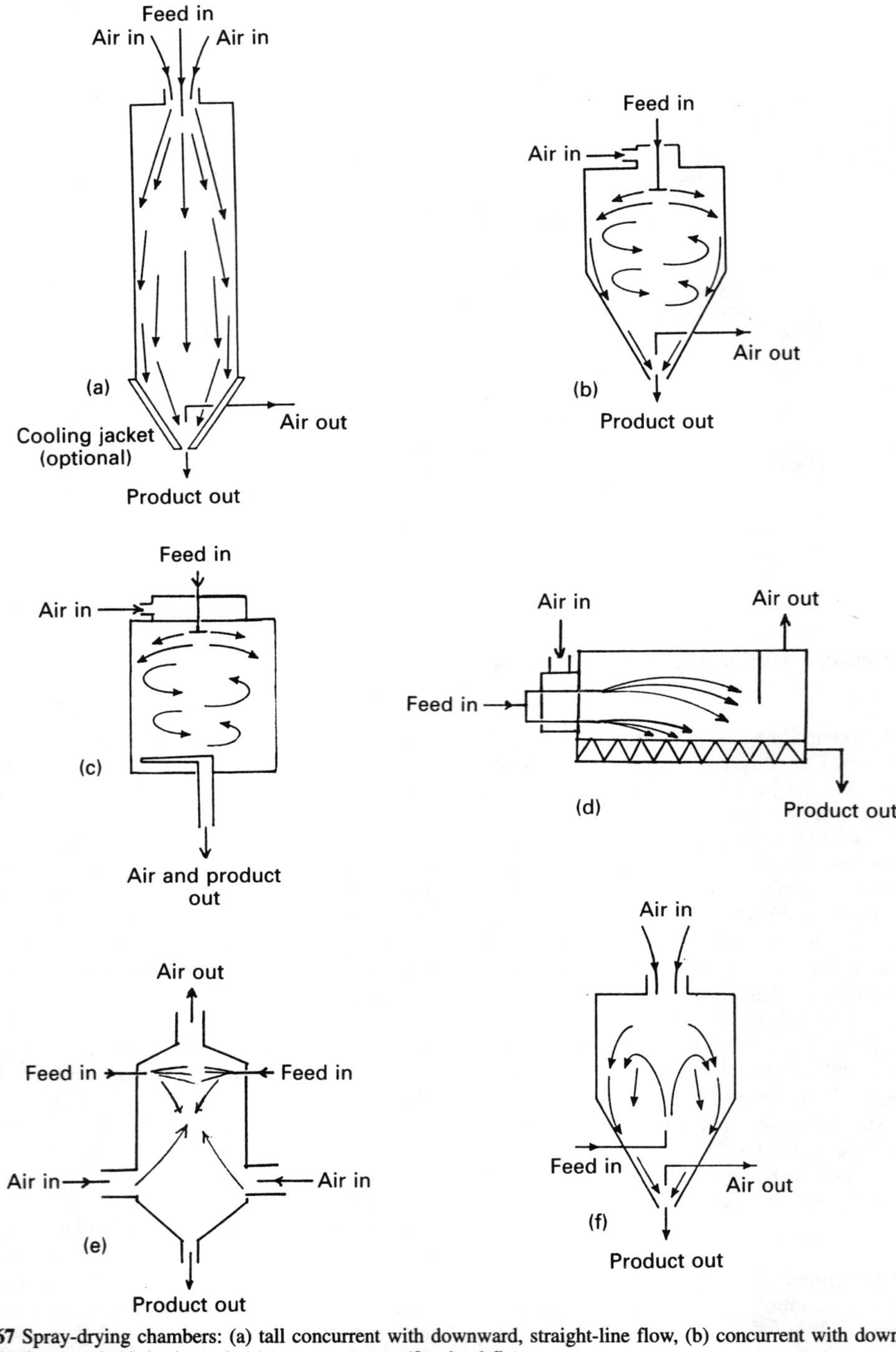

Figure 2.67 Spray-drying chambers: (a) tall concurrent with downward, straight-line flow, (b) concurrent with downward, rotary flow, (c) flat-bottomed, (d) horizontal, (e) countercurrent, (f) mixed-flow

chamber by means of a rotating pneumatic discharger. This design has a reduced, installed height requirement as compared with the cone-bottom type. The powder discharger ensures a well-controlled residence time. Horizontal concurrent chambers, featuring nozzle atomizers, are also available (*Figure 2.67(d)*) but are not widely used now. Countercurrent chamber designs (*Figure 2.67(e)*) are not often used for food applications because of the risk of heat damaging the product. An exception was the BIRS process, introduced over twenty years ago. In this process, the feed was introduced at the top of a very tall tower. Low humidity air at a temperature of not more than 30°C was introduced at the bottom of the tower and rose slowly upwards. Thus very gentle drying took place, limiting heat damage. A mixed-flow chamber is shown in *Figure 2.67(f)*. Such chambers find limited use for dehydration of foods which are not very heat sensitive. Spray-drying chambers featuring multistage drying are also available. Examples include the integrated fluidized-bed drier (IFB) and the Filtermat drier. The properties of the dried powder, i.e. its particle size, bulk density and reconstitution characteristics are influenced by the type of atomizer and the chamber design. The selection of the most appropriate equipment to yield a product of specified properties is usually made after pilot-scale practical trials, although some information will be available in the literature.

Deposition of powder particles on the inner surface of the drying-chamber wall can occur in spray drying. This is most likely to happen when drying foods with a high sugar or fat content. Powders from such feeds may be thermoplastic and become tacky when hot. Such a build-up of powder is undesirable, as powder remaining on the wall may become seriously heat damaged. It can also lead to a fire or explosion hazard. Powder deposits may be loosened by hammers striking the outer surface of the chamber wall at regular intervals. Such hammers may be operated electromagnetically or pneumatically and are usually located around the outside of the cylindrical section of the chamber. Brushes, chains or air brooms sweeping the inner surface of the wall of the conical section of the chamber may also be used to disturb powder deposits. Cooling the chamber wall by reducing its insulation or passing cool air through a jacket around the wall may help. The bulk of the powder is removed from the bottom of the drying chamber through a valve or airlock. The fine powder leaves the chamber with the exhaust air and is recovered by an air/powder separator (Brennan, 1971; Van Arsdel et al, 1973; Green, 1984; Filkova and Mujumdar, 1987; Pisecky, 1987; Brennan et al, 1990; Masters, 1985; Brennan, 1993). *See also* Air broom, Air/powder separator, BIRS process, Centrifugal atomizer, Explosion hazard, Filtermat spray drier, Fire hazard, Integrated fluidized-bed (IFB) spray drier, Jet-spray drier, Pressure nozzle, Spray drying, Two-fluid-nozzle.

Spray-forming devices

Devices for the conversion of liquids and slurries into sprays for the purpose of spray drying. The spray of liquid is mixed with heated air in a drying chamber and a dry powder is formed. These devices are also known as atomizers. There are three types of such devices, i.e. pressure nozzles (also known as pressure jets), centrifugal atomizers (also known as wheel atomizers) and two-fluid nozzles (also known as pneumatic nozzles). *See also* Centrifugal atomizers, Pressure nozzles, Spray drying, Two-fluid nozzles.

Stainless steel

A ferrous alloy in which the principal alloying element is chromium. When chromium is present in amounts greater than 12%, stainless steel possesses a high degree of resistance to corrosion. This property is attributed to the formation of a thin film of chromium oxide on the surface of the metal. There are three main types of stainless steel – ferritic, martensitic and austenitic. The last of these contains nickel, in addition to the chromium, and has good mechanical properties as well as corrosion resistance. It is austenitic stainless steel which is most widely used in the construction of food-processing equipment. A common type contains 18% chromium and 8% nickel known as 18/8 austenitic steel. In order to prevent a fault developing near welded joints, known as weld decay, the carbon content of the steel is kept at very low levels, less than 0.15%. In addition, small quantities of titanium or niobium may be included to further protect against weld decay. The corrosion resistance of stainless steels is mainly dependent on the oxide film formed at the surface which is maintained under oxidizing conditions. Occasionally, changes in the environment can cause fluctuations in this passivity and lead to breakdown, perhaps due to hydrogen being released in the system. The inclusion of 2–4% molybdenum considerably improves corrosion resistance, particularly against localized corrosion and pitting caused by halides. Stainless steel can handle a wide range of chemicals satisfactorily and has found wide application in the food and pharmaceutical industries, including drying equipment (*see* Materials of construction). It has a hard, smooth surface which can be cleaned relatively easily (John, 1993).

Standard atmosphere (atm)

An SI unit of pressure equivalent to 0.101 MN m^{-2}.

Steam blanching *see* Blanching.

Steam ejector

This device is used to create low pressures in evaporators, vacuum driers and freeze driers. High pressure steam is allowed to expand through a jet and thence through a converging-diverging section (Venturi) at high velocity. This generates a zone of low pressure in the suction chamber which is connected to the unit to be evacuated. Both condensible and non-condensible gases are entrained in the steam and discharged with it (*Figure*

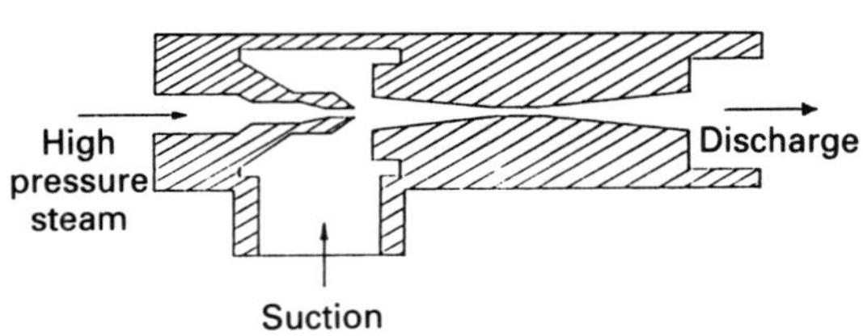

Figure 2.68 Principle of steam ejector (Toledo, 1991)

2.68). Such ejectors are most suitable for generating relatively low pressure, below 2.0 kN m^{-2}. To achieve such low pressures, a number of ejectors are used in series. In *Figure 2.8* (page 32) a four-stage steam ejector system is shown connected to a cabinet freeze drier. The first two stages, known as augmentors, discharge into a water-cooled condenser. This removes most of the condensible gases. The interior of the condenser is at a pressure below atmospheric and the condensate is removed by a 10.5 m-high barometric leg or by an extractor pump. Additional booster ejectors are included to cope with the initial rapid pump-down necessary to avoid thawing of the frozen food in the cabinet. As compared with vacuum pumps capable of producing very low pressures, steam ejectors are of relatively simple design and easy to maintain. However, very large volumes of water are required to use as steam in the ejector and cooling medium in the condenser (Cotson and Smith, 1963; Mellor, 1978; Toledo, 1991). *See also* Barometric leg, Cabinet freeze drier, Vacuum pump.

Steam peeling
Holding some vegetables and fruits in high-pressure steam for short periods is an effective way of loosening their skins. The skin and the tissue beneath it become superheated. When the pressure is suddenly reduced, the moisture in the tissue flashes off and forces the skin off the fruit or vegetable. High-pressure water sprays complete the separation of the skin and tissue. Alternatively, revolving brushes or rubber fingers may be used to remove the skin. Continuous steam peelers operate at pressures of about 600 kN m^{-2}, with holding times of about 2 min. Higher pressures, up to 1.70 MN m^{-2}, are used in batch peelers, with holding times of 30 s (Luh and Woodroof, 1973; Woodroof and Luh, 1973; Arthey and Dennis, 1991).

Steam-swept atomizer *see* Centrifugal atomizer.

Steam tables
Tables in which the thermodynamic properties of water and steam are presented over a range of temperatures and pressures. The properties include specific internal energy, specific enthalpy and specific entropy. *See also* Thermodynamic tables (Haywood, 1968).

Stoke (St)
A metric unit of kinematic viscosity equivalent to 10^{-4} $m^2\ s^{-1}$.

Streamline flow of fluid *see* Laminar flow of fluid.

Structural changes during dehydration
Animal and vegetable tissue undergo some degreè of shrinkage during drying by most methods, with the exception of freeze drying. Colloidal materials also shrink. This shrinkage may lead to various kinds of damage including warping, cracking and crushing of tissue. Very often the shrinkage is not uniform throughout a piece and the end product may be severely distorted. Some food products are more prone to damage than others. For example, pasta products are prone to checking or cracking and have to be dried slowly to minimize such damage. Very often in the early stage of drying, at low rates, the amount of shrinkage bears a simple relationship to the amount of moisture removed. Towards the end of drying, the rate of shrinkage is reduced and the final shape and size of the material is fixed before drying is complete. The rate of drying, particularly in the early stages, can influence the amount and type of shrinkage which occurs. If drying is slow, moisture gradients within the piece will be low and the material will shrink down fully on to a solid core. This will yield a product with a high bulk density which may be slow to rehydrate. If drying is rapid, the surface will dry out quickly while the centre is still wet. The outer layers will be placed under tension which may cause them to become rigid, forming a supporting structure with dimensions close to those of the original wet material. As drying proceeds, stresses will develop within the piece which will cause the tissues to split and rupture. The end product will have an open, porous structure, with a low bulk density. This should rehydrate rapidly but may have poor water-holding properties and a soft texture.

Wang and Brennan (1994a) studied changes in structure in rectangular slabs of potato drying from one large face only in air at 40, 50, 60 and 70°C using microscopy. *Figure 2.69* shows the structural changes in the vertical section during drying at different time intervals. The column on the left-hand side shows the structural changes after 1, 3 and 6 hours drying in air at 40°C, the column on the right-hand side shows the changes in air at 70°C. It can be seen that shrinkage occurs first at the surface and then gradually moves down into the pieces as drying proceeds. The cells assume an elongated shape. *Figure 2.70* shows the structural changes in the potato in horizontal section, drying in air at 40°C. The column on the left-hand side shows the changes from the surface to the bottom after 1 hour of drying time while the right-hand column shows the changes after 6 hours. *Figure 2.71* shows the corresponding changes in air at 70°C. The surface of the sample has been damaged after 1 hour, while the inner parts are intact. As drying proceeds, cracks occur in the inner structure.

The changes in density of the potato pieces during drying are shown in *Figure 2.72*. In the early stages of drying, the density increased as moisture content decreased. At low moisture content there was a small decrease in density. At this stage, the volume of the slab stayed constant while moisture was removed. Marousis and Saravacos (1990) obtained similar results when drying starch materials. The density of the potato at any given moisture content decreased with increase in air temperature. This ties in with the microscopic observations. A simple mathematical model was developed which gave a good prediction of changes in density with

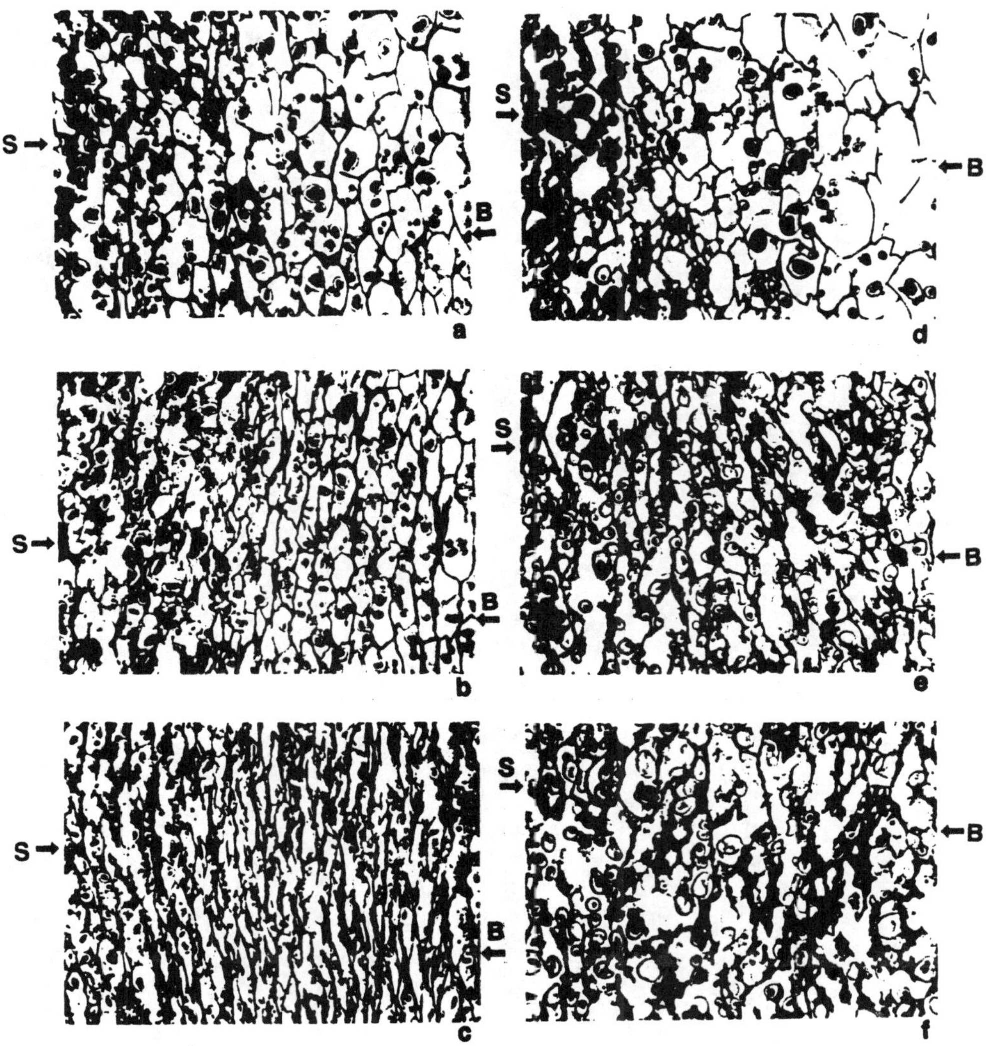

Figure 2.69 Photomicrograph of parenchyma cells in the vertical section during drying at different drying times and at two temperatures. S – surface of sample; B – bottom of sample; a – after 1 h drying in air at 40°C; b – after 3.5 h drying in air at 40°C; c – after 6 h drying in air at 40°C; d – after 1 h drying at 70°C; e – after 3 h drying at 70°C; f – after 6 h drying at 70°C

moisture content. Changes in the three dimensions of the slab during drying were monitored. Linear relationships between these dimensions and moisture content applied over most of the drying curve.

The dimension(s), usually thickness or drying surface area, of solid pieces, drying in heated air, feature in most equations representing both constant and falling-rate periods of drying (see equations on pages 8–11). If such dimensions change during drying then the pattern of drying may be expected to change also. However, any change in dimensions which occurs will depend on the type of shrinkage. At low drying rates, the external dimensions will change and it would seem desirable to take such changes into account in any model representing drying rate. However, at high drying rates, there may only be small changes in external dimensions while the tissues split and rupture internally. This type of shrinkage is also likely to influence the pattern of drying but is much more difficult to take into account in a mathematical model. Few drying models take shrinkage into account. Gorling (1958) took changes in thickness of potato slices into account when presenting drying rate data. On the other hand, Jason (1958), when studying the drying of fish muscle, concluded that the effective value of the diffusion coefficient was independent of shrinkage. Wang and Brennan (1994b) developed a model for

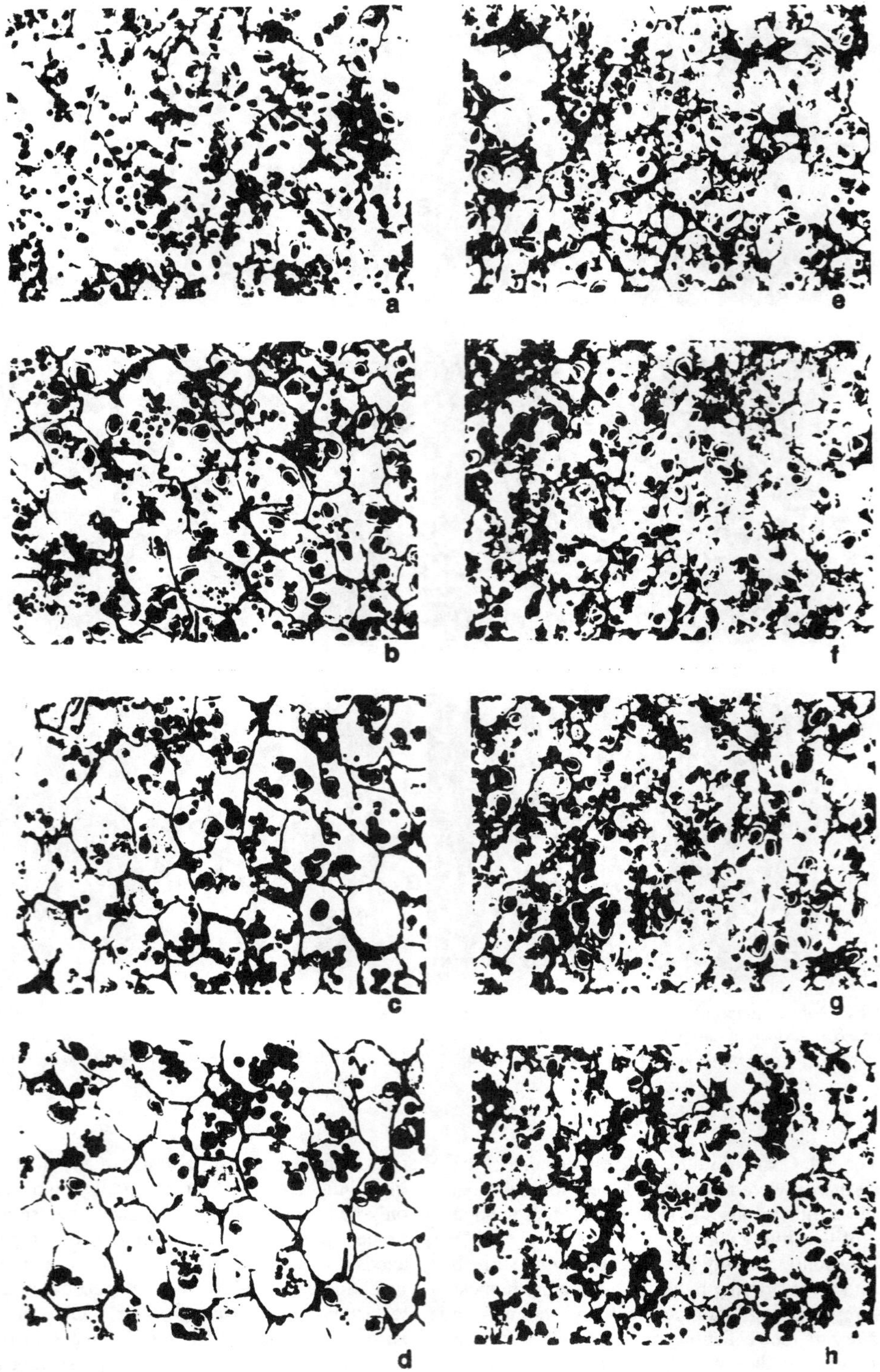

Figure 2.70 Photomicrograph of parenchyma cells in potato in the horizontal section at different drying times in air at 40°C. Left-hand column after 1 h drying. Right-hand column after 6 h drying. a – surface of sample after 1 h; b – 1.3 mm below surface after 1 h; c – 2.6 mm below surface after 1 h; d – 3.9 mm below surface after 1 h; e – surface of sample after 6 h; f – 0.8 mm below surface after 6 h; g – 1.6 mm below surface after 6 h; h – 2.4 mm below surface after 6 h

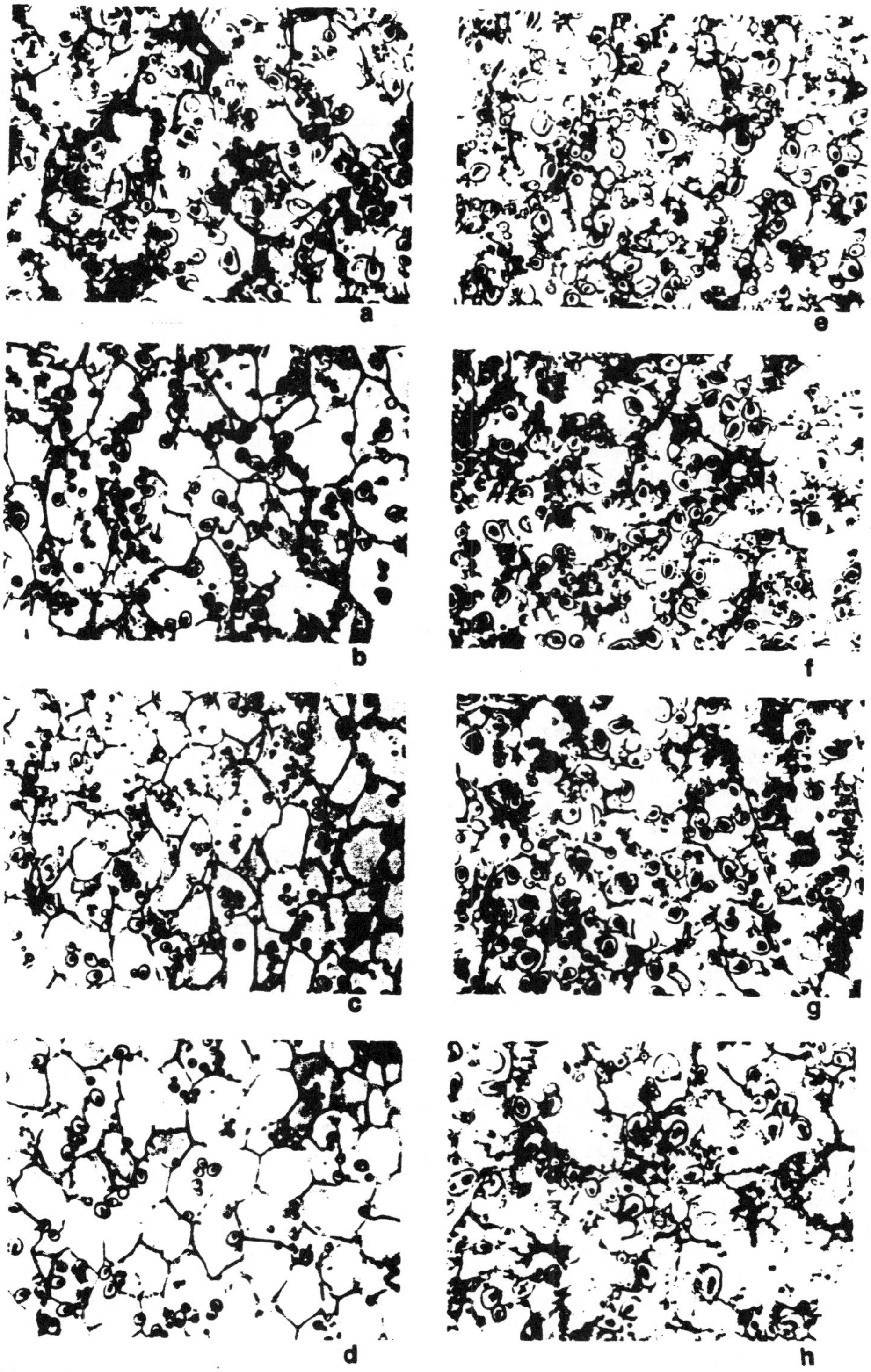

Figure 2.71 Photomicrograph of parenchyma cells in potato in the horizontal section at different drying times in air at 70°C. Left-hand column after 1 h drying. Right-hand column after 6 h drying. a – surface of sample after 1 h; b – 1.3 mm below surface after 1 h; c – 2.6 mm below surface after 1 h; d – 3.6 mm below surface after 1 h; e – surface of sample after 6 h; f – 1.0 mm below surface after 6 h; g – 2.0 mm below surface after 6 h; h – 3.0 mm below surface after 6 h

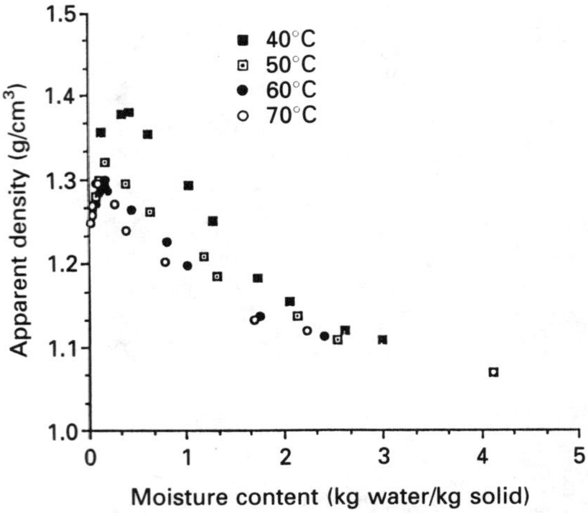

Figure 2.72 Effect of moisture content on the density of potato during drying in air at four different temperatures

the prediction of temperature and moisture distribution within a slab of potato during dehydration in heated air. They found that the relationship between experimental and predicted values of moisture content and temperature was closer when a model incorporating a shrinkage element was used as compared to a model which did not take shrinkage into account. In freeze drying, most of the water is first converted to ice by freezing and subsequently to vapour by sublimation. The water vapour permeates to the surface through the dry solid. There is no movement of liquid water within the solid. Consequently there is no shrinkage and the dry solid has a rigid porous structure (Van Arsdel et al, 1973a; Brennan et al, 1990). *See also* Freeze drying.

Sulphiting *see* Sulphuring.

Sulphuring (sulphiting)
The process of adding SO_2 to most fruits and some vegetables prior to dehydration. The main purpose of this is to inhibit non-enzymic browning during processing and storage of the dried food. This type of discoloration is discussed elsewhere in this text, under Water activity and food quality. In addition to inhibiting non-enzymic browning, SO_2 treatment can reduce the destruction of carotene and ascorbic acid during blanching and drying. The traditional method of sulphuring fruits is to expose them to the fumes from burning sulphur in a chamber known as the sulphur house. Such houses may be brick-built structures, wood-framed cells or plastic tents. It is difficult to control the amount of SO_2 in the dried food by this method. In recent years, injection of bottled SO_2 gas into chambers containing the fruit has been practised instead of burning the sulphur. This method facilitates control over the amount of SO_2 absorbed by the fruit, causes less air pollution and can be less expensive as compared with the use of burning sulphur. An alternative method of introducing SO_2 is to dip the food pieces in a sulphite solution, usually a mixture of sodium sulphite and sodium bisulphite, for a predetermined time, or the food may be sprayed with a similar solution. This method is preferred for some fruits and most vegetables. Many factors influence the amount of SO_2 absorbed by the food including time of sulphuring, SO_2 concentration, temperature, type of fruit or vegetable and its size, condition and maturity. Excessive sulphuring may adversely affect the flavour of the product. The optimum level of SO_2 in the dried material varies from product to product and depends on the storage temperature and time. Most countries have set limits on the SO_2 content of dried fruit and vegetable products (Roberts and McWeeny, 1972; Luh and Woodroof, 1975; Woodroof and Luh, 1975; Wedzicha, 1986; Salunkhe et al, 1991). *See* Water activity and food quality.

Sun drying *see* Solar drying.

Surface tension
At the interface between a liquid and a gas there is an imbalance in the forces of attraction between molecules. Molecules on the surface of the liquid will be subjected to a net attraction by the molecules in the bulk of the liquid. This results in the surface being in a state of tension. Thus, there will be a tendency for the surface area to contract. The forces acting on the surface causing this tendency are known as surface tension forces. One definition of surface tension is the force per unit length acting on a surface, at right angles to one side of a line drawn on the surface. Common units used are $MN\,m^{-1}$ or $dyn\,cm^{-1}$. A second definition of surface tension is the work required per unit area to increase the surface area under constant temperature conditions (Lewis, 1987).

Système International d'Unités *see* SI units.

Sweat box
A wooden box in which sun-dried fruits are held to equilibrate (sweat), to produce a product with a uniform moisture content. *See* Sweating.

Sweating
After sun drying some fruits are placed in wooden boxes (sweat boxes) or in bins or in heaped piles on a clean floor and held for periods of time which range from a couple of days to a few weeks to equilibrate. Moisture will migrate within the pieces to their surfaces. Pieces with a relatively low moisture content will absorb moisture from those with a higher moisture content and a more uniform product will result (Cruess, 1948; Woodroof and Luh, 1975).

T

Thermal conductivity

This is a measure of the rate at which heat moves through a substance. In the context of this work, it mostly applies to the transfer of heat through a solid material. The rate of heat transfer along a piece of solid material is proportional to the cross-sectional area of the piece, normal to the direction of flow of the heat, and the temperature gradient per unit length in the direction of flow. Thus the rate of heat transfer may be represented by the relationship:

$$\frac{dq}{dt} = kA\,\frac{(\theta_1 - \theta_2)}{l}$$

where q is the heat transferred, t is time, A is the cross-sectional area, θ_1 and θ_2 are the temperatures at each end of the flow path and l is the length of the flow path. k in this equation is the thermal conductivity of the solid. Common units are $W\,m^{-1}\,K^{-1}$. The thermal conductivity of a food material depends on its composition and temperature. Mathematical models have been developed to calculate the thermal conductivity of complex foods (Miles et al, 1983; Toledo, 1991; Singh, 1992; Wang and Brennan, 1992). Data on the thermal conductivities of food materials is available in the literature. Some of this data is comprehensively reported, i.e. the composition of the food and the test conditions used in determining its thermal conductivity are given. In other cases, not so much information is provided and the data is less useful. Polley et al (1980) published a compilation of the thermal properties of foods. In this, values of the thermal conductivities of fresh foods (fish, meat, poultry, fruits and vegetables) ranged from 0.24 to $0.61\,W\,m^{-1}\,K^{-1}$, and for frozen products from 1.10 to $1.62\,W\,m^{-1}\,K^{-1}$. They present very little data on dried foods. Other sources of data on the thermal conductivities of foods are Heldman and Singh, 1981 and Holmes and Woodburn, 1981. Approximate thermal conductivity values for some non-food materials, in $W\,m^{-1}\,K^{-1}$, are: water at 0°C 0.57, ice at 0°C 2.30, air at 0°C 2.42×10^{-2}, copper at 0°C 403.0, aluminium at 0°C 202.0, cast iron at 0°C 55.0, stainless steel at 0°C 8.0–16.0, glass at 0°C 0.1–1.0, building brick at 20°C 0.69 (Lewis, 1987; Toledo, 1991; Singh and Heldman, 1993).

Thermal diffusivity

The ratio of thermal conductivity to the product of the specific heat and density of a material, i.e:

$$a = \frac{k}{c\rho}$$

where a is the thermal diffusivity, c the thermal conductivity and ρ the density. Common units are $m^2\,s^{-1}$. It is a measure of the rate at which the temperature of a material will change when it is heated or cooled. A high value of thermal diffusivity identifies a material whose temperature will change rapidly. The term features in equations representing unsteady-state heat transfer, e.g. for unsteady-state heat transfer by conduction:

$$\frac{d\theta}{dt} = a\,\frac{d^2\theta}{dx^2}$$

where θ is temperature, t is time and x is the distance over which heat transfer is taking place, in one direction only.

The thermal diffusivity of water at 30°C is $1.48 \times 10^{-7}\,m^2\,s^{-1}$ and ice at 0°C is 11.82 in the same units. Singh, 1982 published a list of thermal diffusivities of fresh fruit, vegetables, meat and fish. Values ranged from 1.05 to $1.91 \times 10^{-7}\,m^2\,s^{-1}$.

Thermal efficiency

The thermal efficiency, E_t, of a drying chamber is the ratio of the heat used in evaporating moisture to the total heat supplied to the chamber, usually expressed as a percentage. In the case of a hot-air drier, if the drying chamber is assumed to be perfectly insulated and leakproof, the thermal efficiency may be expressed as

$$E_t = \frac{Q_1}{Q_1 + Q_2} \times 100$$

where Q_1 is the energy used in the evaporation of water and Q_2 is the energy lost in the exhaust air. Thermal efficiency may also be expressed in terms of temperatures. For a perfectly insulated and leakproof chamber E_t may be expressed as:

$$E_t = \frac{\theta_1 - \theta_2}{\theta_1 - \theta_0} \times 100$$

where θ_1 and θ_2 are the temperatures of the air entering and leaving the chamber respectively and θ_0 is the temperature of the air entering the heater. If radiation and other losses are to be taken into account, thermal efficiency may be expressed as:

$$E_t = \frac{\text{Rate of evaporation of water} \times L_E}{\text{Flow rate of air} \times C_H \times (\theta_1 - \theta_0)} \times 100$$

where L_E is the latent heat of evaporation of water at θ_E, θ_E the evaporating temperature and C_H the humid heat of the air passing through the heater. This relationship does not take account of sensible heat changes in the feed and product. Another measure of efficiency of utilization of energy, known as fuel efficiency or economy, is defined on page 59. *See also* Energy used in drying, Economy, Fuel efficiency.

Thermal energy *see* Internal energy.

Thermal wheel
A rotating heat exchanger (*Figure 2.73*) suitable for recovering sensible heat from large volumes of gas. It is used to recover heat from the exhaust air from drying chambers (Strumillo and Lopez-Cacicedo, 1987; Fellows, 1988). *See also* Energy used in drying, Heat exchangers.

Thermodynamic tables
Tables in which the thermodynamic properties of liquids and gases, over a range of temperatures and pressures, are presented. Properties include specific internal energy, specific enthalpy and specific entropy. Tables are available for water and steam, air and refrigerants such as ammonia, carbon dioxide and methyl chloride. Tables containing data on water and steam only are also known as steam tables (Haywood, 1968).

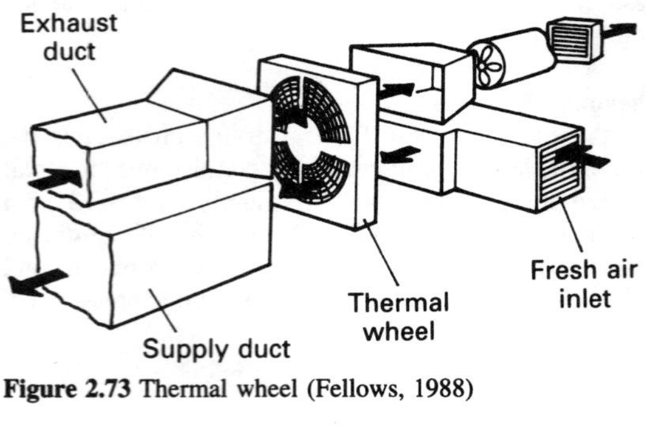

Figure 2.73 Thermal wheel (Fellows, 1988)

Tonne (t)
A metric unit of mass equivalent to 10^3 kg.

Toroidal-bed drier
A fluidized-bed drier in which the heated air enters the base of the drying chamber, at high velocity, through blades or louvres which impart a rotary motion to the air (*Figure 2.74*). This creates a compact rotating bed of particles. High rates of heat and mass transfer are attained within this bed, resulting in high initial rates of drying. The air is usually heated directly by gas. The depth of the bed of particles varies from a few millimetres to more than 50 mm. The drier can accommodate a wide range of particle sizes and shapes. This equipment can also be used for roasting, toasting and expanding particulate foods (Grikitis, 1988; Brennan, 1989). *See also* Fluidized bed drier.

Torr
A metric unit of pressure equivalent to $133.0\,\mathrm{N\,m^{-2}}$.

Tray drier *see* Cabinet drier.

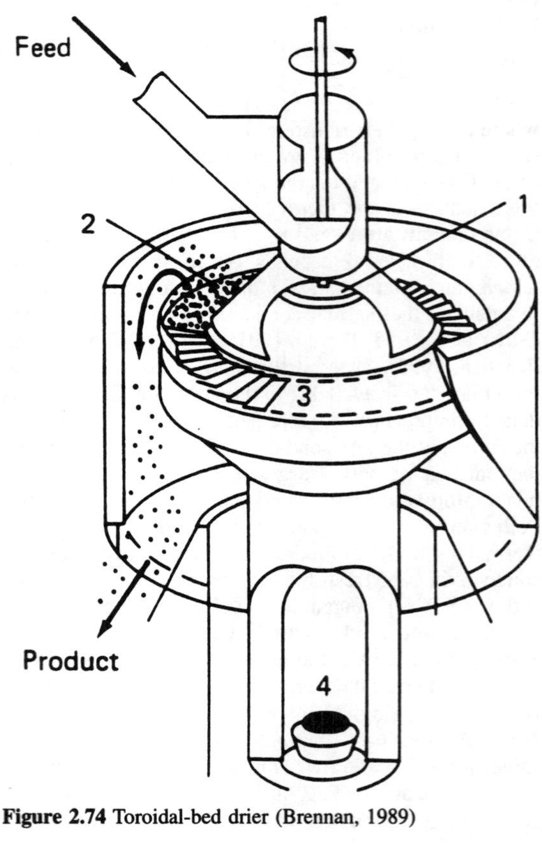

Figure 2.74 Toroidal-bed drier (Brennan, 1989)

Trehalose (alpha-D-glucopyranosyl-alpha-D-glucopyranoside)
This sugar is a disaccharide consisting of two glucose molecules linked by their reducing carbons. It is a non-reducing sugar which is chemically inert and biologically non-toxic. It has one quarter to one half the sweetness of sucrose. It is found in large amounts in certain organisms from both the plant and animal kingdoms. These organisms, known as anhydrobionts, have the ability to avoid damage during drying. They can lose up to 99% of their total water content, remain inert in the dry state, but recover their activity when rehydrated. Mushrooms and baker's yeast are examples of such organisms. Trehalose may be present in mushrooms in amounts up to 15% of their dry weight. It is also known as mushroom sugar. Mushrooms are the main source of trehalose in the western diet. It is also present in other fungi, bacteria, insects and plants.

The addition of trehalose to biologically active materials prior to drying has been shown to increase the retention of their activity as compared with samples dried under similar conditions without the addition of the sugar. The activity of monoclonal antibodies has been retained in this way even after several years of storage at room temperature in the dried state. So also has the activity of highly unstable enzymes. Trehalose has also been added to foods prior to drying. Most of this work has been done with liquid or liquidized foods. Blended fresh egg with added trehalose, dried at 30–50°C, yielded an odourless yellow-orange powder which could be stored at room temperature. On rehydration the product was of high quality, indistinguishable from fresh egg. Purées of banana, strawberry, mango, avocado, apple and raspberry with added trehalose were dried. When the powders were reconstituted, even after prolonged storage, they recovered the colour and flavour of the original fruit. Herbs dipped in trehalose when dried and reconstituted retained their odour and flavour very well. Analysis of the volatiles released from reconstituted mango powder containing trehalose, showed that there was an increase in number and intensity of volatile aromatic esters, compared to powder without added trehalose. There was also a decrease in the concentrations of furfural and α-humulene. Such compounds form in the Maillard reactions which cause undesirable colour and flavour development in dried foods. It would appear that trehalose inhibits the reactions between proteins and sugars in the fruit. Some work has been done on food pieces. Banana slices soaked in trehalose solution and then dried retained their colour and texture better than untreated slices. Therefore, there could be many more applications for trehalose in food dehydration. An even more promising prospect is the development of genetically engineered foods that can synthesize their own trehalose.

The mechanism whereby trehalose has this protective effect is not fully understood. One explanation proposed is that the hydroxyl groups in this sugar replace the water molecules that are usually bonded to proteins and other macromolecules. Another theory is that, as trehalose dries out of solution, it passes through a stage when it is in the form of a glass. It has been shown that pure trehalose in this form becomes unstable and crystallizes when stored for a few days. However, when such a glass contains some protein, it is stable and does not crystallize. The good odour and flavour retention in dried foods containing trehalose may be due to the volatile aromatic esters being trapped within the stable trehalose glass. Neither of these theories fully explains the protective role of trehalose in food dehydration and further research is needed to establish a better explanation (Roser, 1991; Roser and Colaco, 1993). *See also* Glass transition, Maillard reactions.

Tubular reverse osmosis or ultrafiltration membrane *see* Membrane concentration.

Tunnel drier
A hot-air drier which operates on a continuous or semi-continuous basis at high throughputs. The drying chamber consists of an insulated tunnel up to 25 m long with a square or rectangular cross-section, typically 2 m × 2 m. The prepared food is placed on trays made of slatted wood, perforated metal plate or wire mesh. These trays are assembled, one above the other, on trolleys. The trolleys enter the 'wet end' of the tunnel and exit from the 'dry end'. Some tunnels are equipped with a moving chain, or similar device, which pulls the trolleys through the tunnel at a uniform rate. In other tunnels, the trolleys run on rails. As a trolley loaded with fresh material is introduced at the 'wet end', one carrying the dried product is displaced out at the 'dry end'. Air movement within the tunnel is mainly between the trays. Air velocities are typically in the range 2–6 m s^{-1}. Tunnels featuring throughflow of air are also available. The food material and air may move concurrent or countercurrent to each other or more complex flow patterns may prevail.

In a *concurrent* tunnel, air and fresh feed enter at the 'wet end' and exit at the 'dry end' (*Figure 2.75(a)*). The characteristic features of this design of tunnel are: (i) High rates of evaporation attainable at the 'wet end' of the tunnel, as relatively high air temperatures may be used with little risk of heat damaging the food. The surface temperature of the food is kept relatively low by evaporative cooling in the early stages of drying. In the case of sliced or diced vegetables, these high initial drying rates tend to produce a dry product with an open structure and hence low bulk density. (ii) As the food moves down the tunnel, it comes into contact with cooler, moister air and the drying rate falls off. There is little risk of heat damaging the product near the 'dry end' of the tunnel. (iii) Low moisture content is difficult to achieve because of the poor drying conditions at the 'dry end' of the tunnel.

In a *countercurrent* tunnel the fresh feed enters at the 'wet end' of the tunnel while the air enters at the 'dry

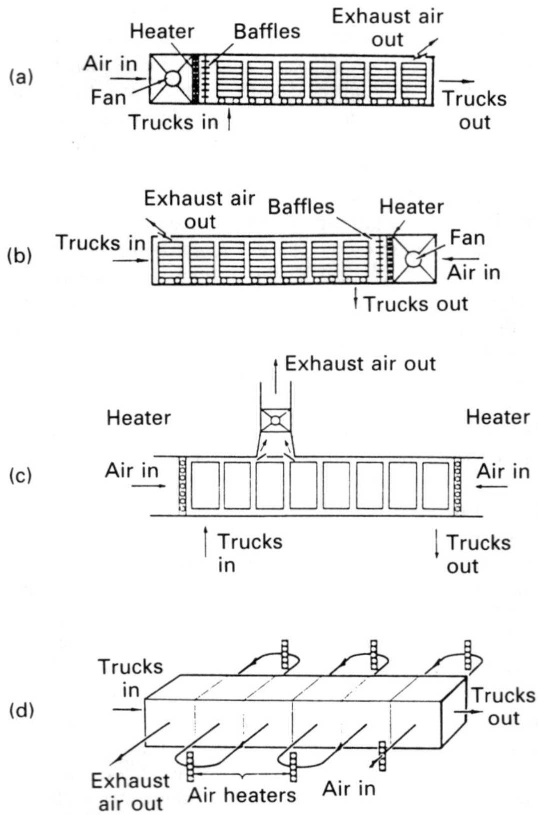

Figure 2.75 Tunnel driers: (a) concurrent, (b) countercurrent, (c) two-stage, centre exhaust, (d) crossflow, adapted from Brennan et al (1990)

end' (*Figure 2.75(b)*). The features of this design of tunnel are: (i) Relatively low rates of drying occur at the 'wet end' of the tunnel as the fresh feed first encounters cool, moist air. Overloading of the drier with wet feed can result in the food being exposed for an extended period to warm, moist conditions which may lead to spoilage. In the case of vegetable pieces, severe shrinkage can occur, yielding a product with a high bulk density, which is slow to reabsorb water on reconstitution. (ii) At the 'dry end' of the tunnel, drying conditions can be severe. Low moisture content is attainable but the risk of heat damaging the product is high. (iii) Countercurrent tunnels are usually more thermally efficient than concurrent designs.

Where a single tunnel does not satisfy the requirements for a particular product, two or more tunnels may be used in series. A common arrangement is to pass the food first through a concurrent and then through a countercurrent tunnel. In this way, advantage can be taken of the high initial rates of drying, characteristic of the concurrent design, while, in the countercurrent second stage, relatively high final drying rates and low moisture content are attainable. The concurrent tunnel is usually shorter than the countercurrent one where the drying rates are lower. The air entering the second tunnel is usually at a lower temperature than that used in the first tunnel, to minimize heat damage to the product. In the *centre-exhaust* tunnel (*Figure 2.75(c)*), two-stage drying is achieved in one unit. The characteristic features of this type of tunnel are: (i) Shorter drying times and increased throughput attainable as compared with single stage unit of similar size. (ii) Better control over the drying conditions at different stages of drying. (iii) Difficulty may be encountered in achieving a uniform flow of air where the two streams meet inside the tunnel. (iv) The capital cost of the two-stage unit is higher than for a single-stage tunnel.

In the *cross-flow* tunnel, the air travels at right angles to the movement of the food. Often the direction of flow alternates at different stages of drying (*Figure 2.75(d)*). The characteristics of this design of tunnel are: (i) Because of the frequent change in direction of the air flow, more uniform drying is possible, as compared to designs in which the air flows parallel to the food. (ii) If separate heaters are provided at different stages of drying, as shown in *Figure 2.75(d)*, good control may be exercised over the drying conditions throughout the drying cycle. (iii) Such units are more expensive to purchase and maintain, as compared with simpler systems (Van Arsdel et al, 1973; Williams-Gardner, 1976; Charm, 1979; Brennan et al, 1990).

Tunnel freeze drier

This is comprised of a cylindrical tunnel, typically 1.8–2.4 m in diameter, made up of sections each capable of handling 3–4 tonnes of frozen material per day. The number of these sections determines the total capacity of the tunnel. The tunnel is fitted with entry and exit vacuum-locks, separated from the main body of the tunnel by large gate valves. Fixed heater plates are cantilevered in pairs from both sides of the tunnel. The frozen food material in trays or ribbed dishes is carried through the tunnel either on a trolley running on ground rails or suspended from an overhead rail. The trays pass between the heater plates. In some designs, the trays do not touch the plates and heating is by radiation from above and below the trays. In other units, the trays, when stationary, can be lowered on to the heater shelves, permitting heat transfer by conduction from below. The vacuum and heating systems are designed to suit the loads at the various stages in the tunnel. Both vapour and heat loads decrease as drying proceeds. A tunnel freeze drier consisting of five sections, including the vacuum locks, is shown in *Figure 2.76*. Vapour restriction plates are fitted within the main body of the tunnel. These closely fit the inside wall of the tunnel but allow the trolleys to pass through. Thus, the tunnel consists of five zones, i.e. the two vacuum locks and three zones in the main body of the tunnel. Each of these zones is equipped with a separate vacuum system. Each fresh trolleyload of frozen food is introduced into the entry vacuum lock. This is sealed and pumped down rapidly to the required

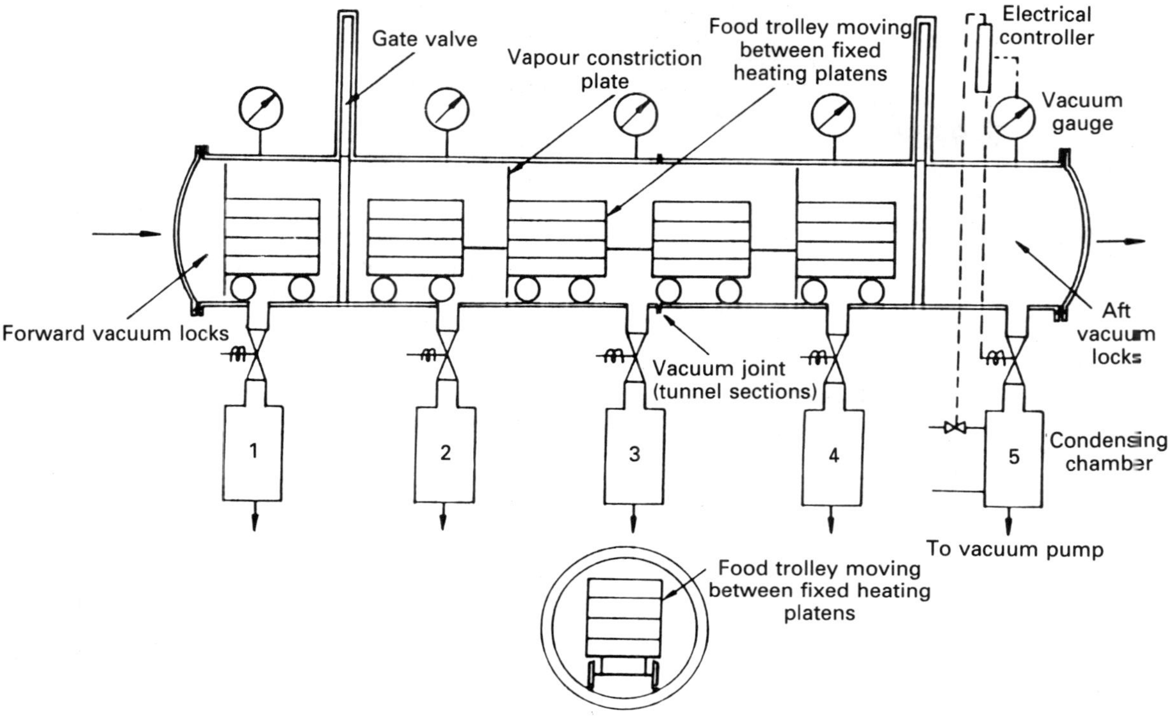

Figure 2.76 Tunnel freeze drier (Mellor, 1974)

pressure. The gate valve is then opened and the trolley passes into the main body of the tunnel. The gate valve is closed and the vacuum in the entry lock is broken. The door of the lock is opened to receive the next trolley. A similar sequence of steps occurs when a trolley of dried food is removed through the exit vacuum lock. The dry product is usually discharged into an air-conditioned chamber for packaging.

Other designs of continuous freeze drier have been reported, in which granulated frozen material is carried by screws, belts or vibrating decks through vacuum chambers. In one design described by Lorentzen (1975), circular heater plates are mounted one above the other in a vacuum chamber. Frozen granules are fed on to the top plate alternately from each of two entrance locks. A central, vertical shaft rotates within the chamber. This carries arms equipped with shovels. These shovels mix the granules and move them outwards towards the edge of the plate from where they drop on to the plate below, which has a larger diameter. On this second plate the shovels again mix the granules but move them towards the centre of the plate where there is a discharge port on to a third plate which has the same diameter as the first one. Thus the material is carried downwards from plate to plate until it drops from the bottom plate into each of two exit locks alternately and is removed from the vacuum chamber (Lorentzen, 1975; Mellor, 1978; Dalgleish, 1990). *See also* Cabinet freeze drier, Freeze drying, Multicabinet freeze drier, Vacuum-spray freeze drier.

Turbulent flow of fluid
A pattern of flow of a fluid in a pipe or over a surface in which eddies occur and there is mixing in a direction at right angles to the main direction of flow of the fluid. A Reynolds number (R_e) of 4000 or more is indicative of turbulent flow, while a value of 2000 or less describes laminar flow. *See* Laminar flow of fluid, Reynolds number.

Twin-drum drier *see* Drum drier.

Two-fluid nozzle
This is one type of device used to convert liquids and slurries into sprays for the purpose of spray drying. It functions by converting the energy in a high velocity stream of gas into surface energy. Two such nozzles are shown in *Figure 2.77.* As the gas exits from the annular opening, a Venturi effect is created and the liquid is drawn out through the central circular orifice and atomized. The size of the droplets formed may be varied by changing the gas-liquid ratio. A typical empirical equation for the prediction of droplet size from such an atomizer is:

$$D = \frac{585 \times 10^3 \sqrt{\gamma}}{V_{RL} \sqrt{\rho}} + 597 \left(\frac{\mu}{\sqrt{\gamma\rho}} \right)^{0.45} \left(\frac{1000\ V_L}{V_G} \right)^{1.5}$$

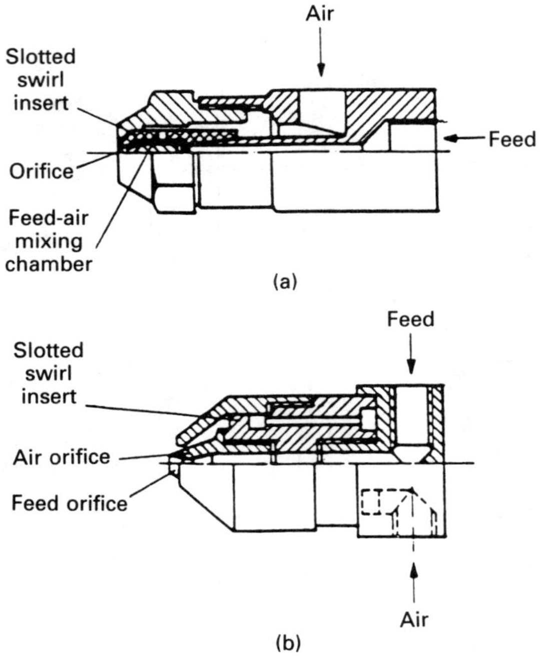

Figure 2.77 Two-fluid (pneumatic) nozzles: (a) internal mixing, (b) external mixing (Masters, 1991)

where D is the Sauter mean droplet diameter, γ, ρ and μ are the liquid surface tension ($N\,m^{-1}$), density ($kg\,m^{-3}$) and viscosity (cp), respectively, V_L and V_G are the volumetric flow rates of liquid and gas, respectively, ($m^3\,s^{-1}$) and V_{RL} is the relative velocity of the two fluids ($m\,s^{-1}$) (Filkova and Mujumdar (1987)). Such nozzles can handle up to $100\,l\,h^{-1}$ of liquid. It takes about $0.5\,m^3$ of compressed air, at a pressure within the range 0.15–1.0 MPa, to atomize 1 kg liquid. They are quite efficient at producing small, uniform droplets of low-viscosity liquid, but when handling high-viscosity material the droplet sizes are less uniform. They are subject to blocking and abrasion by insoluble solid particles in the feed. Because of the large requirement for compressed air, operating costs are relatively high. The application of two-fluid nozzles for food spray drying is limited (Van Arsdel et al, 1973; Coulson et al, 1978; Green, 1984; Masters, 1985; Filkova and Mujumdar, 1987; Pisecky, 1987; Brennan et al, 1990). *See also* Centrifugal atomizers, Pressure nozzles, Spray drying.

Two-phase freezing

A method of freezing in which the freezing medium is a combination of two phases. These are a solid and a gas where solid CO_2 is used, a liquid and a gas where liquid nitrogen is used. *See also* Freezing methods.

U

Ultrafiltration *see* Membrane concentration.

Ultrasonic cleaning *see* Dry cleaning of raw materials.

Unbound moisture
The water present in a wet solid in greater quantity than the amount which is in equilibrium with the saturation humidity at the relevant temperature, i.e. water which exerts a vapour pressure equal to that of pure water at the same temperature. *See also* Bound moisture.

Units of measurement *see* British units and/or SI units, Conversion factors.

V

Vacuum
A space which exhibits a pressure of less than one atmosphere. It is normally created by evacuation with a vacuum pump or a water or steam ejector. The extent of the vacuum is usually expressed in terms of pressure less than atmospheric, i.e. absolute pressure = atmospheric pressure – vacuum. Common units are Pa ($N\,m^{-2}$), inHg or mmHg.

Vacuum band drier
A continuous vacuum drier for liquids and slurries. A continuous belt, moving clockwise, passes over a heating and cooling drum inside a vacuum chamber (*Figure 2.78*). The belt may be a band of stainless steel or very fine stainless steel wire mesh or made up of hinged stainless steel plates. The feed is supplied from outside the cabinet through a valve and applied to the belt surface in the form of a thin layer by a roller. Additional heating is provided by radiant heaters or heated platens in contact with the belt. After passing over the cooling roller, which solidifies thermoplastic material facilitating its removal from the belt, the product is scraped off the belt. It is removed from the vacuum chamber through a rotary valve, a sealed screw or into two receivers working in parallel. While one receiver is under vacuum and receiving product, the other is being emptied. The empty receiver is then evacuated ready to receive product. The chamber is evacuated via a vapour outlet by means of a condenser and vacuum pump. Conditions in the vacuum chamber are similar to those in vacuum shelf driers, i.e. pressure in the range 5–30 $kN\,m^{-2}$, absolute, and belt temperature in the range 35–80°C. The equivalent of 2–3 kg steam is required to evaporate 1 kg water. A number of belts may be located in one vacuum cabinet. As is the case with the vacuum shelf drier, this is a relatively expensive method of drying and only used for very heat-sensitive material which can bear the high drying costs. Both types of vacuum drier may be used for vacuum puff drying (Nonhebel and Moss, 1971; Brennan, 1989; Brennan et al, 1990). *See also* Double-cone vacuum drier, Vacuum shelf drier, Vacuum puff drying.

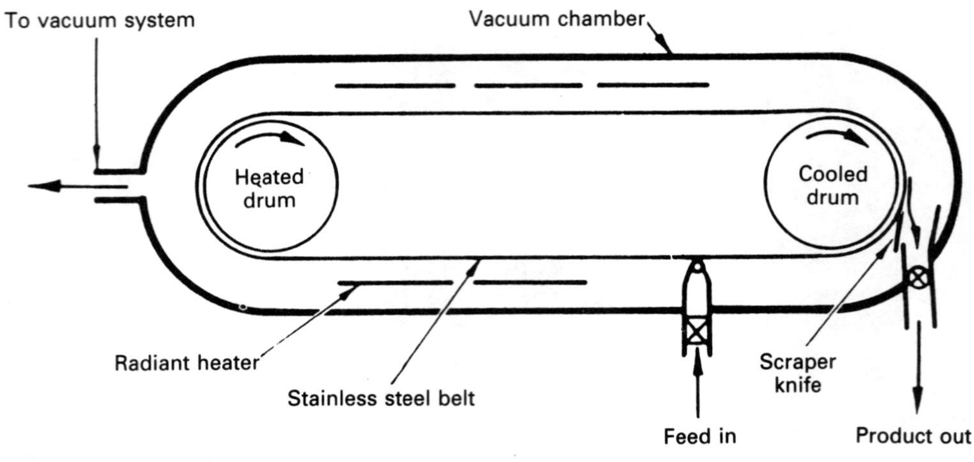

Figure 2.78 Principle of vacuum band drier (Brennan et al, 1990)

Vacuum cabinet drier *see* Vacuum shelf drier.

Vacuum contact plate (VCP) drying
Drying in a vacuum cabinet fitted with heater shelves, which are operated hydraulically so that they can be brought into contact with both surfaces of the layer of material being dried. *See also* Vacuum shelf drier.

Vacuum drum drier *see* Drum drier.

Vacuum evaporation *see* Evaporation, Evaporator.

Vacuum puff drying
A method of vacuum drying concentrated liquids which can promote relatively rapid drying. A small amount of a foaming agent, similar to that used for foam-mat drying, may be added to the concentrate. Some concentrates do not require a foaming agent. The concentrate is applied in a thin layer on to trays, in the case of vacuum shelf drying, or on to the belt in a vacuum band drier. On the application of a vacuum to the chamber, bubbles of water vapour and entrapped air form within the concentrate and expand, and the concentrate froths up to form a foam. By careful control of pressure, temperature and viscosity of the concentrate, it may be made to expand to occupy a space twenty times that of the original material. When heat is applied, the foam dries rapidly to form a porous dry product, which is then cooled and removed from the tray or belt. Because of the porous nature of the product, it rapidly reconstitutes when added to water (Van Arsdel et al, 1973; Salunkhe et al, 1991). *See also* Foam-mat drying, Vacuum band drier, Vacuum shelf drier.

Vacuum pump
The low pressure required in vacuum evaporation, vacuum drying and freeze drying may be attained by the use of vacuum pumps. Such pumps are classified into two categories, low-vacuum and high-vacuum types. Because they operate at pressures which are less than atmospheric, vacuum pumps must be capable of handling very large volumes of air. An example of a low vacuum pump is shown in *Figure 2.79*. This is a rotating vane, liquid-sealed pump and is capable of attaining pressures down to 20 kN m^{-2}, absolute. A rotor (1) revolves in a circular casing (3). The rotor consists of a series of blades projecting from a hollow, cylindrical hub. The blades do not touch the inner surface of the casing. They form a series of chambers. The casing contains a liquid (4), usually water. In position (5) the chamber is empty. As the rotor turns, the converging casing forces the water into the chamber until it is full at (6). As the water recedes from the rotor chamber (7), it is replaced by air entering through the inlet port (2) which is connected to the pump inlet. As the rotor continues to turn, the chamber refills with water and the air that filled the chamber is forced out through port (2) to the pump discharge. The piston pump may be used as a low-vacuum pump. In designs intended for vacuum duties, the clearance at the end of the cylinder and around the valves is relatively small and the valves are lightweight.

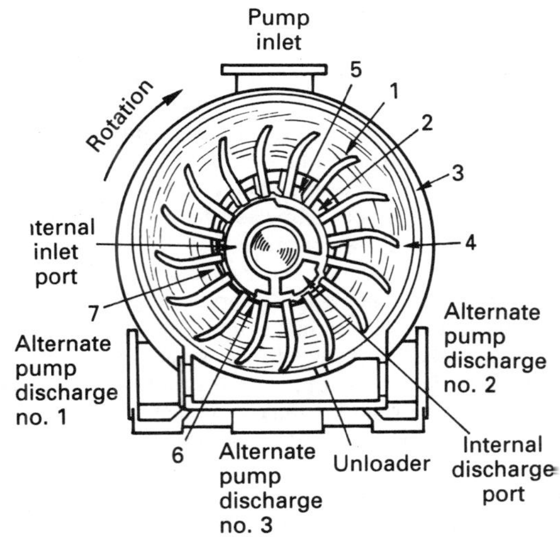

Figure 2.79 Rotary vacuum pump (Hall et al, 1986). See text for detail

Reciprocating piston pumps are not capable of reaching very low pressures, such as those required for freeze drying. Rotary pumps are more suited to such duties. In *Figure 2.80*, two types of rotary oil-sealed pump, suitable for very low-pressure work, are shown in series. The first in the series is a rotary piston pump. A shaft carrying an eccentrically mounted cam rotates in a cylindrical cavity. This cam turns in a cylindrical

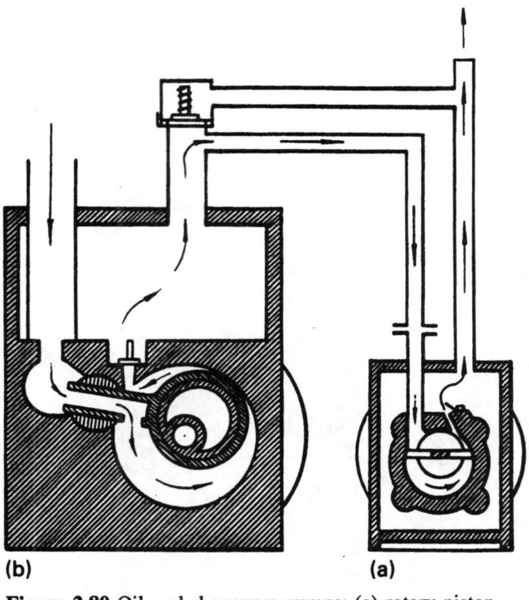

Figure 2.80 Oil sealed vacuum pumps: (a) rotary piston pump, (b) rotary vane pump (Cotson and Smith, 1963)

member which touches the inner surface of the cavity as it rotates. The member is attached to a hollow piston of rectangular section equipped with an inlet port. As the cam rotates, the piston oscillates in the inlet passage. As it turns anti-clockwise the space behind the cam increases. The air enters this space while the gas in front of the cam is compressed and expelled through the exhaust valve. Oil both lubricates the pump and acts as a seal between the high-pressure side and the low-pressure side in the pump. The second pump in the series in *Figure 2.80* is known as a rotary vane pump. A rotor is eccentrically mounted in a cylindrical cavity and as it rotates it makes contact with the inner surface of the cavity at the top of its cycle. Two spring-loaded vanes mounted on the rotor also make contact with the inner surface of the cavity. Air entering through the inlet port is confined between the vanes and as the rotor turns it is compressed and expelled through the exhaust valve. Oil again acts as a seal between the high- and low-pressure sides of the pump. For freeze-drying applications two oil-sealed pumps are often used in series, the second stage being gas-ballasted. The Roots pump (or blower), *Figure 2.81*, is also used to create low pressures. Two interlocking lobes rotate within a cavity. The movement of the lobes is synchronized by gears. There is a small, fixed clearance between the lobes and between them and the inner surface of the cavity. The lobes rotate at up to 3000 rpm. Such pumps are capable of handling very large volumes of air. In freeze-drying plants, Roots pumps are used in series with oil-sealed vacuum pumps. When oil-sealed pumps or Roots pumps are used for freeze drying, they are located downstream of a refrigerated condenser. Barometric condensers and steam ejectors are also used to create low pressures in evaporators and driers (Cotson and Smith, 1963; Hall, Farrall and Rippen, 1986). *See also* Barometric condenser, Evaporation, Freeze drying, Steam ejector, Vacuum cabinet drier, Vacuum band drier.

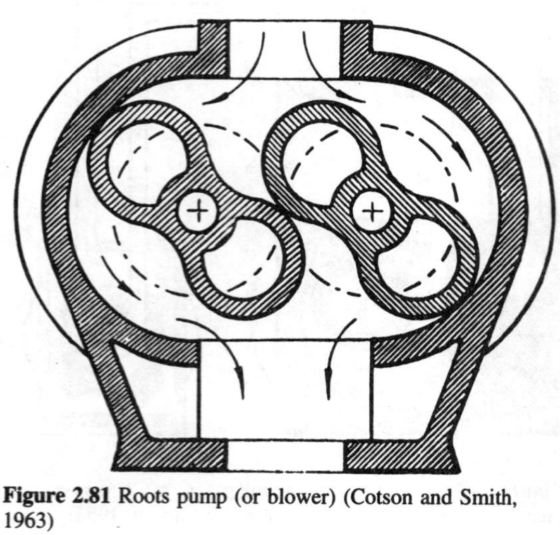

Figure 2.81 Roots pump (or blower) (Cotson and Smith, 1963)

Vacuum shelf (cabinet, tray) drier

In its simplest form, this drier consists of a vacuum-tight chamber fitted with one or more access doors. The chamber is usually cylindrical in shape and positioned horizontally. It is equipped with hollow shelves, heated internally with steam at above or below ambient pressure, depending on the shelf temperature required. Alternatively, heated water, or some other thermal fluid, may be circulated through the selves. Typically shelves measure 2.0 × 1.5 m. The chamber may contain up to 24 shelves. A partial vacuum is drawn on the chamber via a vapour outlet which leads to a condenser and a vacuum pump in series. The condenser may be water cooled, or, if very low pressures are required, cooled by refrigerant. The food, which may be a solid, liquid or paste, is deposited in thin layers on metal trays. These rest on the heated selves in the chamber. Shelves and trays must be designed and manufactured to ensure good thermal contact between them. After loading the food into the chamber, the door is closed and the pressure reduced to the desired level. This may range from just below atmospheric pressure to as low as 0.1 kN m^{-2}, absolute. Very low pressures are used when freeze drying. For vacuum drying, chamber pressure is usually in the range 5–30 kN m^{-2}, absolute. Heat is applied to the plates to raise them to the appropriate temperature which is usually in the range 35–80°C. The equivalent of 1.2–2.5 kg steam is required to evaporate 1 kg water. While most of the heat is supplied to the food by conduction from the heated plates on which the trays rest, some heat will reach it by radiation from the selves above and other hot surfaces in the cabinet. The initial drying rate is high and constant. This is followed by an extended falling-rate period. In addition to resistance to movement of moisture within the solid, poor thermal contact between the food and the tray may contribute to the reduction in drying rate. Shrinkage of the tissues in the food may cause it to move away from the tray surface in places. Drying may occur from those parts of the food surface which are not in contact with the tray. A dry layer may form which acts as a further barrier to heat transfer. It is important that the plate temperature be well controlled, so as not to cause heat damage to the dry areas of the surface of the food. Vacuum cabinets have been used which are fitted with hydraulically operated heated shelves so that they can be brought into contact with both surfaces of the food. This method of drying is known as the vacuum contact plate (VCP) method and is reported to have led to a reduction in the drying time in some cases, as compared to fixed plates. However, the escape of water vapour from the drying surfaces may be restricted by the close contact between the plates and the food surfaces and this may counteract the beneficial effect of this type of cabinet. Such cabinets are not widely used for vacuum drying but are used in freeze drying.

In one design of vacuum shelf drier each tray is sealed and heated independently so that it can be withdrawn from the cabinet without interrupting the drying cycle in the other trays. The trays are jacketed and may be plugged into a heating system at the back of the cabinet. This design is more flexible and capable of a higher production capacity as compared with the conventional single-compartment design. Vacuum shelf and vacuum band driers have relatively high capital and maintenance costs. This method of drying is only used for very heat-sensitive materials which can bear the high costs of drying. Both types of vacuum drier can be used for vacuum puff drying (Nonhebel and Moss, 1971; Williams-Gardner, 1976; Green, 1984; Brennan et al, 1990). *See also* Double-cone vacuum drier, Freeze drying, Vacuum band drier, Vacuum puff drying.

Vacuum-spray freeze drier
A type of freeze drier suitable for drying pre-concentrated liquid foods. The concentrated liquid food is sprayed into a cylindrical vacuum chamber, surrounded by a refrigerated coil. The droplets, typically 150 μm in diameter, evaporatively freeze, losing about 15% of moisture (wwb) in doing so. The frozen droplets fall on to a belt, which carries them through a vacuum tunnel, where energy is applied by radiant heaters and drying is completed. The dry product is removed from the vacuum chamber intermittently, through a vacuum-lock (Mellor, 1978; Liapis, 1987). *See also* Cabinet freeze drier, Freeze-drying, Multicabinet freeze drier, Tunnel freeze drier.

Vacuum steam blancher *see* Blanching.

Vacuum tray drier *see* Vacuum shelf drier.

Venturi
A converging-diverging section in a tube or pipe which restricts the flow of a gas or vapour. The pressure drop across the restriction is related to the flow rate of the gas or vapour through it and measurement of this pressure drop provides a means of determining the flow rate of the gas or vapour. This is the basis of a Venturi meter. In a steam ejector high-pressure steam expands through a jet and then through a Venturi at high velocity. This creates a low-pressure zone in a suction chamber which is attached to a vessel which is to be evacuated, see *Figure 2.68*. Steam ejectors are used to create low

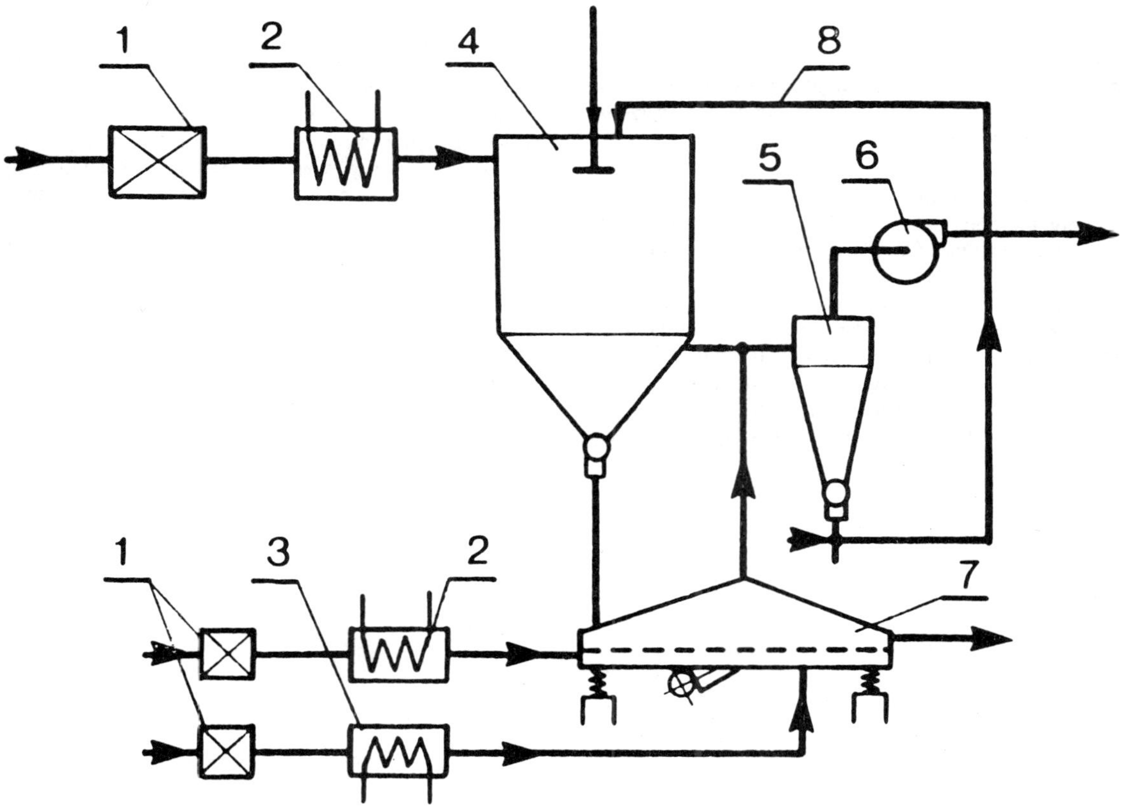

Figure 2.82 Two-stage spray-drying system with vibrated fluidized bed as second stage: (1) air filter, (2) air heater, (3) air cooler, (4) spray-drier, (5) cyclone, (6) exhaust fan, (7) vibrated fluidized-bed drier, (8) return line for fine powder, adapted from Filkova and Mujumdar (1987)

pressures in evaporators, vacuum driers and freeze driers. *See also* Steam ejector.

Vibrated fluidized-bed drier (vibro-fluidizer)
For certain applications, a vibrating fluidized-bed drier may have advantages over the stationary type. A vibrating bed is capable of handling particles with a wider size distribution and with sticky properties as compared with a stationary bed. It is also less likely to cause damage to particles by abrasion and the break-up of agglomerates of particles. Such driers usually have a long, narrow shape with a relatively shallow bed of particles. They vibrate in a frequency range of 5–25 Hz and a half amplitude of a few millimetres. Usually, the air velocity required to keep the particles moving is much less than that in a stationary bed, even less than the incipient velocity in the stationary bed. Vibrated beds are often used as after-driers in spray-drying systems (*Figure 2.82*) and also in rewetting plants for producing instant products. Such beds may be divided into sections so that multi-stage drying and cooling may be carried out in a single unit (Hovmand, 1987; Masters, 1991). *See also* Fluidized-bed driers, Rewetting process.

Vibro-fluidizer *see* Vibrated fluidized-bed drier.

Viscosity
Internal friction within a fluid, giving rise to a resistance to flow. *See* Dynamic viscosity, Kinematic viscosity and Apparent viscosity.

Volatile loss/retention during dehydration
The organic compounds which are responsible for the odour and flavour of foods are relatively volatile and have evaporation temperatures well below that of water. Consequently, they are likely to be lost to some extent during dehydration. The extent of loss of such volatiles will depend on the drying conditions. If the initial drying rate is high, so that a dry layer forms quickly at the drying surface, the loss of volatiles is likely to be less than at a low initial drying rate. The dry layer is virtually impermeable to the volatile flavour compounds. Kerkhof and Schoeber (1974), in their study of the drying behaviour of droplets in spray driers, also considered the retention of aroma compounds. From their calculations they concluded that aroma retention increases with (1) increase in the soluble solids content of the feed, (2) increase in feed temperature, particularly when its soluble solids content is high, (3) increase in air outlet temperature, (4) decrease in air outlet humidity. Under optimum conditions, up to 80% of aroma retention was possible.

Volatile retention in freeze drying is usually quite high and can exceed 90%. This may seem surprising, in view of the very low pressure used in this method of drying. During freeze drying, water and volatiles evaporate together until a certain moisture content is reached. From that point on, the loss of volatiles stops while that of water continues. Re-humidification of the dried material to a moisture content above that at which volatile loss ceased usually causes evaporation of the volatile to recommence. Some authors state that the critical moisture content above which volatile loss ceases is the BET monolayer moisture content. Two mechanisms have been suggested in the literature as accounting for volatile retention during freeze drying. These are (1) surface adsorption of the volatile on the dry layer of the freeze-drying material and (2) entrapment of the volatile in micro-regions formed during the freezing and subsequently stabilized in the freeze-dried material. Much research has been carried out to establish which of these theories is correct. The bulk of the published evidence seems to favour the entrapment mechanism. However, adsorption may also play a part in volatile retention (Karel, 1974; Kerkhof and Schoeber, 1974; Bartholomai et al, 1975a, 1975b, 1976; Mellor, 1978). *See also* Freeze drying, Monolayer moisture content, Spray drying.

Vortex-bed drier
A spouted-bed drier in which the drying air is introduced tangentially in the drying chamber. *See also* Spouted-bed drier (*Figure 2.64*).

W

Waste disposal
Food dehydration plants, in common with other types of food-processing plants, generate large quantities of waste. Examples include: peelings and trimmings that result from the preparation of fruits and vegetables for dehydration, wash water and blanching water; residues left after the extraction of solubles from ground roasted coffee or tea leaves; bones, fat and blood from meat-processing; whey and buttermilk from processing dairy products; water containing detergents and sanitizers used in cleaning processing plant; human sewage from toilet facilities. Some solid waste may be disposed of directly (see below). Liquid wastes and those containing suspended solids generally have to be treated for disposal. The treatment may be done wholly or partially on-site or in municipal sewage treatment plants. Waste treatment is a very expensive business and so the amount of waste produced should be kept to a minimum and, wherever possible, some use should be made of waste products.

The strength of food waste may be specified in terms of the amount of suspended solids it contains and its biochemical oxygen demand (BOD). The latter is a measure of the quantity of oxygen required for the oxidation of organic matter in water by the microorganisms present, over a specified time under specified conditions. Food wastes may contain large amounts of organic matter and hence have high BOD values. If discharged without treatment into rivers or streams such wastes would lead to pollution. Treatment of food wastes may involve some or all of the following stages:

Screening
The first stage involves the removal of relatively large solid particles by passing the waste through screens. The screens may be stationary, vibrating or rotary and the aperture size is selected to suit the size of the solid particles.

Settling
The liquid stream from the screens may pass slowly through a settling tank in which the solids settle to the bottom and are removed as a sludge. The 'clarified' liquid is removed through an overflow near the top of the tank. If the effluent contains fats and oils it may pass through additional settling tanks where the fats rise to the surface and are skimmed off. In some systems compressed air is injected into the bottom of the tank and the bubbles formed help to carry grease and fine solids to the surface.

Coagulation and flocculation
To remove finer particles a coagulant solution may be mixed into the effluent. Lime and ferrous sulphate are used for this purpose. The agent overcomes the repulsive forces between the small particles causing them to form small groups. Further gentle mixing causes these groups to form larger flocs. These readily settle out and are removed as sludge.

Following the physical and chemical treatment outlined above the effluent may require biological treatment before it is discharged into streams or rivers. In such treatment, dissolved or finely divided organic matter is broken down by microorganisms, under controlled conditions, to produce a sludge and a liquid stream which can be discharged safely without the risk of pollution. Biological treatment may be aerobic or anaerobic.

Aerobic biological treatment
A number of different methods are used to treat effluent aerobically. These include: trickling filters, activated sludge systems, oxidation ditches, lagoons and spray irrigation.

Trickling filter
This consists of a cylindrical concrete tank packed with broken rocks or stones. The effluent is sprayed over the top of the bed and percolates down through it. A slime of biologically active organisms (zoogloeal slime) forms on the surface of the rocks. These organisms use the organic material in the effluent as food and remove it from the liquid. The slime may contain bacteria, fungi, protozoa, worms, fly larvae and rotifers. The treated liquid leaves through drains located beneath the bed of rocks.

Although this liquid contains very little dissolved organic matter, it will carry in it suspended material released from the slime layer. It therefore passes through settling tanks where this is removed as sludge. The liquid may then be suitable for direct discharge into waterways. The sludge from these tanks may be mixed with that obtained by physical and/or chemical treatment and further treated anaerobically. There are many variations on this basic design of filter. Wood or plastic packing materials may be used for the bed. These can lead to relatively high rates of flow through the filter. Two filters may be used in series (alternating double filtration, ADF). A moving film system is also available. In this, a number of circular discs rotate partially submerged in effluent in a tank. The slime layer builds up on these discs.

Activated sludge system
In this system the effluent after preliminary treatment is fed to a reaction vessel through which air is bubbled. A mixed population of microorganisms, in the form of a floc, is maintained suspended in the vessel. The finely divided organic matter in the effluent agglomerates in the flocs of microorganisms and some of the organic matter in solution is also adsorbed. This material is utilized by the microorganisms which convert it to water, carbon dioxide and additional biomass. After 6–8 hours in the vessel, the suspension is discharged into a settling tank where the suspended matter sediments and 'clean' liquid is removed via an overflow. This liquid may then be discharged directly into waterways. Part of the sludge is recycled to maintain the active microbial suspension in the vessel. The ratio of food to recycled microorganisms can be varied to suit the type of effluent being treated. The more readily degradable the waste, the higher this ratio may be. In the extended aeration process, this ratio is very low and aeration continues for 1–2 days. Relatively small amounts of sludge are produced by this method. This sludge is relatively inert and may need no further treatment before disposal on land.

Lagoons (ponds)
Large, shallow lagoons or ponds 1.0–1.5 m deep may also be used for effluent treatment. These normally rely on natural surface aeration, hence the limit on depth. Aerobic conditions prevail in the bulk of the liquid, but some anaerobic activity may occur in the sludge at the bottom. Mechanical aeration may be used in large lagoons. Lagoons are useful when sufficient land is available and/or when the effluent supply is intermittent (seasonal) and the cost of more sophisticated equipment is prohibitive.

Oxidation ditches
The effluent is continuously circulated around a lined, annular trough 1–2 m deep, dug in the ground. Circulation is achieved by surface agitators. They behave very much like activated sludge vessels, with microbial flocs responsible for the breakdown of the organic matter so that part of the sludge is recycled. The residence time of the liquid in the ditch is 1–4 days.

Spray irrigation
Screened or otherwise pretreated effluent may be sprayed over land carrying grass crops. If this is done carefully, pollution of rivers and streams can be avoided and the spray can irrigate crops. The land needs to be relatively porous.

Anaerobic biological treatment
Anaerobic microorganisms may be used to break down organic matter in waste. The equipment costs are less and the amount of sludge produced smaller than by aerobic treatment. However, such processes are slow and relatively large reactors are required. Odorous liquids may be produced which need further treatment before discharge into waterways. Wastes with high concentrations of less degradable solids may be treated anaerobically. Sludges produced during aerobic treatment of waste may be further treated anaerobically before final disposal.

Lagoons
Anaerobic treatment may be accomplished in lagoons deeper than those used for aerobic treatment. The microbial activity takes place in the sludge at the bottom of the pond.

Anaerobic reactors (digesters)
These are large, enclosed tanks equipped with external heat exchangers. The digesting sludge is recirculated through these exchangers to maintain its temperature, usually in the range 30–35°C. The main activity occurs in the middle third of the tank. Methane and carbon dioxide gases are collected from the top and sludge from the bottom. The liquid is taken off from the top third of the tank. Residence times may be 10–30 days. The liquid from such digesters usually requires further aerobic treatment. The sludge may be de-watered before disposal. Residence times can be reduced to 10–20 days by treatment in two stages.

Disposal of solids and sludges
Solids separated in the early stages of waste treatment and sludges produced in the later stages have to be disposed of. Where possible and economic, materials should be recovered from solids and sludges. Sludges contain large amounts of water and removing some of this facilitates disposal of the solids. Sludges may be chemically treated to assist the de-watering process. This involves the addition of inorganic ions of iron or aluminium or organic polyelectrolytes which improves the settling rates of solids. Heat treatment, to 200°C at 1.0–2.0 MN m^{-2}, may also facilitate settling. Pre-treated sludges may be fed to gravity thickeners where slow-speed stirring gently compresses the settled sludge. The

liquid removed is returned to the earlier stages of waste treatment. Flotation may also be employed to de-water sludges. Sludges from gravity thickeners and flotation tanks may be further concentrated on moving-belt filters. Concentrated sludges may then be finally disposed of by dumping or land-filling, by incineration or by composting (Brennan et al, 1990; Hang, 1992). *See also* Water treatment.

Water activity

Water activity of a solution or wet solid, a_w, is defined by the relationship:

$$a_w = \frac{p_w}{P_w}$$

where p_w is the water vapour pressure exhibited by the solution or wet solid and P_w is the vapour pressure of pure water at the same temperature. This expression also describes the relative humidity of an air-water vapour mixture. Thus, if a solution or wet solid is in equilibrium with its surrounding atmosphere, the water activity of the solution or wet solid will be numerically equal to the relative humidity of the atmosphere. Usually, water activity is expressed as a fraction and relative humidity as a percentage (*see* Equilibrium relative humidity, Equilibrium moisture content, Relative humidity). When the water activity is less than 1, the moisture is restrained in some way within the solid. It may be adsorbed on the surface of the solid, held within capillaries or by solution effects. The relationship between the water activity and moisture content of a wet solid at a fixed temperature, expressed graphically, is known as a sorption isotherm. *See also* Adsorption, Bound moisture, Sorption, Sorption isotherm, Water activity and food quality.

Water activity and food quality

The level of available moisture in foods can influence their chemical stability and the activity of the enzymes and microorganisms which they can support. These effects, in turn, can influence the quality and shelf life of the foods. Some of these effects are represented in a general way in *Figure 2.83* and further information is shown in Table 2.9.

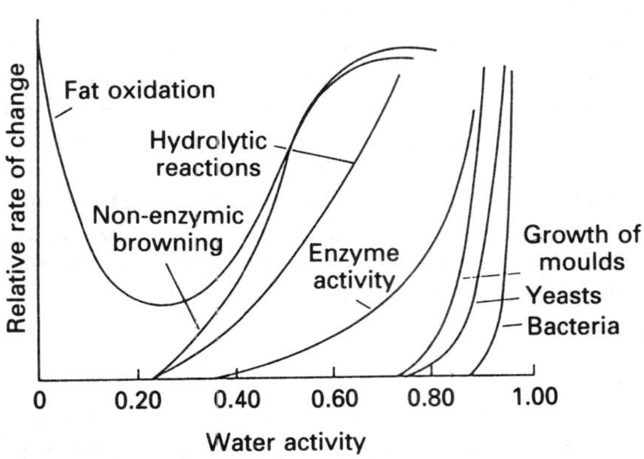

Figure 2.83 Influence of water activity on the stability of foods (Brennan et al, 1990)

Chemical stability

Water in foods acts as a mobilizer and solvent for reacting species. It is also more directly involved in hydrolytic reactions. In recent years, techniques such as nuclear magnetic resonance (NMR) and electron spin resonance (ESR) have been used to study the mobility of moisture in foods. At a moisture content just above the monomolecular value, a small level of mobility is detectable. This contributes to chemical reactions in low-moisture foods. At higher levels of moisture content, a more significant increase in mobility occurs. This corresponds to the point where the solute begins to dissolve. Beyond this point, rapid increases in rates of reaction, e.g. non-enzymic browning, occur. Different compounds become mobilized at different levels of moisture content. The initial form of the substance, e.g. amorphous, crystalline, gelatinized, may influence the moisture content at which mobility is detected. *Oxidation* occurs in foods, particularly those that contain polyunsaturated fats. The reaction involves free radicals and is catalysed by traces of heavy metals. Once the radicals are formed, an autocatalytic reaction can start, with reaction products catalysing the reaction so that the rate increases with time. Oxidation leads to rancidity, due to the formation of volatile carbonyl compounds, and to polymerization due to a recombining of lipid radicals. Cross-linking may occur in proteins, which causes toughening. Aldehydes and ketones produced by oxidation may cause browning. Ascorbic acid is destroyed by oxidation.

The influence of water activity on oxidation is complex. Water can increase the rate of oxidation by increasing the mobility of reactants and bringing catalysts into solution. As the solid matrix swells, new surfaces for catalysts are exposed. However, water may also slow up oxidation by hydrating or diluting heavy metal catalysts or precipitating them as hydroxides. It may also slow down peroxide decomposition by hydrogen bonding with hydroperoxides. Water may also encourage radical recombination which could interrupt the oxidation reaction chain. The net result of all these influences is that, in many foods, the rate of oxidation reaches a minimum in the water activity range 0.2–0.4 (*Figure 2.83*) which usually corresponds with the monomolecular moisture content of the foods. Apart from the adjustment of water activity, e.g. by dehydration, oxidation may be reduced by storing at low temperature, by reducing access to oxygen through vacuum or gas packaging and/or the selective use of antioxidants. Antioxidants are of two main types, i.e. free radical scavengers such as butylated hydroxyanisole (BHA) and butylated hydroxytoluene (BHT), or metal chelators such as citric acid and citrates. It is important to comply with legislation governing the use of such antioxidants. *Non-enzymic browning* of foods may occur

Table 2.9 The importance of water activity in foods, from Fellows (1988)

a_w	*Phenomenon*	*Examples*
1.00		Highly perishable fresh foods
0.95	*Pseudomonas, Bacillus, Clostridium perfringens* and some yeasts inhibited	Foods with 40% sucrose or 7% salt, cooked sausages, bread
0.90	Lower limit for bacterial growth (general). *Salmonella, Vibrio parahaemolyticus, Clostridium botulinum, Lactobacillus*, and some yeasts and fungi inhibited	Foods with 55% sucrose, 12% salt, cured ham, medium-age cheese. Intermediate-moisture foods (a_w = 0.90–0.55)
0.85	Many yeasts inhibited	Foods with 65% sucrose, 15% salt, salami, mature cheese, margarine
0.80	Lower limit for most enzyme activity and growth of most fungi; *Staphylococcus aureus* inhibited	Flour, rice (15–17% water), fruit cake, sweetened condensed milk, fruit syrups, fondant
0.75	Lower limit for halophilic bacteria	Marzipan (15–17% water), jams
0.70	Lower limit for growth of most xerophilic fungi	
0.65	Maximum velocity of Maillard reactions	Rolled oats (10% water), fudge, molasses, nuts
0.60	Lower limit for growth of osmophilic or xerophilic yeasts and fungi	Dried fruits (15–20% water), toffees, caramels (8% water), honey
0.55	Deoxyribonucleic acid becomes disordered (lower limit for life to continue)	
0.50		Dried foods (a_w = 0–0.55), spices, noodles
0.40	Minimum oxidation velocity	Whole egg powder (5% water)
0.30		Crackers, bread crusts (3–5% water)
0.25	Maximum heat resistance of bacterial spores	
0.20		Whole milk powder (2–3% water), dried vegetables (5% water), cornflakes (5% water)

as a result of heating, concentrating or dehydrating them. This phenomenon is due to a group of chemical changes involving aldehydic or ketonic compounds and amino compounds which react to produce coloured polymeric products. The carbonyl compounds may be carbohydrates or the products arising from the oxidation of lipids. The amino groups come from proteins or amines. The Maillard reaction is usually described as involving a reducing sugar and an amino-acid. In addition to the colour change, volatile flavour compounds may be produced. In some products these are desirable, e.g. flavours produced during the baking of bread or the roasting of coffee. In other products they may be regarded as undesirable, cooked flavours. The main factors which influence non-enzymic browning are temperature, pH and a_w. Rate of browning increases as temperature increases. At pH values below 5.0 the rate of browning is low but at pH above 6.0 it increases rapidly. Rate of browning is usually most rapid in the a_w range 0.4–0.6 (*Figure 2.83*). Below that range reaction is slow due to lack of mobility of the water. Above that range the reaction slows down, probably due to the dilution of the reactants. Non-enzymic browning may be reduced by storing foods at low temperature and/or by adjusting the pH or a_w. The addition of SO_2 by exposing the food to the fumes of burning sulphur, or by dipping in a sulphite solution prior to dehydration can also retard this type of browning. It is important to comply with legislation governing the use of these additives.

Nutritional quality

The nutritional quality of foods may also be influenced by water activity. Nutritionally valuable amino acids may be involved as reactants in non-enzymic browning and hence become unavailable to the consumer. Lysine is the amino acid that has been most widely studied and there is considerable evidence that a loss of this nutrient occurs in foods with high water activity, both in foods that exhibit browning and in some that do not. This loss increases with the increase in temperature. Destruction of ascorbic acid increases rapidly at a_w values above 0.65. There is also some evidence to suggest that other vitamins, such as thiamin and riboflavin, may also be destroyed at high a_w values.

Enzyme activity

Enzymes may be described as proteinaceous compounds which catalyse organic reactions in a specific way. Reaction rates may be increased 4–5-fold and sometimes even higher. This catalysis involves two stages.

First, the enzyme binds to the substrate, usually as a covalent complex or an acyl-enzyme intermediate. Secondly, the complex breaks down into a reaction product and the enzyme. Water plays a role in both stages by increasing the mobility of the substrate and the reaction products. At moisture content below the monomolecular value there is little enzyme activity as there is insufficient moisture available to support such activity. Above the monomolecular region, activity increases somewhat and at a_w values above 0.8 it accelerates rapidly (*Figure 2.83*). The enzymes affected in this way include: *amylases* which catalyse the breakdown of starches; *invertases* which catalyse the breakdown of sucrose into glucose and fructose; *lipases* which catalyse the hydrolysis of triglycerides into glycerol, fatty acids and mono- and di-glycerides; *proteases* which catalyse the breakdown of proteins into amino acids via a number of intermediate products and others with more specific effects such as *lipoxidases, peroxidases, phenoloxidases, phospholipidase and lecithinase.*

Microbiological activity
The growth pattern of microorganisms features a lag phase, a phase in which growth accelerates, a phase in which a high, steady growth rate prevails, a phase in which the rate of growth is offset by the rate of death and the population remains constant – known as the stationary phase – and a final phase in which the rate of death exceeds the rate of growth and the population reduces. a_w has been shown to affect each of the four main growth phases and can influence the rate of growth, the extent of growth, i.e. the size of the stationary phase population, and the presence or absence of growth. The most widely quoted criterion is the lowest value of a_w at which a particular microorganism will grow. However, less extreme effects may be important, e.g. at a_w values above the minimum that will support growth, the population of a pathogenic microorganism may be too low to produce a toxic product or an infectious dose. A reduction in growth rate due to reduced a_w may allow synergistic inhibition using other physical or chemical agents, e.g. the addition of salt or reduction in pH. On the other hand, a reduction in a_w can increase the heat resistance of bacterial vegetative cells and spores. Non-halophilic bacteria will not normally grow at an a_w less than 0.90. In the case of halophilic (salt-loving) bacteria, water relations are quite different. Not only can they tolerate high concentrations of salt and grow at low levels of a_w, they are unable to grow at high a_w levels. Extreme halophiles can grow at a_w as low as 0.75. The lower limit of a_w for the growth of most yeasts is 0.85 and moulds 0.80. Osmophilic (sugar-loving) yeasts and moulds may grow down to an a_w of 0.60. Xerophilic (dry-loving) moulds may grow down to an a_w of 0.70. The maximum heat resistance of bacterial vegetative cells and spores occurs at a_w values of 0.75 and 0.25 respectively. Many other factors influence the activity of microorganisms including temperature, pH, availability of oxygen and nutrients and these may affect their behaviour at different levels of a_w. In Table 2.10 some interactions of a_w, pH and temperature in selected foods are presented (Morse, 1974; Eichner, 1975; Labuza, 1975; Duckworth et al, 1976; Williams, 1976; Troller and Christian, 1978; Duckworth, 1981; Simatos et al, 1981; Kirk, 1981; Labuza and Saltmarch, 1981; Karel, 1985; Fellows, 1988; Troller, 1989). *See also* Antioxidant, Sulphuring.

Water blanching *see* Blanching.

Water treatment
Very large volumes of water are used in food dehydration plants for a variety of purposes. Water is used in wet cleaning and blanching of raw materials. It may be incorporated into the feed before drying, e.g. in making up ice-cream mix or oil-in-water emulsions used as

Table 2.10 Interaction of a_w, pH and temperature in selected foods, from Fellows (1988)

Food	*pH*	a_w	*Shelf life*	*Notes*
Fresh meat	>4.5	>0.95	days	Preserve by chilling
Cooked meat	>4.5	0.95	weeks	Ambient storage when packaged
Dry sausage	>4.5	<0.90	months	Preserved by salt and low a_w
Fresh vegetables	>4.5	>0.95	weeks	'Stable' while respiring
Pickles	<4.5	0.90	months	Low pH maintained by packaging
Bread	>4.5	>0.95	days	
Fruit cake	>4.5	<0.90	weeks	Preserved by heat and low a_w
Milk	>4.5	>0.95	days	Preserved by chilling
Yogurt	<4.5	<0.95	weeks	Preserved by low pH and chilling
Dried milk	>4.5	<0.90	months	Preserved by low a_w

coffee/tea whiteners. Comminuted fruits or vegetables may have to be diluted before drying, although this should be limited as far as possible to minimize the cost of drying. There are many other more specialized uses for water, e.g. in the extraction of solubles from ground roasted coffee or tea leaves. For all applications where water is in direct contact with the food it has to be of high potable quality, i.e. clean, free from colour, taint, odour or toxic materials and of acceptable microbiological standard. There may be other requirements, e.g. the hardness/softness of the water may need to be specified as this can influence product characteristics such as the texture of some vegetables. Water is used for cleaning and sanitizing equipment and utensils. For this purpose, water of high potable quality is also required. It may also have to be softened to avoid unsightly deposits forming on metal surfaces or scale build-up on heat transfer surfaces. Water is used as a heating and/or cooling medium in heat exchangers and condensers. In these applications it does not come in direct contact with food. Consequently, it does not have to be of such high quality. It should still be of good microbiological standard and free of toxic substances, in case of leaks in the equipment which could result in the water getting into the product. It may also need to be softened to avoid scale build-up in the equipment. However, the requirements regarding colour and taste need not be so stringent. Water is also used to generate steam. This may also need to be softened to avoid scaling in the boiler. A food-processing plant may have its own water supply. This may be surface water collected in lakes or reservoirs or ground water from wells or springs. Such waters may need considerable treatment. Water may be purchased from municipal authorities. This can be fully treated and of potable standard. Alternatively, it may be untreated or partially treated and require further treatment before use. Whatever the source of water, it is an expensive commodity and so its use should be monitored and controlled and it should be recycled whenever possible.

Some or all of the following steps may be involved in treatment of the water supply to a dehydration plant:

Removal of insoluble solids
Relatively large, insoluble solid particles, 20 µm–5.0 mm in diameter may be removed by passing the water through a screen of suitable aperture size. This is often the first stage in the treatment process. The next stage may involve sedimentation or settling. The water flows upwards or horizontally across large settling tanks or basins and the solid particles settle out under the influence of gravity. The 'clear' liquid is discharged as overflow. To facilitate the settling of very fine particles they may be coagulated and flocculated. The addition of coagulating agents such as salts of aluminium or iron, with vigorous mixing, overcomes the repulsion between small particles arising out of electrostatic charges on their surfaces, and causes them to agglomerate into small groups. Further gentle mixing of these agglomerates causes them to form even larger groups or flocs of particles which readily settle out in the settling basins. The overflow from these tanks will contain only relatively small quantities of very fine particles. These may be removed by passage through sand filters. In some low-rate sand filters a zoogloeal slime forms on top of the bed of sand. A high concentration of microorganisms develops in this slime and this oxidizes organic matter. In such filters a very large proportion of the microorganisms present in the water is removed and so are nitrogen compounds and colouring matter, as well as insoluble solid particles. In other high-rate sand filters, no layer of slime forms and the sand bed simply removes insoluble solid particles and some microorganisms. In some systems a high-rate filter may precede a low-rate one.

Removal of microorganisms
Water may contain a wide range of microorganisms in large numbers. This is particularly the case with surface waters. Such microorganisms may cause discoloration and tainting of food. Pathogenic microorganisms may be a health hazard to the consumer. Microbiological standards for drinking water have been set by the World Health Organisation (WHO 1971). Water used in processing, which comes in direct contact with food, should be of comparable quality. Treatment with chlorine or its derivatives is the most commonly used method of sterilizing water. These will react with organic nitrogenous matter in the water. Sufficient chlorine must be added to ensure that there is enough residual-free chlorine to sterilize the water. Ozone is another effective sterilizing agent. Equipment for the production of ozone for the purposes of sterilizing water is now in use in many countries. Heat treatment, ultrasonic treatment and UV radiation are also used to sterilize water. Fully automated UV plants are now in operation in some countries.

Removal of organic matter and materials that cause colour, taint and odour in water
The combination of coagulation, settling and filtration can remove most of these undesirable contaminants from water. The use of coagulant aids, such as activated silica, can help to remove colours, as can chlorination. Ozone treatment can help to remove taints. Activated carbon is very effective in removing colour, taint and odour from water. This can be mixed in with the water which is then filtered through sand-beds, or the water may be filtered through a bed of charcoal.

Softening of water
Calcium and magnesium salts in solution in water give rise to hardness. Temporary hardness is due to the presence of calcium and magnesium bicarbonates. Permanent hardness is the result of calcium and magnesium sulphates, chlorides and nitrates being in solution in the water. Water hardness can influence the texture of some vegetables and cause the build-up of

scale on heat transfer surfaces in heat exchangers and condensers. The use of hard water for washing requires more soap and alkaline detergents as compared with soft water. The hardness of water may be reduced, i.e. the water softened, by precipitation of the salts giving rise to the hardness. Hydrated lime can reduce temporary hardness, while sodium carbonate can reduce both temporary and permanent hardness. The fine precipitate is coagulated and removed by filtration. After precipitation, the water may be passed through a bed of granulated zoolite. The calcium, magnesium, iron and manganese ions are replaced by sodium ions (ion exchange). This can produce water with zero hardness. Organic ion exchange materials can remove minerals and organic matter from water. These may be polystyrene resins or acrylic polymers.

Removal of dissolved gases
Water may contain oxygen, nitrogen, carbon dioxide and hydrogen sulphide in solution. Such gases may cause problems in boilers and heat exchangers where they form films at the heat transfer surfaces and reduce the rate of heat transfer. Oxygen, carbon dioxide and hydrogen sulphide can also react with iron, causing corrosion in equipment and pipelines. Carbon dioxide can be removed by boiling the water. The addition of sodium sulphite or hydrazine can inhibit oxygen corrosion. Chlorination can counteract the effect of hydrogen sulphide (WHO, 1971; Barnes and Wilson, 1983; Brennan et al, 1990).

Watt (W)
An SI unit of power equal to 1 joule per second.

Wet-bulb temperature
The steady state temperature attained by a small amount of water evaporating in an air stream under adiabatic conditions, in such a way as to produce no significant change in the properties of the air. It is measured by means of a thermometer whose bulb is covered by a wick which is kept saturated with water. There is a balance between the rate of evaporation of water from the wick and the rate of heat transfer from the air, thus:

$$H_w - H = \frac{h_c}{L_w k}(\theta - \theta_w)$$

where H = absolute humidity of the air; H_w = the saturation humidity of the air at θ_w (dwb); L_w = the latent heat of vaporization at θ_w; h_c = heat transfer coefficient; k = mass transfer coefficient. Other temperature sensing elements may be used similarly to measure θ_w.

For air-water vapour mixtures wet-bulb temperature and adiabatic saturation temperature are approximately equal.

Wet basis *see* Wet weight basis.

Wet cleaning of raw materials
Wet cleaning techniques are generally more effective than dry cleaning methods, particularly in removing contaminants which are strongly attached to the food particles. However, wet cleaning requires very large volumes of water of good quality, produces very large volumes of effluent and can accelerate chemical and microbiological spoilage, unless careful control is exercised over washing times and hold-up times, prior to drying. Wet cleaning may involve one or more of the following methods:

Soaking
This is useful for foods such as root vegetables which may be heavily contaminated with earth or other contaminants. The vegetables or fruits are held in tanks of water. The water may be stirred with guarded impeller stirrers, the food may be gently stirred with slow-moving paddles or tumbled gently in a perforated drum which is partially submerged in the water. Alternatively, compressed air may be introduced into the tank to agitate the water. This latter method causes less mechanical damage to the food and is used with soft, easily bruised materials. Heavy contaminants sink to the bottom of the tank and are removed through a grill. Light contaminants rise to the top and are removed over a weir or through a pipe. Heated water may be used, provided it does not speed up spoilage. Suitable detergents are added to the water to assist in removing pesticide, herbicide and fertilizer residues. Chlorinated water may be used where it does not adversely affect the quality of the product.

Spraying
Directing water sprays on to the surface of the food is a widely used method of removing soil. The pressure of the sprays should be as high as possible without damaging the food. In belt washers, the food is conveyed on a perforated belt beneath the sprays. Roller conveyors may be used for near-spherical items, so that they may spin, thus exposing all the surface to the sprays. In drum-washers, the food is tumbled gently in a cylinder made of metal rods (rod washer) or wires. The food is retained in the cylinder while the contaminants pass out between the rods or wires. Water is sprayed over the food from a pipe located centrally in the cylinder (*Figure 2.84*).

Flotation
The food suspended in water is made to flow down an inclined chute, passing over a number of adjustable weirs. Heavy contaminants will be trapped behind the weirs. The food and light contaminants flow over the weirs and then over a vibrating screen, while the light contaminants are washed away with water sprays. In *froth flotation*, food particles such as peas are introduced into a mineral oil-in-water emulsion. Air is blown through the liquid and a froth formed on the top. The peas sink to the bottom and light debris rises to the top

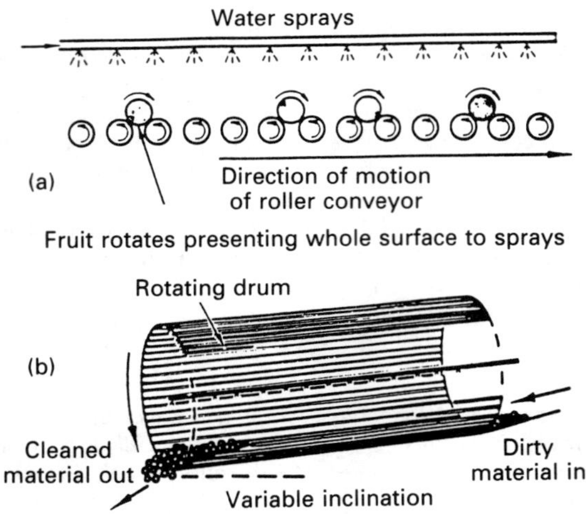

Figure 2.84 Spray washers: (a) belt-type washer, (b) drum washer (Brennan et al, 1990)

and floats on the froth. This is skimmed off. The peas are washed before proceeding to the next stage of the process.

Ultrasonic cleaning
The use of ultrasonic waves to loosen soil adhering to foods has been investigated experimentally, but is not yet used commercially to any significant extent. Ultrasound is extensively used for this purpose in engineering and medical fields.

When wet cleaning methods are being used, it is important to monitor the microbiological quality of the water at all stages and to maintain the plant in a good sanitary condition. Water should be conserved as much as possible. A waste disposal system capable of dealing with the effluent from the washers must be provided (Fellows, 1988; Brennan et al, 1990; Arthey and Dennis, 1991). *See* Waste disposal, Water treatment.

Wet collector *see* Wet scrubber.

Wet scrubber (wet collector)
This is a device used to remove fine powder particles from an air stream, usually exhaust air from a spray-drying chamber. Scrubbers are seldom used alone but in combination with other types of air/powder separators, e.g. a cyclone separator may be used to remove relatively large particles while a scrubber is used to clean up the air leaving the cyclone. In a wet scrubber, the powder particles are contacted with a liquid which washes them out of the air stream. Contact may be achieved by passing the air containing the powder through a spray of the liquid or through a plate or packing over which a head of the liquid is maintained. In another design, known as the Venturi type, the air

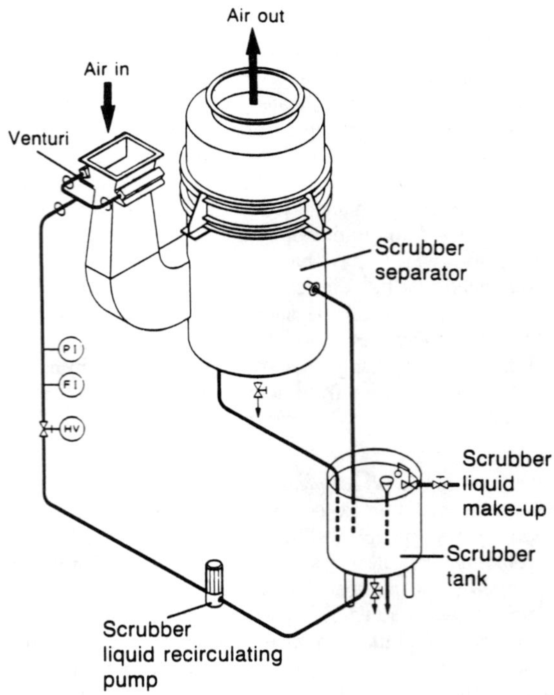

Figure 2.85 Venturi scrubber (Masters, 1991)

containing the powder passes through a Venturi where it is accelerated. At the throat of the Venturi, the wash liquid is injected into the air stream. The liquid is atomized into drops which wash the powder from the air. The liquid is separated from the cleaned air by means of a separating chamber or wet cyclone separator (*Figure 2.85*). The Venturi type of scrubber is often preferred for food applications as it is relatively easy to clean and maintain in a hygienic condition as compared with other designs. A typical collection efficiency curve for this type of scrubber is shown in *Figure 2.86*. It can remove up to 99% of all particles with diameters of 5 μm or more. Wet scrubbers are mainly used to clean up air to be

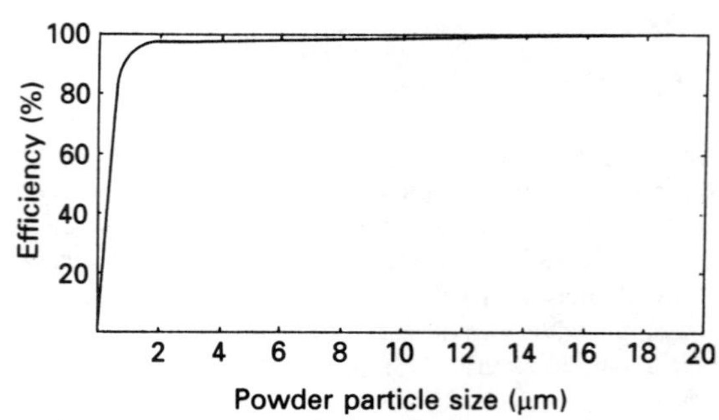

Figure 2.86 Grade efficiency curve for Venturi scrubber (Masters, 1991)

discharged to atmosphere. For this application water is used as the wash liquid. When recovery of the powder is required, some of the incoming liquid feed to the drier can be used as wash liquid. This wash liquid may be recirculated a number of times through the scrubber to increase its concentration and then introduced back into the feed entering the drier. The wash liquid may need to be concentrated in an evaporator before being returned to the drier. Up to 60% of the heat in the exhaust air may be recovered in a wet scrubber (Coulson et al, 1978; Green, 1984; Masters, 1991) *See also* Bag filter, Cyclone separator, Electrostatic precipitator, Energy used in drying, Spray drying.

Wettability of dried powders *see* Reconstitutability of dried powders.

Wet weight basis (wwb)
A method of expressing the moisture content of a wet material based on a unit mass of the wet material. If M is the mass of water and S the mass of dry solids in a sample, then the moisture content (wwb) may be expressed as $M/M + S$, e.g. kg moisture per kg of wet material or as a percentage $(M/M + S) \times 100$. *See also* Dry weight basis.

Wheel atomizer *see* Centrifugal atomizer.

Part 3

A Guide to Dehydrated Foods

Note: In this section moisture content is expressed in percentage wet weight basis in all cases.

A

Abattoir by-products *see* under Blood, Gelatine, Offal.

Acerola (West Indian cherry, ground cherry)
This fruit has been pulped and the pulp dried in a vacuum band drier at a temperature of c. 80°C. Acerola pulp has also been freeze-dried (Salunkhe and Desai, 1984b). *See* Freeze drier, Vacuum band drier.

Apple
Two types of dried apple are manufactured. One contains 10–25% moisture and is sometimes known as evaporated apple. The second contains less than 3% moisture and is often referred to as low-moisture apple. Apple varieties that are firm in texture and yield a white dried product are preferred. The apples are peeled, cored and trimmed. They are then immersed in 1–3% sodium bisulphite solution to prevent browning. They are subsequently cut into appropriate-sized pieces, i.e. rings, slices or wedges. Dehydration to 10–25% moisture is usually carried out in a kiln or tunnel drier. The apple pieces may be further treated with bisulphite solution or exposed to SO_2 fumes during the early stages of drying. In a kiln drier, the fruit pieces need to be turned several times. Air temperature is usually in the range 65–75°C and drying times 14–18 hours, depending on the size of the pieces. If a countercurrent tunnel drier is used air inlet and outlet temperatures of 75 and 55°C, respectively, are maintained and the drying time is typically 8 hours. If a two-stage, concurrent-countercurrent tunnel is used, the air may enter the concurrent section at up to 85°C and the countercurrent section at not more than 75°C. In order to reach the desired level of SO_2 in the product, i.e. 1000 to 3000 ppm, the apple pieces may be dried down to 16–18% moisture and subsequently remoistened to 24% moisture with a solution of bisulphite. This product has a limited shelf life unless refrigerated. It is used mainly for pie filling and other similar applications. To produce low-moisture dried apple, the partially dried pieces may be further reduced in size and subjected to secondary drying in a tunnel or conveyor drier, down to a moisture content of less than 3%. This type of product has a long shelf life if stored or packaged in an appropriate manner. It is free flowing and has a reduced SO_2 content as compared with the partially dried product. Low-moisture dried apple may also be produced directly from fresh apple.

Apple pieces have also been dried by other methods, but not on a large commercial scale. They may be freeze dried to produce a high quality product with good flavour and reconstitution characteristics, but at high cost. They have been reduced to 50% of their fresh weight by osmotic drying. This partially dried product may be rendered shelf stable by a variety of methods. The drying may be completed in heated air or in a vacuum drier (the osmovac process). The osmosed fruit may be frozen (dehydrofreezing) or canned (dehydrocanning). These techniques have been used to produce an apple product for the catering trade. Apple pieces have also been explosion puffed. Apple sauce is pulped apple flesh with added sugar and acid. This has been dried using drum-, vacuum- and freeze-drying techniques. Drum-dried apple purée may be used as an ingredient in dried baby food (Van Arsdel et al, 1973; Woodroof and Luh, 1975; Salunkhe and Desai, 1984a). *See also* Freeze drying, Drum drying, Osmotic drying, Osmovac process, Dehydrocanning, Dehydrofreezing.

Apricot
This fruit is sun dried on a large scale. Ripe fruits are pitted and cut into halves along the suture. They are not peeled. They are then sulphured, either by exposure to sulphur fumes for up to 3 hours or treated with bisulphite to attain 2000–5000 ppm of SO_2 in the end product. They are exposed to the sun, cup upwards, for up to four days, after which they are held in well-ventilated sheds until the moisture content falls to about 18%. They are then stored in boxes or bins for several weeks to attain a uniform moisture content (sweating).

Apricots have also been dehydrated in countercurrent tunnel driers with air inlet temperatures of 65–70°C. However, the quality of the dried product was not as good as the sun-dried fruit. Improvements in quality were achieved by steam blanching the fruit prior to dehydration or at an intermediate stage in the drying process. Partial sun drying of the fruit for 3–5 hours followed by dehydration in a tunnel drier also resulted in an improvement in quality compared with fruit dried only in the tunnel. The moisture content of dried apricot may be further reduced from 18% to less than 5% in a vacuum drier operated at a temperature of 65°C. Apricot paste or purée may be dried to a similar moisture content by drum drying, vacuum drying or freeze drying. Bolin et al (1973) describe the production of apricot 'cloth'. Apricot concentrate, treated with 0.5% sodium bisulphite, was dried on a drum drier at a temperature of 130°C down to 12% moisture content. The dried material had a bright orange colour and good flavour. It could be cut into strips and sold in that form. Several layers of the 'cloth' could be pressed and shaped into bars or apricot halves. Low-moisture content apricot products, either on their own or mixed with other dried fruits, are used in dry baby foods, as fillings for pies or tarts or stewed as desserts (Van Arsdel et al, 1973b; Woodroof and Luh, 1975; Salunkhe et al, 1991). *See also* Drum drier, Freeze drying, Solar drying, Tunnel drier, Vacuum drier.

Aubergine *see* Eggplant.

Avocado
No record of dehydration of avocado was found, with the exception of a freeze-dried avocado guacamole salad which had a shelf life of 48 weeks at 18°C (Salunkhe and Desai, 1984b). *See* Freeze-drying.

B

Banana

Sun-dried banana products are produced on a limited scale. Peeled, whole, ripe bananas may be sun dried but the product is dark in colour, although it has a pleasant flavour. A better quality product is produced if the ripe fruit is sliced lengthways, treated with fumes of sulphur or bisulphite solution and sun dried. Sliced or diced bananas, sulphur-treated, may be dried in cabinet or tunnel driers. The dried fruit pieces are usually milled to a powder. The quality of this type of product is not very good. Banana flesh in piece form has also been vacuum dried down to less than 4% moisture content. Shelf temperature in the range 65–75°C was used. This product was of better quality than the air dried fruit. Pieces of banana flesh may be partially dried osmotically, in 70° Brix sugar solution for up to 10 hours, and then dried down to 2.5% moisture content or less in a vacuum shelf drier, in about 6 hours (osmovac process). Sliced or diced banana flesh may be freeze dried to produce a high quality, porous, dried product, which can be milled to a powder.

Banana pulp or purée may be dried by a variety of methods. It may be drum dried to produce a flake or powder product. The fruit is steam blanched and may be treated with SO_2 to retain the light colour. The final moisture content of the product may be as low as 2.5%. This form of dried banana is used in dry baby food formulations. Banana purée may be vacuum puff dried or freeze dried. It may also be dried using the foam mat technique. In the latter method, glyceryl monostearate may be used as a foaming agent and SO_2 added to slow up oxidation and retain the light colour in the product. Banana purée may be spray dried continuously if additives, such as 15% malto-dextrin, are used to facilitate the handling of the hot, dry powder. The purée, which usually contains 25% total solids, may need to be diluted down to 18–20% total prior to atomization. It must also be well homogenized. Centrifugal atomizers are used. Banana purée, without additives, may be spray dried on a batch principle. The drying has to be stopped at intervals to clean out the drying chamber (Van Arsdel et al, 1973b; Woodroof and Luh, 1975; Masters, 1991). *See also* Cabinet drier, Drum drier, Foam-mat drying, Freeze drying, Solar drying, Tunnel drier, Vacuum shelf drier.

Bean, green

Green beans may be dehydrated by methods similar to those used for green peas. The beans are size-graded, sorted for quality and have the pod ends snipped off. They are then washed and cut transversely into pieces about 25 mm long. Alternatively, they may be cut transversely into longer pieces, about 40 mm, and then cut along their length to give a French-style product. They are blanched in steam at atmospheric pressure for 3–6 minutes and cooled quickly with water sprays. Treatment with sodium sulphite and bisulphite follows. This may be done by dipping or spraying. The dried beans should contain 400–500 ppm of SO_2. The beans may be dried in cabinet, tunnel, conveyor or fluidized-bed driers. In recent years multistage, fluidized driers are most widely used, in particular for the shorter pieces. A relatively high air temperature, up to 90°C, is used in the early stages of drying with lower air temperatures, 50–75°C, being used near the end of drying. It is important that the pods and beans are dried down to less than 5% moisture content. Cooked beans may also be dehydrated. The blanching time is extended until the beans are well cooked. Tunnel or conveyor driers are used for cooked beans, as they are more prone to mechanical damage, as compared with the uncooked beans. Beans have been dehydrofrozen satisfactorily and an expensive, but high quality, product may be made by freeze drying. Beans may be puréed and the purée drum dried and milled to a flaky product. Dried green beans are marketed in consumer and catering packs and incorporated into dried soup mixes and complete meals (Creuss, 1948; Luh and Woodroof, 1975). *See also* Conveyor drier, Dehydrofreezing, Fluidized-bed drier, Freeze drying, Tunnel drier.

Beef *see* under Meat.

Beet (beetroot)
Beets are washed by strong water sprays and peeled. Abrasion, steam and lye peeling methods have been used. After inspection, the peeled beets are cut into dice, slices or strips as required. The pieces may be blanched in steam at atmospheric pressure for 6 minutes. Some sources suggest that blanching is unnecessary. The red pigment of beetroot, betanin, is water soluble. During preparation of the beetroot pieces, contact with water should be kept to a minimum to limit leaching losses. Beetroot pieces may be dried in cabinet, tunnel, conveyor or fluidized-bed driers. Initial air inlet temperature may be in the range 90–100°C, reducing to less than 70°C towards the end of drying. The material may be removed from the drier at a moisture content of 10% and further drying, down to 5% moisture content or lower, accomplished in a bin drier. Diced beats have also been explosion puffed at an intermediate stage of drying to improve the reconstitution characteristics of the dried product (Van Arsdel et al, 1973b; Luh and Woodroof, 1975). *See also* Cabinet drier, Conveyor drier, Explosion puffing, Fluidized-bed drier, Tunnel drier.

Bell pepper (capsicum)
Large, thick-walled peppers with a high solids content are used for dehydration. They are washed, inspected and graded for colour and size. The stem, seeds and placenta are removed in a coring machine and the peppers are cut in half. The halves are diced and treated with sulphite-bisulphite solution, by dipping or spraying, to attain 1000–2500 ppm of SO_2 in the dried product. The diced peppers are dried in cabinet, tunnel or conveyor driers. Air inlet temperatures of up to 80°C have been used initially, falling to below 65°C towards the end of drying. They may be dried to 10% moisture content in the continuous drier and further reduced to 5% moisture content in a bin drier. Diced bell peppers have also been dehydrofrozen (Van Arsdel et al, 1973b; Luh and Woodroof, 1975). *See* Cabinet drier, Conveyor drier, Dehydrofreezing, Tunnel drier.

Blood
Blood from slaughterhouses can be converted into a number of dried products, as shown in *Figure 3.1* (Masters, 1991). Some of these products may be suitable for human consumption. In the manufacture of blood products for human use, only disease-free animals must be bled and strictly hygienic conditions maintained throughout the process. *Dried coagulated whole blood powder* may be prepared by drum or spray drying. The blood is coagulated with steam, screw-pressed and heat-sterilized in an autoclave. It is then drum dried or spray dried. This product may be used as an ingredient in animal feed or as a fertilizer. *Red cell blood* is the name given to dried blood corpuscles. Citrate is added to the blood to prevent it coagulating and the corpuscles separated from the plasma by centrifugation. They are then spray dried. The product contains 56% moisture and 90–92% protein. It is used in sausages and puddings. It may be mixed with other waste products from the slaughterhouse, the mixture sterilized and used for animal feeding. *Plasma powder* is manufactured by concentrating and spray-drying the corpuscle-free plasma fraction from the centrifuge. The product has a moisture content of 5%. This product has some medical applications. *Blood albumin powder* is made from plasma from which the fibrin has been removed. This is achieved by adding an excess of calcium ions and agitation. The plasma is then screened, concentrated and spray dried. The product has a moisture content of 7–9% and contains 70–75% protein. It is used as a whipping agent in confectionery and bakery products and also has non-food uses including lithograph coatings, textile printing and adhesives. *Soluble blood powder* is made from defibrinated whole blood. This is concentrated to 30–35% solids content and spray dried using relatively low air inlet and outlet temperatures. The powder is readily soluble in cold water. It is used in sausages and puddings. Non-food applications include adhesives and asphalt emulsions (Masters, 1991). *See also* Drum drier, Spray drying.

Breakfast cereals
Drying is an important stage in the manufacture of most types of breakfast cereal. For example, the following steps are involved in the manufacture of traditional cornflakes. The ingredients, cornmeal grits, sugar, malt syrup, salt and water are mixed and cooked, usually by direct steam injection, for 2 hours at 105–125 kN m^{-2}. The cooked product will contain 28–32% moisture. It is dumped on to a belt conveyor to cool. It is then delumped by passing over flatbed vibrating screens or through rotating screens. The next state is drying. Single- or multistage conveyor driers are usually used. Inlet air has a temperature of 80–110°C and a relative humidity of 8–15%. The target moisture content is 10–14%. This is a suitable value for the subsequent flaking operation. The dried cornmeal is cooled on a belt cooler and held in large bins (tempered) for some hours to allow the moisture to migrate, so as to attain a uniform moisture content. It is then flaked by passing between large metal rollers which are water-cooled to maintain the temperature at their surfaces in the range of 43–50°C. The flakes are then toasted in an oven. A common type of oven is a rotating cylinder, similar to a rotary drier, through which air at 275–330°C is passed. The residence time is about 90 seconds and the final moisture content 1.5–3.0%. Wheat and rice flakes are manufactured in a similar way. Flaked cereal products may also be made by extrusion cooking. The pellets leaving the extruder have a moisture content in the range 18–24% and do not require drying prior to flaking and toasting. Puffing of cereal products may be accomplished in puffing guns, similar to those discussed under explosion puffing, or in flaking ovens as described above. The toroidal-bed drier could also be used to puff

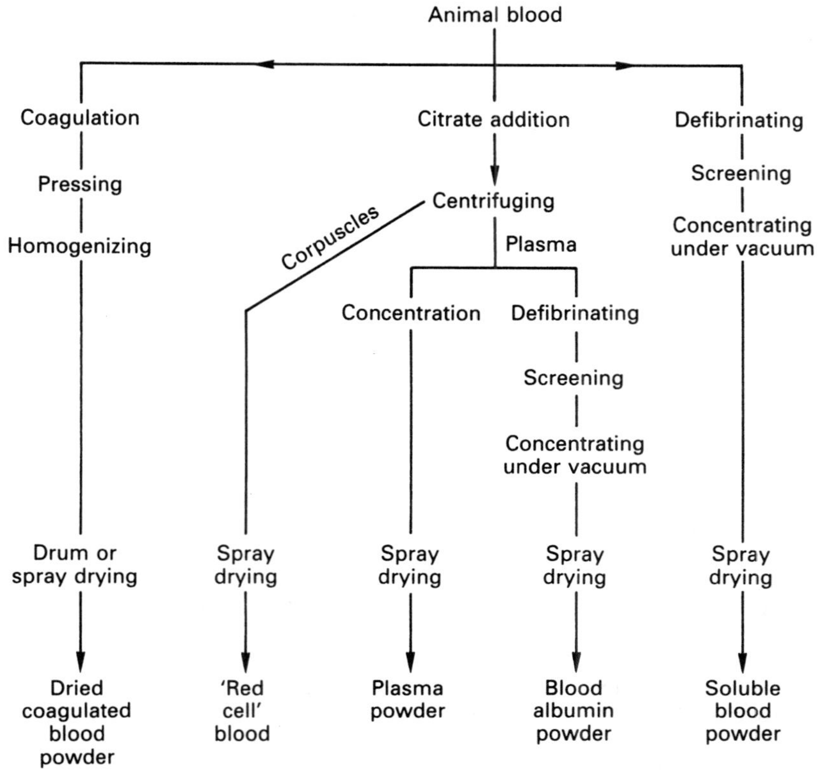

Figure 3.1 Stages in the processing of animal blood by spray-drying (Masters, 1991)

cereal products. Microwave heating has been used to a limited extent as an alternative to heated air for drying breakfast cereal products (Fast, 1990). *See also* Conveyor drier, Explosion puffing, Rotary drier, Toroidal-bed drier.

Brewers' yeast *see* Yeast.

Brussels sprout
This vegetable may be dehydrated by washing the sprouts, blanching them in steam at atmospheric pressure for 4–5 minutes and treating them with sulphite-bisulphite solution to attain an SO_2 content of 1000–2000 ppm in the dried product. They may then be dried in cabinet or tunnel driers using air temperatures of about 60°C (Creuss, 1958). *See also* Cabinet drier, Tunnel drier.

Butter *see* under High fat powder.

Buttermilk
This is the by-product from the churning of cream in buttermaking. Sweet buttermilk is obtained when fresh cream is churned and acid or sour buttermilk comes from ripened cream. Sweet buttermilk typically contains 9.0% total solids including 4.40% lactose, 0.01% lactic acid, 3.5% protein, 0.35% fat and 0.73% ash. Acid buttermilk contains 8.4% total solids including 3.40% lactose, 0.50% lactic acid, 3.30% protein, 0.50% fat and 0.65% ash. Both sweet and acid buttermilk may be spray dried in equipment similar to that used for skim milk. A common system used is a spray drier and fluidized-bed cooler. Air inlet and outlet temperatures used are in the range 175–190°C and 75–90°C respectively. Moisture content of the powder is c. 4.0% and bulk density in the range 0.75–0.83 g ml^{-1}. Sweet buttermilk can be concentrated to 45–50% total solids prior to spray drying without difficulty. However, there are problems with concentrating acid buttermilk since the casein in it coagulates. The viscosity increases and burn-on can occur in the evaporator. Relatively low temperatures have to be maintained in the evaporator and the maximum total solids content that can be attained normally is 25%.

Buttermilk may also be drum dried, but this is seldom done commercially. A double-drum drier is most suitable for this application.

Dried buttermilk is used as an ingredient in animal feed (Masters, 1991). *See also* Drum drier, Skim milk, Spray drying.

C

Cabbage

Varieties of cabbage best suited to dehydration should have a relatively high solids content, up to 10%, and a strong green colour. The cabbage should be delivered as quickly as possible to the processing plant after harvesting, as the outer leaves bleach quickly when exposed to the sun. In the plant, the outer leaves which are discoloured, withered or damaged are removed by hand. The heads are cored mechanically by means of a rotating knife or manually with the aid of a coring tool. The whole or half heads are washed with water sprays and then shredded or diced. The shredded or diced material is spread on to trays which are then conveyed through a steam blancher. Blanching time, in steam at atmospheric pressure, is in the range 1–2 minutes. At an intermediate stage in the blanching operation, the cabbage is sprayed with a solution of sodium sulphite and bisulphite, sufficient to give an SO_2 content of 1000–2000 ppm in the dried product. The prepared cabbage may be dried in cabinet, tunnel or conveyor driers. A two-stage concurrent-countercurrent tunnel is frequently used. Air inlet temperature to the first stage may be as high as 75°C, but in the second stage air at about 60°C is used. A drying time of 6 hours is required to reduce the moisture content of the cabbage down to 7%. It may be dried further down to 4% moisture content in a bin drier. Dehydrated cabbage may be incorporated into dry soup or sauce mixes, used in canned soups, salads and many other foods (Cruess, 1958; Van Arsdel et al, 1973b). *See also* Cabinet drier, Conveyor drier, Tunnel drier.

Carrot

Relatively large carrots, high in solids but free from woody fibre, are most suitable for dehydration. The carrots are dry cleaned, then wet cleaned with high pressure water sprays. They are then peeled, in lye or with steam, and sorted on the basis of size. The green tops are cut off in automatic topping machines. The carrots are than sliced and/or diced and the pieces blanched in steam at atmospheric pressure for 6–8 minutes. They are then treated with sodium sulphite-bisulphite solution which is applied by spraying. The dried carrots should contain 500–1000 ppm of SO_2. As an alternative to sulphiting, the blanched carrot pieces may be sprayed or otherwise coated with a 2% starch suspension at 80°C. This has the effect of extending the shelf life of the dried pieces, as compared with the blanched, sulphited product, by reducing oxidation. The pieces are then dried in a single-stage countercurrent tunnel drier or a two-stage concurrent-countercurrent tunnel drier or on a conveyor drier or fluidized-bed drier. In single countercurrent tunnels air inlet temperature of not more than 70°C is used. In two-stage tunnels, higher air inlet temperatures, up to 95°C, can be used in the first stage and lower temperatures, 65°C, in the second stage. In single-stage driers it may take 7–9 hours to reach a moisture content of 10%. In two-stage driers it may take 5–6 hours to achieve a similar moisture content. Two- or three-stage conveyor driers may be used for carrot drying. Air inlet temperatures up to 105°C may be used in the early stages and up to 85°C in the later stages. Drying times may be as little as 1 hour in such driers. Drying of the partially dried carrot pieces, down to 6–7% moisture content, is usually done in a bin drier. However, in multistage conveyor or fluidized-bed driers, drying may be completed in the one unit. In recent years fluidized beds have been used increasingly for drying diced carrot. The explosion puffing technique has been successfully applied to carrots at an intermediate stage in drying to accelerate the drying process. Carrot pieces have been partially dried by osmosis, using salt solutions, and then frozen (dehydrofreezing) or the drying completed in a vacuum drier (osmovac process). High quality but expensive dried carrot pieces can be produced by freeze drying. Cooked and puréed carrot may be drum dried to give a flaked product. Very finely divided, cooked carrot may be spray dried to a fine powder. Dried carrot products are sold in consumer and catering packs and are used as an ingredient in dried soup and sauce mixes, dried baby foods and complete meals (Van Arsdel et al, 1973b; Luh and Woodroof, 1975; Holdsworth, 1983). *See also* Conveyor drier, Drum drier, Explosion puffing, Fluidized-bed drier, Freeze drying, Spray drying, Tunnel drier.

Casein

Casein can be produced in two ways. Acid casein is produced by the use of hydrochloric acid. Pasteurized skim milk is heated to 40–46°C. Sufficient hydrochloric acid to reduce the pH of the milk to 4.6 is diluted with four parts of water and added to the milk with thorough mixing. After a short holding period in a vat, the whey is drained off. The curd is washed twice with cold water, pressed and milled. Acid casein may also be produced by inoculating skim milk with acid-producing bacteria and incubating it at 20–30°C until the acidity reaches 0.64%. The curd is stirred and heated to 50–65°C. The whey is drained, the curd washed twice with cold water and pressed for 10–15 hours. It is then milled ready for drying. Rennet casein is made by adding sufficient rennet and calcium chloride to skim milk to cause it to clot in 20–30 minutes. Agitation is commenced 2–5 minutes after coagulation starts, the temperature is raised to 55–70°C and the curd cooked for 30 minutes. The whey is then drained off. Acid or rennet whey may be dried in a cabinet or tunnel drier at an air temperature of 50–65°C to a moisture content of 4%. The dried casein is milled and screened to an appropriate particle size. Casein is used in the manufacture of paint, plastics and for paper coating. It is also used in coffee/tea whiteners, whipping powders and imitation milks (Hall and Hendrick, 1966). *See also* Cabinet drier, Sodium caseinate, Tunnel drier.

Cassava

Cassava chips, raw or partially cooked, may be sun dried. The dried pieces may be ground to a flour. Another dried cassava product is produced by grating peeled, washed roots and allowing the mash to ferment for 2–5 days. The water is squeezed from the fermented mass and the solid remaining is partly gelatinized and dried in a heated metal pan to a moisture content of 12–13%. In West Africa this product is known as gari (Salunkhe and Desai, 1984d). *See* Solar drying.

Cauliflower

This vegetable is not dehydrated commercially to any significant extent. Dried cauliflower may be prepared by washing the heads, breaking them up or cutting them into smaller pieces, blanching them in steam at atmospheric pressure for 4–5 minutes and treating them with sulphite-bisulphite solution to attain an SO_2 content of 1000–2000 ppm in the dried product. They may then be dried in cabinet or tunnel driers using relatively low air temperatures, c. 60°C (Creuss, 1958). *See* Cabinet drier, Tunnel drier.

Celery

Dehydrated celery in the form of stalk dice, stalk and leaf flakes and powder is produced commercially. The whole celery heads are prewashed. The stalks are removed by cutters. The leafy portion of the celery is cut to size and the leaves separated from the stalk by elutriation with air. A secondary washing and inspection takes place at this stage. The stalks are diced and the leaves are sliced. A spray of sulphite-bisulphite solution is applied to the dice, to attain an SO_2 content of 500–1000 ppm in the dried product. A two-stage concurrent-countercurrent tunnel drier is a suitable type of drier for celery. An air inlet temperature of 80–85°C may be used in the first stage and 55°C in the second. A number of techniques to improve the quality and reconstitution characteristics of dried celery have been reported. These include impregnating the diced celery with sucrose, dextrose or glucose prior to drying and explosion puffing at an intermediate stage in drying. Dried celery products may be used as ingredients in dried soup mixes or meals, canned soups and sauces, poultry stuffings and salads (Van Arsdel et al, 1973b; Luh and Woodroof, 1975). *See also* Explosion puffing, Tunnel drier.

Cereal grains

Grains such as wheat, corn (maize) and rice, intended for milling, usually need to have their moisture content adjusted to enable them to be stored and/or to condition them for the milling process. Traditionally, such grains were dried in the field to a stable moisture content. However, when they are combine-harvested, the moisture content at harvesting is usually too high for storage stability and so drying is necessary. Drying in heated air is employed. The initial moisture content is normally not more than 30% and drying down to 13–16.5% is the usual practice, to ensure storage stability. However, different grains, or indeed different varieties of one grain, will have optimum moisture contents for milling. Such moisture contents are usually in the range 12.5–14.5%. Under severe drying conditions, i.e. high air temperature and/or rapid drying, stresses may be set up in grains which cause them to crack or check. This may result in breakdown of the grains during handling and storage and may also adversely affect their milling. Foster (1964) described the 'dryeration process' in which corn (maize) is dried down to within 2% of its target moisture content, then held in a storage bin for 4–8 hours before completing the drying with a through flow of ambient air. Corn dried in this way was less brittle than grain dried to the same moisture content in one continuous operation. In the literature various values for maximum temperatures to be attained by different grains during drying are to be found. The maximum temperature for corn is of the order of 65°C while for rice it is 50°C. Many types of driers have been used for grain drying. Both direct and indirect heating of the drying air is used. Fluidized-bed, pneumatic and rotary driers have been used for drying of grains. Large capacity driers, designed specifically for drying grains, are also in use. One such drier, known as a column drier, consists of a tower with vertical columns of grain retained between screens on two or more sides of a central duct through which air flows. The grain moves downwards under gravity and the air moves horizontally through the grains and is exhausted to the atmosphere. Horizontal column driers are also available. In these, two columns of varying length are arranged around a central air duct. Some such driers may be stacked one above the other in modules. One module may be used to cool the grain after drying. For smaller scale operations, bin driers are used. Heated air passes up through the bed of grains. If only a small amount of moisture is to be removed, 1–4%, ambient air may suffice. Bins fitted with agitators to move the grain about during drying are also available. Van Arsdel et al (1973b) describe 'layer drying' in bins. A layer of grain is put into the bin and drying commenced. When this layer is almost dry, a second layer is put on top. This procedure is continued until the bin is full. The application of infrared and microwave heating for drying grains has been investigated, but no commercial applications have been reported (Matz, 1970; Van Arsdel et al, 1973b). *See also* Fluidized-bed drier, Pneumatic drier, Rotary drier.

Cheese

Cheese powder may be produced by spray drying. The cheese is ground and mixed with water to form a smooth liquid which can be atomized. Wall deposition can be a problem, due to the fat content of the cheese. It is common practice to use a drying chamber with an integrated fluidized bed or to complete the drying on a separate fluidized bed. A second fluidized bed may be used to cool the product. Air inlet and outlet temperatures used are in the range 170–180°C and 70–85°C respectively. Strong odours emanating from the plant can be a problem. A semi-closed-cycle spray drier may be used or the exhaust air may be passed through a deodorizing unit (Van Arsdel et al, 1973b; Masters, 1991). *See also* Integrated fluidized-bed spray drier, Semi-closed-cycle spray-drying system, Spray-drying.

Cherry

Sun drying of cherries is practised to a very limited extent. However, they can be dried successfully in a countercurrent tunnel drier. If whole fruits are to be dried, they are dipped in boiling 0.5% NaOH solution to crack the skin. They are then washed in water. If the fruit is pitted before drying the NaOH dip is not required. The fruit is then sulphured by exposure to sulphur fumes or treatment with bisulphite. Air is supplied to the tunnel drier at 75°C and drying time is 8–12 hours. Cherries have been vacuum dried successfully. Osmotic drying followed by vacuum drying (osmovac process) has also been applied to cherries (Van Arsdel et al, 1973b; Woodroof and Luh, 1975). *See also* Osmotic drying, Tunnel drier, Vacuum shelf drier.

Chicken meat *see* Meat.

Chilli pepper

These may be dried whole or in slices. The washed pods, sliced if desirable, are dried in cabinet, tunnel or conveyor driers. Air inlet temperatures of up to 80°C have been used but some reports state that the colour and pungency of the product is better if lower temperatures are used. The peppers may be dried down to a moisture content of 7–8% and then ground. Another approach, to further improve colour and pungency, is to dry down to 12–15% moisture content, store at 0°C, then dry down to 7–8% moisture content just before grinding (Van Arsdel et al, 1973b; Luh and Woodroof, 1975). *See* Cabinet drier, Conveyor drier, Tunnel drier.

Coffee extract

Ground roasted coffee beans are extracted with hot water and the extract dried to produce so-called instant coffee. In the preparation of the extract, the green beans are first cleaned, using screens, cyclones and metal detectors. The beans are then blended to give mixtures appropriate to the type of product required. They are then roasted and ground. The extent of the roasting will determine the strength and flavour of the end product, to a large extent. The fineness of the grind will influence the rate of extraction. A fine grind will provide a large surface area for extraction. However, too fine a grind may give rise to difficulties in ensuring that the surface comes into contact with the water in the extractor. Ground particle size is usually in the range 1000–2000 μm. Extraction is carried out in static or moving-bed extractors. In the former case 5–8 cells are used, one being isolated from the cycle at any one time, while being emptied and refilled. Heated water percolates through the other cells. Each cell consists of a tall cylindrical pressure vessel into which the ground coffee is filled. The heated water percolates down through the coffee grinds in each cell and then on to the next cell. Heat exchangers are located between the cells to adjust the temperature of the water/solution. When first introduced into the battery, the grounds are extracted with water at 100°C. As extraction proceeds, the solution temperature is raised to a maximum of about 180°C. Not only are the soluble components of the roasted coffee extracted, but a certain amount of hydrolysis of insoluble carbohydrate material occurs at the higher temperatures, increasing the yield of soluble solids. The extent of this hydrolysis must be controlled to ensure that the flavour of the end product is acceptable. The ratio of water to coffee grounds is usually in the range 7:2–5:1 and the cycle time is ½–1 hour. The extract leaving the battery has a solids content of 20–30%. In a common design of moving-bed extractor, the coffee grounds are conveyed through a tilted trough, operated under pressure, countercurrent to the flow of heated water/solution, by means of a helical conveyor. Extract prepared by either of these methods is cooled and clarified, by filtration or centrifugation, and stored at low temperature, c. 5°C prior to dehydration by spray drying or freeze drying. The extract, containing 20–30% total solids, may be spray dried directly or it may be further concentrated to more than 60% total solids by evaporation prior to drying. A tall, cylindrical chamber is used. The extract is introduced into the top of the chamber through a nozzle atomizer and flows downwards, concurrently with the drying air. This flow pattern minimizes the chances of powder deposition on the chamber wall. The cone-shaped base of the chamber may be fitted with a cooling jacket to facilitate removal of the dry powder. The fines from the cyclone may be recycled into the wet zone of the drying chamber. Air inlet and outlet temperatures used are in the range 250–300°C and 105–115°C respectively. The moisture content of the powder is 2–3%. This powder may be agglomerated by the rewetting process to improve its reconstitution characteristics.

The production of freeze-dried instant coffee has increased significantly in recent years. The coffee extract may be frozen in trays in a blast freezer. Extract containing 30% total solids is frozen in 2 cm thick layers, in air at –40°C. Belt freezers are also used for large-scale production. One type consists of a stainless steel belt, cooled from below with sprays of brine or ethylene glycol-water mixture. A slab 1.25 cm thick is frozen at –40°C in 20–30 minutes. Another type of continuous freezer consists of a Teflon/rubber belt which carries the extract through a blast freezer. Extract may also be frozen on a refrigerated cylinder, the frozen product being removed as chips from the cylinder surface by a blade. The rate of freezing can affect the properties of the freeze-dried product, in particular its colour. Slow freezing produces a darker product than does quick freezing. The frozen extract is broken down into granules by a mill or rotary cutter, in a refrigerated chamber. The granules are then dried in a cabinet, tunnel or continuous moving-bed freeze drier. The maximum temperature to which the freeze-dried granules are exposed is 40°C. In batch driers, drying times of c. 8 hours are common. In continuous moving bed freeze driers drying times of as little as 10 minutes are possible.

Instant coffee can also be produced in a vacuum band drier, operating at a maximum temperature of c. 60°C. Coffee extract has also been foam dried experimentally, but not commercially.

Instant coffee substitutes, based on mixtures of coffee, chicory, barley, rye, wheat and sugar beet, are also manufactured. In general, the principles of the processes involved are similar to those used for instant coffee, although some differences in processing conditions exist. Chicory extract is rather more difficult to dry than coffee extract as it contains more carbohydrates and tends to be sticky when hot (Van Arsdel et al, 1973b; Sivetz and Desrosier, 1979; Masters, 1991; Clarke and Macrae, 1987; Brennan et al, 1990). *See also* Foam-mat drying, Freeze drying, Spray drying, Vacuum band drier.

Coffee whitener *see* Whiteners for coffee or tea.

Corn (maize)
Young, tender corn should be used for dehydration. After dehusking, the kernels are blanched, usually in steam at atmospheric pressure, for 2 minutes. The kernels are treated by spraying or dipping, using a solution of sulphite and bisulphite at a pH of 6.7 to attain an SO_2 content of 2000 ppm in the dried product. The kernels may then be dried in a cabinet, tunnel, conveyor or fluidized-bed drier. In a two-stage, concurrent-counter-current tunnel, air inlet temperature to the first stage could be as high as 85°C and to the second about 75°C. The drying may be completed in a bin drier. In recent years fluidized-bed driers have been more widely used for corn drying. After drying, pieces of cob, kernel or silk may be separated by elutriation with air (Van Arsdel et al, 1973b; Luh and Woodroof, 1975). *See also* Cabinet drier, Cereal grains, Tunnel drier, Conveyor drier, Fluidized-bed drier.

Corn flakes *see* Breakfast cereals.

Corn gluten
In the manufacture of corn starch gluten is separated from the kernels. This may be mixed with ground hull and steep water and spray dried (Masters, 1991). *See also* Corn starch, Spray drying.

Corn starch (cornflour)
To produce corn starch, cleaned corn kernels are soaked in tanks of slightly acidified water for up to 40 hours. The water is acidified with sulphuric acid to prevent fermentation during the soaking period. The steep water is then drained off and may be concentrated and spray dried. The softened kernels are milled to free the germ and loosen the hulls. Germ separators remove the germ from which oil is extracted. The kernel remains are ground and hull pieces are removed by screens. The starch is separated from the gluten by centrifugation. A corn gluten suspension is formed by mixing the hull and gluten with steep water and this may be spray dried. The starch may be dried in one of two ways. It may be suspended in water and spray dried. Large capacity driers are used, usually equipped with rotary atomizers. Alternatively, the starch may be gelatinized in a continuous steam cooker, operating at a steam pressure of 0.50–0.55 MN m^{-2}, prior to spray drying. Because of the high viscosity of the gelatinized material, high pressure nozzle atomizers are required to atomize it and the solids content of the feed to the atomizer should not exceed 10%. Gelatinized starch with up to 16% solids content can be satisfactorily atomized by two-fluid nozzles, using steam as the propellant. Spray driers for gelatinized starch are of relatively small capacity. Up to three nozzles are located about halfway up a tall drying chamber and the spray is directed upwards, counter-current to the descending airflow. Gelatinized starch powder is highly soluble in cold water while the

ungelatinized material is not. Starch powders can be agglomerated by the rewetting method to improve their reconstitution properties (Masters, 1991). *See also* Potato starch, Spray drying.

Corn steep water (liquor)
The steep water used in the extraction of corn starch may be concentrated and spray dried. The product contains proteins, carbohydrates and minerals. One use for the powder is in culture media for moulds producing antibiotics (Masters, 1991). *See* Corn starch, Spray drying.

Cranberry
Dehydrated cranberries in the form of whole berries or flakes are produced commercially. The washed berries are perforated, if dehydrated whole berries are to be produced, or sliced, if the end product is to be in the form of flakes. The berries or slices are dried in cabinet or tunnel driers. If a countercurrent tunnel is used, an air inlet temperature of 75°C is suitable. Cross-flow tunnels are also used. The moisture content of the berries may be reduced to 8–10% in a tunnel and further reduced to 5% or less in a bin drier.

Drum-dried cranberry flakes may also be produced. The berries are washed and cooked. The cooked berries are pulped and the pulp drum dried down to 5% moisture content. The product may be marketed in the form of flakes, or the flakes may be ground to a powder (Van Arsdel et al, 1973b). *See also* Cabinet drier, Tunnel drier.

Cream *see* High fat powder.

D

Date
This fruit is mainly sun dried. When harvested from the palm some varieties may be astringent. They are incubated at 30–37°C for a few days. During this ripening period, the sucrose in the date is converted to invert sugar as a result of the activity of enzymes, and they lose their astringency. When picked from the palm the fruits may already have a sufficiently low moisture content to obviate further dehydration. If not, they are exposed to the sun for a sufficient length of time to produce a moisture content of less than 24%. If a lower moisture content is required the partially dried dates may be sliced or diced and dried in cabinet or tunnel driers down to less than 12% moisture content. Alternatively, the partially dried fruits may be vacuum dried down to less than 3.5% moisture content. This dried fruit may be further subdivided into a granular or powdered product (Van Arsdel et al, 1973b; Woodroof and Luh, 1975; Salunkhe and Desai, 1984b). *See also* Cabinet drier, Solar drying, Tunnel drier, Vacuum shelf drier.

Durian
This fruit may be pulped and drum dried, as reported by Salunkhe and Desai (1984). *See* Drum drier.

E

Egg

A number of different dried egg products are available, including spray-dried egg white, pan-dried egg white, spray-dried whole egg and egg yolk. Whole egg and precooked scrambled egg have also been freeze dried. Dried egg products with added carbohydrate are also produced. Other additives, such as whipping aids and emulsifying agents, may be incorporated into eggs prior to drying. Anti-caking agents may be added to the dried egg products. Egg yolk and egg white contain small amounts of glucose. This reacts with several of the egg proteins to produce brown coloured compounds and undesirable odours during the storage of dried egg products, particularly dried egg white. There is also evidence that an interaction occurs between cephalin, a phospholipid, and glucose which causes discoloration and off-flavour development in dried whole egg and egg yolk products. To improve the shelf life of dried egg products, the glucose may be removed by a fermentation process prior to drying. The glucose is converted to acid during the fermentation. Fermentation is usually employed when producing dried egg white but is not so widely practised when drying whole egg or egg yolk. Fermentation may be brought about by bacteria, yeasts or enzymes. *Streptococcus lactis* has been used to desugar egg white. Strains of *Saccharomyces* have been used for egg white and yolk. The most commonly used enzyme system is comprised of glucose oxidase and catalase.

To prepare them for dehydration, eggs are candled to detect internal faults. Dirty eggs are washed with detergent sprays, air-dried and candled again. They are then broken and the whites and yolks separated by hand or machine. An experienced operator can break and separate up to 800 eggs per hour. Modern machines can handle up to 18 000 eggs per hour. Efficient separation of yolk from white is essential if a dried white product with good whipping characteristics is required. Residual egg white adhering to the shells can be recovered by centrifugation. Egg white is fermented, filtered to remove chalazae and foam and pasteurized. Stabilizing salts are added prior to pasteurization which is carried out at a temperature in the range 56–59°C for several minutes. The white, containing 10–12% solids, is fed to the spray drier, which may have a vertical or horizontal concurrent drying chamber. An air inlet temperature in the range 145–200°C is used. Both jet and centrifugal atomizers are used. The dried product has a moisture content of 7–9% and a low density. To increase the density of the powder and improve the thermal efficiency of the drying operation, egg white may be concentrated by vacuum evaporation to 20% solids. However, the functional properties of the powder may be adversely affected during evaporation. Ultrafiltration and reverse osmosis have been used commercially to concentrate egg white to 20% total solids prior to drying, with very little adverse effect on functional properties. To produce whole egg or egg yolk powder the liquid egg is homogenized, fermented if required, filtered and pasteurized at 64–66°C for 2–4 minutes. Whole egg is concentrated to 25–27% total solids, egg yolk is concentrated to 45–48%. The egg is then dried in a spray drier of similar design to that used for egg white, using similar air inlet temperatures. The dry powder has a moisture content of 2–4%. In the early days of spray-drying egg, secondary drying was commonly employed. The product from the spray drier, with a moisture content of 3–5%, was further dried down to less than 2% moisture content in a vertical pneumatic drier (airlift drier). Modern spray driers are capable of producing powders with 2% moisture content, so secondary drying is not necessary. It is important that dried egg powder be removed quickly from the drying chamber and cooled. Fluidized-bed coolers may be used. Egg powders may be agglomerated by the rewetting process, to improve their reconstitution properties. The flow properties of spray-dried egg yolk and whole egg may be improved by the addition of anti-caking agents, if permitted by relevant regulations.

Desirable functional properties of spray-dried egg products depend on their application. Good whipping (foaming or aeration) ability is necessary if they are to be used in the making of sponge cakes, meringues and other open-structured confections. Dried egg white is most commonly used in such products. High drying temperatures and severe mechanical action (homogenizing, mixing, pumping) should be avoided as they can reduce whipping ability. Efficient removal of egg yolk from white is also necessary. Removal of glucose by fermentation, prior to drying, yields a dried egg white which can withstand relatively high temperatures without losing its whipping ability. The whipping ability of whole egg and egg yolk is not as susceptible to mechanical damage as is egg white. However, exposure to high temperatures will reduce the foaming capacity of the dried product. The addition of a carbohydrate such as sucrose to whole egg or yolk, prior to drying, can help to preserve the whipping properties of the dried product. There is an optimum level of such additives which will safeguard the whipping powers without adversely affecting the flavour of the powder. Other additives may be used to improve the whipping ability of dried egg white. These include sodium lauryl sulphate, triethyl citrate, triacetin and sodium desoxycholate. The amount added is about 0.1%, based on the egg white solids. It is important to comply with relevant legal regulations on the use of such additives. The emulsifying properties of dried egg products are important when they are used in salad cream, mayonnaise and other emulsified products. Egg yolk has the best emulsifying properties, followed by whole egg and egg white. The emulsifying power of dried egg products may decrease during storage, particularly at high temperature. Emulsifying agents added before drying can improve both the emulsifying and whipping properties of the dried product. Examples of such emulsifiers include glyceryl monostearate, glyceryl dioleate and sorbitan esters of fatty acids. Too high an emulsifier content may adversely affect the texture of baked goods, which may have other emulsifiers added with other ingredients. Again, compliance with legislative requirements on the use of such additives is essential. The ability of dried egg products to coagulate and bind particles of food together or to thicken foods is another important functional property when they are used in the making of cakes, custards and puddings. Severe drying conditions and high storage temperatures can adversely affect the coagulating power of dried egg products. Removal of glucose prior to drying can help to retain the binding properties of such products. Egg white may be pan dried. This term describes the drying of layers of egg white in trays in hot-air cabinets or in water-jacketed pans. A temperature of about 54°C is used, as the egg white coagulates rapidly above that temperature. Drying is stopped when a moisture content of 12–16% is reached. The product is in the form of flakes, up to 12 mm long. Smaller particles are known as granules. Pan-dried egg white may be ground to a fine powder. Whole egg, with added emulsifiers, has been freeze dried commercially. The product is used as an ingredient in baked goods. Precooked scrambled eggs have also been freeze dried. Vacuum puff drying, foam mat drying and foam spray drying of egg products have been investigated, but no record of their commercial use has been found (Van Arsdel et al, 1973b; Stadleman and Cottrel, 1986; Masters, 1991). *See also* Freeze drying, Spray drying.

Eggplant (aubergine, garden egg)
Eggplants may be sun dried or dehydrated in cabinet or tunnel driers. Plants intended for drying should have a high solids and anthocyanin content and low levels of phenols and glycoalkaloid (Salunkhe and Desai, 1984d).

Egg white *see* Egg.

Egg yolk *see* Egg.

Enzymes
There are many applications for enzymes in the food and related industries: amylases in milling, baking, brewing and papermaking; glucose oxidase in carbonated soft drinks; lactase in ice cream; lipase in detergents; pectinases in fruit juice clarification; proteases in meat tenderizing and brewing. Enzymes in dry powder form are convenient to handle and store. Spray drying is the most common method of drying enzymes. The influence of water activity on enzymes is discussed elsewhere in this text under water activity and food quality. The retention of the activity of enzymes during drying is a most important consideration. Enzymes vary in their sensitivity to heat. Some can be spray dried under conventional drying conditions, e.g. air inlet and outlet temperatures of around 150°C and 70°C, respectively. Others are very heat sensitive and much lower temperatures must be used. The presence of inorganic salts in the feed to the drier can reduce activity loss. These may be present naturally or can be deliberately added, if the end use permits. The source of the enzyme may also influence its heat sensitivity, e.g. pepsin obtained from cattle is more heat resistant than that from pig or sheep. For some applications where contamination with microorganisms has to be minimized, an aseptic spray drier may be used. Enzymes may also be freeze dried with good retention of activity (Masters, 1991). *See also* Aseptic spray drier, Spray drying, Water activity and food quality.

F

Fig

Sun drying of figs is widely practised. White varieties are picked when ripe but some dark varieties are allowed to partially dry on the tree and are collected after falling to the ground. Some varieties of fig are dipped in a solution containing 1% salt and 1% hydrate of lime before drying. This treatment removes the hairs from the fruit, softens the skin and improves its colour. Again, some varieties are sulphured before drying whereas others are not. The figs are not exposed to the sun for long periods of time. 30–50% of the drying is carried out in the shade with the trays of fruit arranged in stacks. Figs may also be dehydrated in counter-current tunnel driers. Maximum air inlet air temperature used is 75°C. The fruits are usually cut in half and drying time is 8–12 hours, to reach a moisture content of 5%. Sliced or diced figs have also been vacuum-dried to less than 3.5% moisture content (Van Arsdel et al, 1973b; Woodroof and Luh, 1975). *See also* Solar drying, Tunnel drier, Vacuum shelf drier.

Fish

For centuries man has sun dried fish to preserve it. Whole, small fish or gutted, washed larger fish, whole or in fillets, were supported on wooden frames or racks and exposed to the sun. The fish pieces had to be turned regularly to expose new surfaces to the sun's rays. Several days were required to reach an acceptably low moisture content of about 15%. In more recent times, an improvement in quality and a decrease in drying time was achieved by covering the fish with a transparent material, glass or plastic. The 'greenhouse effect' resulted in higher temperatures being reached, as compared with drying in the open, and there was less contamination of the dried product. Sun or solar drying of fish is still practised in warm climates but mainly on a village basis rather than commercially. Two other traditional methods of preserving fish which are widely practised commercially today are salting and smoking. Salting is a form of osmotic drying whereby the introduction of salt into the fish flesh lowers its water activity, thereby increasing its shelf life. Some loss of water by evaporation may also occur during or after salting, depending on the method employed. Smoking involves exposing fish to smoke from burning wood. The smoke imparts a characteristic taste, smell and colour to the fish and in addition has antioxidant, bactericidal and bacteriostatic effects, prolonging the shelf life of the product. Some moisture loss occurs by evaporation during or after the smoking process, depending on the method used. Modern smoking kilns may be batch or continuously operated. Continuous smoking tunnels are very similar in design to tunnel driers. Air flows concurrently with the fish through the tunnel. When cold smoking, air temperatures at or below 30°C are used. If higher temperatures are used some cooking of the fish flesh occurs, depending on the temperature reached. Smoking at higher temperatures is known as hot smoking. In cold smoking, the loss in weight due to dehydration is limited to about 10–11%. Fish smoked in this way would have a shelf life of about 1 week at 0°C. If frozen, it would have a shelf life of 4–8 months depending on the type of fish. Fish hot smoked at temperatures up to 120°C for long periods of time may have a sufficiently low moisture content to make them shelf stable without refrigeration. This type of product, often known as jerky, has a tough, chewy texture. Kippers and kipper fillets are the most widely produced smoked fish products. Bloater is the name given to whole ungutted herring which is dry salted and cold smoked and buckling to the same fish which is dry salted but hot smoked. Mackerel, haddock, salmon, trout and many other types of fish may be smoked, as well as cod roes, sprats, eels and oysters. Salted and smoked fish products are also available. Unsalted and unsmoked fish may be dried in heated air in cabinet or tunnel driers. Relatively low air temperatures, about 30°C, are used. Air-dried fish is not widely produced commercially. An exception is stockfish which is dried, unsalted cod. This product is produced in Iceland, Norway and Belgium and is popular in West African countries. Fish steaks and fillets, prawns, shrimps and oysters have been dried in vacuum cabinet driers. Again, this is not widely practised commercially. Freeze-dried fish products are available. In the early days of freeze drying, the accelerated freeze drying (AFD) methods were used for fish steaks. In modern freeze driers, the heater plates are stationary and heat is applied by conduction from below and radiation from above the product. Because of the high costs of freeze drying, it is only applied to relatively expensive fish products such as prawns and shrimps. These dried products are used in dried soup mixes and ready meals. Many dehydrated fishery by-products are produced commercially (see below) (Van Arsdel et al, 1973; Aitken et al, 1981; Burt, 1988). *See also* Accelerated freeze drying, Cabinet drier, Freeze drying, Solar drying, Tunnel drier.

Fishery by-products *see* Fish flour, Fish hydrolysate, Fish meal, Fish protein concentrate, Fish stickwater, Surimi.

Fish flour
Sardines are cooked and pressed. The press cake is mixed with stickwater and the mixture spray dried to 6–8% moisture content. The product has a strong flavour and contains 8% oil. This may be extracted with ethanol which removes most of the oil and yields a powder which is odourless and tasteless. The product can be used as a protein additive in bread (Masters, 1991). *See also* Fish hydrolysates, Fish meal, Fish stickwater, Spray drying.

Fish hydrolysate (hydrolysed fish product, HFP)
The raw material for hydrolysis is usually white fish fillets or while fish offal. Fatty fish such as sprat, mackerel, herring and anchovy may also be hydrolysed. However, the oil has to be removed to prevent a strong flavour forming in the product. The feed is chopped into small pieces, diluted with some water and fed to a digestion vessel. Enzymes are added at about the 0.5% level and hydrolysis takes place in the temperature range 25–70°C. Temperatures above 45°C will limit undesirable bacterial growth. The pH of the suspension may be controlled by the addition of a buffer. Various enzymes are used including ficin, papain, pancreatin and trypsin. Sometimes mixtures of enzymes are added. When hydrolysis is complete, the broth is filtered and the filtrate pasteurized at 80°C for 15 minutes. This treatment also inactivates the enzymes. The filtrate is then concentrated and dried by drum or spray drying. The dried product has a moisture content of 3–8%. It is used mainly in animal feed. However, it has potential as an ingredient in food for human consumption. If fatty fish is used as raw material, the oil is removed by centrifugation after hydrolysis. This does not always remove the strong flavour of the end product (Windsor and Barlow, 1981 Masters, 1991). *See also* Drum drier, Fish flour, Fish meal, Fish stickwater, Spray drying.

Fish meal
This is made from freshly caught white or fatty fish. The fish is milled, cooked and pressed. The cake from the press is dried. A nutritionally superior fish meal product, sometimes known as whole-fish meal, is manufactured by mixing some concentrated stickwater with the pressed cake prior to drying. Direct and indirect rotary driers are commonly used for drying fish meal. Where direct driers are employed, air inlet temperatures of up to 600°C are used and the air and meal flow concurrently through the drier. The air may be heated directly or indirectly. Where direct air heating is used the fuel should be of good quality, low in sulphur and nitrogen, and burners should be efficient, completely combusting the fuel. Oil and gas are used as fuels. The meal should not reach temperatures above about 80°C and drying times are of the order of 15 minutes. Indirect driers are heated by steam through hollow discs, tubes, coils or a jacket surrounding the cylinder. In some such driers, the cylinder itself is stationary but the discs or coils rotate on a central shaft. Unheated air flows through the cylinder, usually countercurrent to the solid, to remove the water vapour produced. The steam temperature is about 170°C and drying times are of the order of 30 minutes. Another option is to spray dry the press cake using a centrifugal atomizer capable of handling the semi-solid material. The product from rotary driers is milled to a suitable particle size. When fatty fish is used as raw material antioxidants are added to the product to levels of 400–1000 $mg\,kg^{-1}$. If the meal is spray dried, antioxidant may be added to the feed entering the drier. The dried product typically contains 8–12% moisture and 60–65% protein. Whole-fish meal is richer in vitamins and minerals as compared to meal without added stickwater. The powder is hygroscopic and has to be handled and packaged in a low humidity environment. The packaging must provide a good water vapour barrier. Anti-caking agents may be added to the powder. This product has a strong flavour and is only used during early animal feeding, otherwise the meat may have a fish flavour (Windsor and Barlow, 1981; Masters, 1991). *See also* Anti-caking agents, Fish flour, Fish hydrolysate, Fish stickwater, Rotary drier, Spray drying.

Fish protein concentrate (FPC)
An odourless, tasteless protein product may be manufactured by solvent extraction of white fish. Ethanol and propanol have been used as solvents. The process

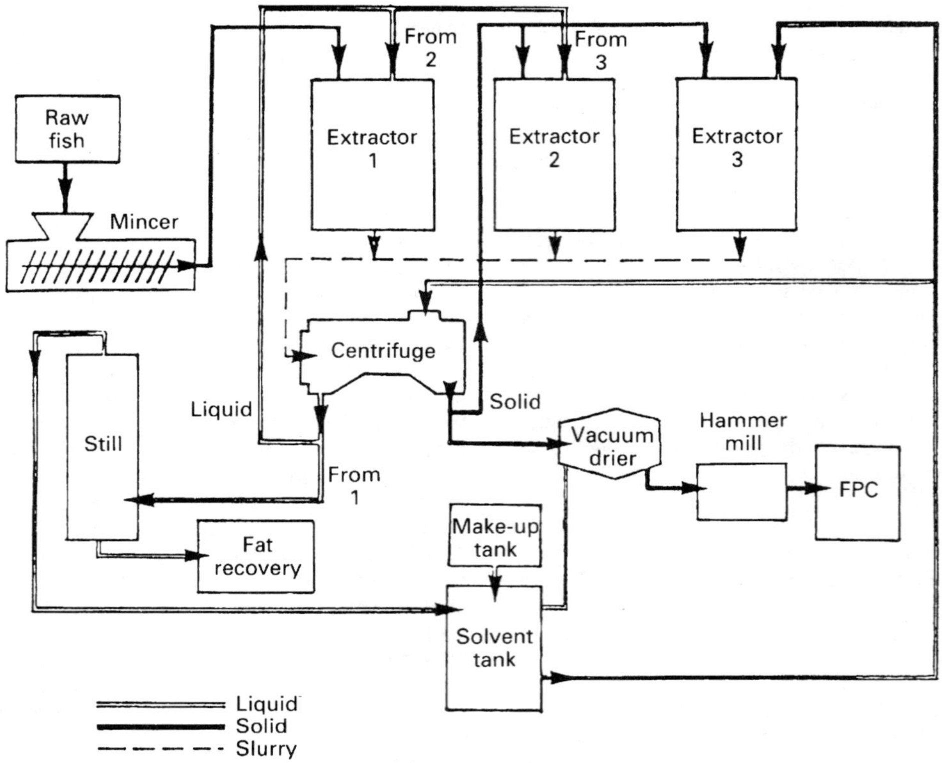

Figure 3.2 Diagram of a fish protein concentrate production plant (after Windsor and Barlow, 1981)

is shown in *Figure 3.2* after Windsor and Barlow (1981).

The fish is minced and extracted in three or more stages. The minced fish is agitated in the first extraction vessel with the solvent for about 50 minutes at ambient temperature. The slurry is separated into a solid and liquid fraction in a centrifuge. The liquid fraction, containing the oil, goes to a still where the solvent is recovered. The solid fraction goes to the second extraction vessel where it is agitated with solvent for 90 minutes at a temperature of 75°C. The slurry from this vessel is again separated in a centrifuge, the liquid goes to the still and the solid to the third extraction vessel. In this third stage, the solid is agitated with fresh solvent for a further 70 minutes at 75°C. A third centrifugal separation ensues. The liquid fraction goes to the still and the solid fraction goes to a vacuum drier where the solvent is evaporated off and goes to a solvent tank. The dry material from the drier is milled as required and packaged. FPC has been incorporated into numerous food products such as bread, biscuits, breakfast cereals, infant foods, pasta, soups and dietetic foods, but only experimentally. It was developed as a protein supplement to be used in countries where cereals and vegetables form a major part of the diet. However, it did not find wide acceptance and is now mainly confined to the health food market in western countries (Windsor and Barlow, 1981). *See also* Fish flour, Fish hydrolysate, Fish meal, Fish stickwater.

Fish stickwater (fish solubles)

This is the term used to describe the aqueous liquid fraction which results when milled, cooked fish, or fish offal, is pressed and the liquid from the press is separated into solid, liquid oil and aqueous liquid fractions by centrifugation. It contains 6–10% total solids made up of salts, proteins and vitamins. It is concentrated to 50% total solids by vacuum evaporation and spray dried. A strong odour is generated which is a problem in urban areas. An environmental spray drier may be used or the exhaust air deodorized by some other means. The dried product contains about 67% protein, 13% lipids and 20% minerals. It contains vitamins B_1, B_2, B_6, B_{12}, nicotinamide and pantothenic acid. It is used as an additive in animal feed (Masters, 1991). *See also* Environmental spray drier, Fish flour, Fish hydrolysate, Fish meal, Spray drier.

Flavouring material

Food flavouring materials are generally highly volatile and, if dried directly, the loss of volatiles would be great and the dry product of very inferior quality. One solution is to mix the flavouring materials with edible, water-soluble gums and carbohydrates prior to spray drying. After drying, the volatile flavouring material is contained within a coating of solid, edible material (encapsulated). The extent to which the volatiles are retained depends on many factors including the volatility of the material, the speed at which the protective film is formed, the completeness of the coating and the size of the liquid droplets. Gums that are permeable to water and water vapour but relatively impermeable to volatile compounds are used. Gum arabic is often used for this purpose. Losses of volatiles in the early stages of drying may be reduced by the use of additives which lower the volatility of the flavouring material (Masters, 1991). *See* Encapsulation.

Fodder yeast *see* Yeast.

G

Garlic
Garlic may be dehydrated in a manner similar to that used for onions. The bulbs bruise easily and must be handled as gently as possible. Each bulb may consist of 6–36 cloves. The bulbs may be exposed to a current of unheated air for a short time to separate the cloves. They are then passed between rubber-covered rollers which break up the bulbs into separate cloves, without crushing them. The husks are separated by screening and aspirating and the cloves are size-graded. Pieces of skin, root stubs and other light debris are floated off as the cloves are washed in water. They are recovered from the wash water by means of a screen. They are then sliced by revolving blades. The slices may be dried in cabinet, tunnel or conveyor driers, under conditions similar to those used for onions, down to about 8% moisture content. They may be further dried down to a moisture content of 6.5% in bin driers. The dried product may be marketed in the form of slices, granules or powder. It may be packed in small packages for consumers or in larger catering packs. Dried garlic products are used as ingredients in dried soup and sauce mixes and in complete meals. They are used in garlic salt, canned soups and meat products, tomato products, salad dressings, marinades and many other foods (Van Arsdel et al, 1973b; Luh and Woodroof, 1975; Salunkhe and Desai, 1984d). *See* Onion.

Gelatine
Defatted bones and hides are extracted to obtain gelatine. The gelatine may be spray dried. A good quality product results if the feed to the drier is relatively dilute, not more than 10% solids content. Hydrolysed gelatine is also spray dried. In this case, a feed solids content of up to 45% can be used (Masters, 1991). *See also* Spray drying.

Glucose
A free-flowing glucose powder can be produced by spray drying. Some of the dried powder is back-mixed with saturated glucose solution to obtain a mixture suitable for spray drying (Masters, 1991). *See also* Spray drying.

Gluten *see* Corn gluten, Wheat gluten.

Grains *see* Cereal grains.

Grape
Grapes are sun dried on a large scale to produce raisins. Prior to drying, the fruits are dipped in an alkali solution to remove the waxy layer from their skins. Various dipping treatments are used. Examples are (a) dipping in 0.2–0.3% NaOH at 93°C for 2 seconds, (b) dipping in 0.3% NaOH + 0.5% KCO_3 + 0.4% olive oil at 76.5°C for 2–3 seconds. Clusters of grapes are placed on wooden or paper trays or plastic sheeting in the vineyard and sun dried down to about 16% moisture content. The clusters need to be turned periodically to ensure uniform drying. Indirect solar drying of grapes, using a solar collector to heat air, which then passes through a drying chamber containing the fruit, is also practised. Bolin and Salunkhe (1982) described a system in which inflated polyethylene hemispheres were used as collectors. Each collector consisted of a dome, 60 m long and 1.5 m wide. The top was translucent and the base black. The tube was connected to a cabinet drier and was inflated by a fan which blew the warm air through the cabinet. The tubes were located between the rows of grapes in the vineyard. Grapes may also be dehydrated in a counter-current tunnel drier. They are dipped in a hot 0.25% NaOH bath, washed and sulphured by exposure to sulphur fumes for about four hours. They are then dried down to 10–15% moisture content in the tunnel. The drying time required is in the range 24–46 hours. The dried product is known as the 'golden bleached' raisin or sultana. The characteristic appearance of this raisin depends mainly on the SO_2 content which is in the range 1500–2000 ppm. Whole grapes may also be dried down to less than 4% moisture content in a vacuum drier.

Raisins produced in this way are used in breakfast cereals, cakes and desserts (Van Arsdel et al, 1973b; Woodroof and Luh, 1975; Salunkhe and Desai, 1984a). *See also* Solar drying, Tunnel drier, Vacuum shelf drier.

Grapefruit

Whole or sliced grapefruits are not dried commercially to any significant extent, much less so than oranges. Grapefruit peel may be dried in a similar manner to orange or lemon peel but, again, very little dried peel is produced commercially.

Grapefruit juice may be concentrated and dried by various methods, similar to those used for orange juice. Spray, foam-mat, vacuum puff and freeze drying techniques may be used to dry the concentrated juice. Nelson and Tressler (1980) discuss the application of foam-mat drying to grapefruit juice concentrate. The production of dried grapefruit powders commercially is on a much smaller scale compared with dried orange powders. *See also* Orange.

Guava

Guava slices or pieces have been dried in heated air. The pieces are blanched in boiling water. They may then be sulphured and dried in air at a temperature in the range 55–76°C down to a moisture content of 4–7%. Drying times ranging from 15 down to 8 hours have been reported. Osmotic drying of guava slices is also possible. The fruit slices are heated in a 70° Brix sugar syrup containing 0.1% potassium metabisulphite at 90° for 3 minutes. They are then cooled to room temperature and allowed to stand in the syrup overnight. The slices are drained. The syrup is concentrated back up to 70° Brix. The fruit is returned to the syrup and allowed to stand in it overnight. The fruit slices are again drained and dried in a hot-air drier, with air temperature in the range 65–80°C, for 7–10 hours. As an alternative to the above process, guava slices may be dried osmotically down to about half of their fresh weight and the drying completed in a vacuum drier operated at a temperature of 60–65°C, down to a final moisture content of 2.0%.

Guava purée has been dried by the foam mat technique. Various foaming agents have been investigated including egg albumin, glycerol monostearate, carboxy-methyl cellulose, soyabean and peanut protein. The foamed purée is then dried on trays or a conveyor in air at 65–70°C, down to a moisture content below 3.0%. Freeze drying has also been applied to guava juice and purée. Sugar is added so that there is 25–40% dissolved sucrose in the mixture. This mixture is frozen rapidly and placed in the freeze drying chamber. The chamber is pumped down quickly, causing the juice to puff up slowly. The porous structure thus produced leads to relatively rapid freeze drying. Heating plate temperature is maintained at 50–55°C and the maximum product temperature is about 40°C (Nelson and Tressler, 1980; Jagtiani et al, 1988). *See also* Foam-mat drying, Freeze drying, Osmotic drying.

H

High fat powder
High fat powders, including those made from butter and cream, may be produced by spray drying. An oil-in-water emulsion is formulated containing 40–82% fat. Where butter or cream is to be used, caseinate and emulsifying agent(s) are included and the pre-mix is homogenized. This emulsion can then be spray dried. Emulsions with such high fat content are generally unstable and easily breakdown. The powder should be handled gently, particularly when warm. Tall chambers with nozzle atomizers have been used. Today, chambers with integrated fluidized beds or belts (Filtermat drier) may be used to cool the powder prior to handling and packaging (Masters, 1991). *See also* Filtermat drier, Homogenization, Integrated fluidized-bed spray drier.

Horseradish
Horseradish roots are washed with high pressure water sprays and brushed to remove dirt. They are not peeled, blanched or sulphited before drying, so as not to inactivate enzymes which hydrolyse glucoside to produce volatile oil, and give the horseradish its pungency. They are sliced with sharp knives to minimize crushing of the tissue. The slices may be dried in cabinet or tunnel driers. In a two-stage, concurrent-countercurrent tunnel drier, air inlet temperature in the first stage is relatively low, about 65°C, and to the second stage about 55°C. The moisture content may be reduced to 8% in the tunnel drier and further reduced to 5% or less in a bin drier (Creuss, 1958; Van Arsdel et al, 1973b; Luh and Woodroof, 1975). *See also* Cabinet drier, Tunnel drier.

I

Ice-cream mix
This is an oil-in-water emulsion consisting of 10–12% fat dispersed in an aqueous solution containing proteins and some inorganic salts in colloidal state and carbohydrates and inorganic salts in solution. A typical dairy mix would contain 12% fat, 10.5% milk solids-not-fat, 13.0% sugar and 0.5% stabilizer. To a basic component of milk, skim milk or water is added fat in the form of cream, butter or butter-oil, solids-not-fat in the form of skim milk powder or condensed milk, sweetening agent in the form of sugar, dextrose or invert sugar and gelatine, alginates or gums as stabilizers. In non-dairy mixes, vegetable fat replaces some or all of the butterfat and caseinate some or all of the skim milk powder. Synthetic emulsifying and stabilizing agents are also used. In the manufacture of ice-cream mix, the basic liquid ingredient is heated to 40–50°C and the dry ingredients added and mixed in. The mix is heated further to c. 70°C and the fat or concentrated milk is added. The mix is then pasteurized in a plate heat exchanger at 80°C for 15–25 seconds. It is then subjected to two-stage homogenization, cooled to less than 5°C and held at that temperature for a period of time to 'age' the mix. This mix may then be spray dried. To reduce problems due to the stickiness of the powder, some of the sugar may be withheld from the mix and added to the dry powder by dry mixing at a later stage. A spray drier with integrated fluidized bed may be used or alternatively drying may be completed in a separate fluidized bed. The moisture content of the powder is 1–2%. To manufacture ice cream, the powder is reconstituted and frozen in a freezer of the scraped surface type, until about 50% of the water is frozen. This partially frozen product is then packaged and freezing is completed in a hardening tunnel operated at –20 to –40°C (Masters, 1991; Rothwell, 1985; Brennan et al, 1990; Brennan, 1992). *See also* Integrated fluidized-bed spray drier, Spray drying.

Instant coffee *see* Coffee extract.

Instant milk powder *see* under Skim milk and Whole milk.

Instant tea *see* Tea extract.

L

Lactose
Crystalline lactose may be manufactured from whey or from the permeate obtained when whey is ultrafiltered. When the starting material is whey, it is concentrated by multistage vacuum evaporation to a total solids concentration of 52–62%. Temperatures less than 70°C are used to avoid denaturation of the whey proteins as much as possible. If permeate is the starting material, temperatures up to 95°C may be used and a solids concentration of 70% is attainable. Multistage falling film evaporators are commonly used. The concentrate is then seeded with fine lactose crystals and cooled to promote crystallization of the lactose. The lactose thus produced is alpha-lactose-monohydrate. The crystals are separated from the liquid by means of a decanting centrifuge. They are then washed and decanted a second time. The crystals are then dried in a rotary or fluidized-bed drier. This product is known as edible lactose. A more refined product, used for pharmaceutical purposes, may be obtained by the inclusion of additional decolorizing and recrystallizing steps (van den Bos, 1982; Rajah and Blendford, 1988). *See also* Evaporation, Fluidized-bed drier, Rotary drier.

Lamb *see* Meat.

Leek
Leeks should be harvested when they have fresh green tops and medium-sized necks. The roots are sliced off and the bulbs and leaves washed. They are then sliced or diced and sulphited, by spraying or dipping, to give an SO_2 content of 500–1000 ppm in the dried product. The pieces are dried in cabinet or tunnel driers. Air inlet temperatures of up to 80°C are used in the early stages of drying reducing to about 55°C towards the end of drying (Van Arsdel et al, 1973b). *See* Cabinet drier, Tunnel drier.

Lemon
Whole lemon fruits or fruit slices are not dried commercially to any significant extent. Lemon peel may be dried in a manner similar to orange peel, i.e. the oil cells are ruptured, the peel is treated with a corn syrup-glycerol solution and dried under vacuum to a moisture content of 8–16%.

Lemon juice is only dried commercially to a limited extent, much less so than orange juice. A number of different drying techniques may be used similar to those used for orange juice. The juice is first concentrated to 25–35% total solids content, usually by multiple effect vacuum evaporation. The volatile aroma compounds are stripped from the juice prior to or during the early stages of evaporation and added back to the concentrate or later to the dried product. This concentrate may be dried by spray, foam-mat, vacuum puff or freeze-drying methods. The dried product has little flavour and so the volatiles, stripped from the juice earlier, may be added to it. Alternatively, encapsulated lemon oil may be used (Nelson and Tressler, 1980). *See also* Orange.

Lime
Whole or sliced lime fruits or lime peel are not dried commercially to any significant extent. Lime juice may be concentrated and dried by various methods similar to those used for orange juice. However, industrial production of dried lime juice products is very limited. *See also* Orange.

M

Macaroni *see* Pasta.

Mango

Sliced and diced mango flesh may be hot-air dried. To give the product a soft pliable texture, the slices or cubes are first held in a 30° Brix sugar solution for up to 2 hours. This is a partial osmotic drying stage. The mango pieces are then sulphured by the traditional method of exposing them to burning sulphur or by holding them in a solution of sodium metabisulphite. The pieces are then dried in a cabinet drier with an air temperature of 65°C, to a moisture content of 18%. Mango pieces have also been partially dried osmotically by immersion in a 65–67° Brix sugar solution for 5–6 hours during which their fresh weight was reduced by about 50%. The drying was completed in a vacuum shelf drier held at a pressure of 0.40 kN m^{-2} absolute and with a shelf temperature of 65°C, down to a moisture content of 2.0%.

Mango purée may also be dried in heated air. It has been successfully spray dried. Because of the sticky and hygroscopic nature of the dried material additives are necessary to facilitate removal of the product from the chamber and the further handling of the powder. Pasteurized, sulphited purée has been successfully spray dried with the addition of 5% liquid glucose and 0.5% tricalcium phosphate. Centrifugal atomization was used and air inlet and outlet temperatures of 165°C and 80–85°C, respectively. The use of a drying chamber with a cooling jacket or a vibrating fluidized-bed cooler would seem to be suited to this application. Mango purée with added skim milk powder has also been successfully spray dried. The skim milk content of the feed ranged from 10 to 40%. Air inlet and outlet temperatures used ranged from 175 to 185°C and 75 to 85°C, respectively. Such powders are suitable for use in products such as ice cream and milk shakes. Mango purée has been dried using the foam mat technique. Polyglycerol stearate was used as the foaming agent. The foam layer was 2.2 mm thick and was cratered. Air temperature was in the range 50–80°C and it took 20 minutes to reach a moisture content of 3%.

Drum drying has been applied to mango purée. A double-drum drier was used with drum clearance of 0.25 mm. The steam pressure was 4.0×10^2 kN m^{-2} and the residence time on the drum surface was 25". The product had a moisture content of 3% and was hygroscopic.

Mango purée has also been dried under vacuum, using a puffing technique. A sulphited mixture of 50% sugar and 50% mango purée was dried in a vacuum shelf drier held at a pressure of 3.4 kN m^{-2} absolute and a shelf temperature of 65°C, down to a moisture content of less than 2%. During the drying, the purée puffed up and the dry product had a porous structure. Freeze-dried mango purée has been prepared by blast freezing it in a layer 12 mm thick, cutting up the layer into pieces measuring 25 × 25 × 12 mm and freeze drying the pieces down to a moisture content of 1.0–2.0% in a cabinet held at a pressure of 0.4 kN m^{-2} absolute.

Mango purée is also converted into fruit bars or 'leathers' which have a chewy texture. A sulphited mixture of purée and sugar of 35° Brix is spread on a tray to a thickness of 6.0 mm and dried in a cabinet dried with air at 80°C for 2.5 hours, by which time its thickness has reduced to 2.0 mm. The thickness of the dried material is increased to 6.0–8.0 mm in stages by adding and drying additional layers (Salunkhe and Desai, 1984a; Jagtiani et al, 1988). *See also* Cabinet drier, Drum drier, Freeze drying, Osmotic drying, Spray drying, Vacuum shelf drier.

Meat

Meat has been sun dried for centuries. Air-dried venison or buffalo meat, mixed with fat, was used by pre-Columbian American Indians. This was known as pemican. Today, many traditional cured and smoked products such as hams and dry sausage involve dehydration, either during or after smoking. Cooked minced meat can be air dried in cabinet, conveyor, fluidized-bed and rotary driers down to a moisture content of 4–6%. The main cause of deterioration of such dried material is oxidation of the fat leading to rancidity. Dried pork is most susceptible to oxidation, lamb and beef less so. Chicken meat is the most stable of dried meats. Antioxidant added during cooking can help to extend the shelf life of dried meat, particularly pork, but care must be taken not to infringe relevant regulations. Packaging in a low oxygen atmosphere, either under vacuum or by the introduction of nitrogen gas, is necessary. In-package desiccants and/or oxygen scavenging agents may be included in catering packages, to further extend the shelf life. The manufacture of air-dried meats is very limited. Cooked minced meat may be dried in vacuum shelf driers. Temperatures of the order of 70°C are used and drying time is about 5 hours. The use of rotary vacuum driers, such as the double-cone vacuum drier for drying meat has also been reported. Packaging requirements are similar to those for air-dried products. Both uncooked and cooked meat may be freeze dried, in the form of steaks, slices, chunks and mince. It was for meat and fish steaks that the accelerated freeze-drying (AFD) method was developed. However, it is little used now. In modern freeze-drying cabinets heater shelves are stationary and heat is supplied by a combination of conduction from below and radiation from above. The moisture content of freeze-dried meat is usually in the range 1.5–3.5%. Because of its porous structure, freeze-dried meat absorbs moisture very quickly and is also highly susceptible to oxidation. Gas packaging in nitrogen is necessary and, again, in-package desiccants and oxygen scavenging agents may be used in large packs. Freeze-dried uncooked meat may have a tough texture due to the denaturating of actomyosin. Browning may occur during storage. This can be limited by suitable packaging as outlined above. Dehydrated meat products, in particulate form, are used as ingredients in dried soup mixes, sauces and ready meals. They are also used as rations for troops, explorers and the like. For this purpose they may be compressed into bars, blocks or cubes.

Various types of meat may be spray dried. Beef is cooked with 15–30% added water. Acetic acid and other additives are added. The cooked meat is milled to a consistency suitable for atomization and spray dried in a concurrent drying chamber. Rotary atomization is used. Air inlet and outlet temperatures are typically 150–200°C and 85–100°C, respectively. Cooked chicken meat may also be spray dried. It is minced and cooked in its own stock to a consistency suitable for atomization. Antioxidant may be added before it is spray dried, if regulations permit. Dried meat powders may be used as ingredients in dried soup, gravy and sauce mixes and in puddings, pies and other processed meat products. Meat may be formulated into intermediate moisture food products. Many such products were developed for military purposes. However, no record was found of an intermediate moisture food for human consumption which is manufactured commercially for retail marketing. Many moist pet foods containing meat are available which are intermediate moisture products (Price and Schweigert, 1971; Van Arsdel et al, 1973b; Pearson and Tauber, 1984; Masters, 1991). *See also* Antioxidant, Cabinet drier, Conveyor drier, Double-cone vacuum drier, Fluidized-bed drier, Freeze drying, In-package dessicant, Intermediate moisture food, Oxygen scavenging agent, Vacuum shelf drier.

Milk *see* Buttermilk, Skim milk, Whole milk.

Mushroom

The most widely cultivated white mushroom is *Agaricus bisporus*, but very little of this is dried commercially. Most of the mushrooms commercially air dried are derived from species of Boletus, which grows wild, *Boletus edulis* in Europe and *Boletus luteus* in South America. The mushrooms may be partially sun dried down to 15–20% moisture content. They are then transported to the drying plant where they are fumigated, dry cleaned and sorted to remove dirt and other contaminants. They are then air dried, using a relatively low air inlet temperature, 50°C, in the early stages of drying rising to 70°C towards the end of the cycle. A high quality dried mushroom may be obtained by freeze drying. Cultivated mushrooms are used. They are sliced, blanched in boiling water or steam at atmospheric pressure for about 2 minutes and/or treated with sulphite-bisulphite solution containing 200 ppm of SO_2 before freezing. They are then freeze dried down to less than 3% moisture content. Blanched, freeze-dried mushrooms have a lighter colour but poorer flavour than unblanched mushrooms. The sulphite treatment does retard darkening of the freeze-dried product during storage (Van Arsdel et al, 1973b; Luh and Woodroof, 1975). *See also* Freeze drying.

Mycelium

When enzymes and antibiotics are produced by fermentation large volumes of mycelium slurry have to be disposed of. This is rich in protein, amino acids and vitamins. When spray dried, this can be used as animal fodder. A suspension of mycelium containing 20–30% solids is fed to the spray drier. High inlet and outlet temperatures are used, 300–500°C and 120–140°C, respectively. There is a considerable explosion hazard when drying this material, and so a semi-closed-cycle spray-drying system is used. There is 5% or less of oxygen in the drying gas. In addition, a very strong odour is created during drying. To overcome this, the gas being vented from the system is deactivated by passing through the flame area prior to being exhausted. This drying system is known as an environmental spray drier. A further feature of this system is that the heat in the vented gas is recovered in a gas-gas heat exchanger. The dry product has a moisture content of 4% or less (Masters, 1991). *See also* Environmental spray drier, Safety aspects of drying, Semi-closed-cycle spray-drying system.

N

Noodles *see* Pasta.

O

Offal

Tissues and glands of slaughtered animals are good sources of enzymes and hormones. Drying of such offal facilitates storage and transport to pharmaceutical plants where purified products are prepared. The offal should be frozen quickly after the butchering of the animal for transport to the drying plant. There it is thawed, milled and finely divided in a colloid mill before spray drying. Relatively low air temperatures are used, not more than 140°C at inlet to chamber (Masters, 1991).

Okra

The pods are washed. Small pods are left intact but large ones are cut lengthways or crossways as required. They are blanched in steam at atmospheric pressure for 2–5 minutes after which they are washed with cold water to remove the gelatinous covering formed during blanching. They are dried in cabinet or tunnel driers using relatively low air temperatures c. 60°C (Creuss, 1958). *See* Cabinet drier, Tunnel drier.

Onion

Sliced onions may be sun dried but mostly they are dehydrated using hot-air drying systems. Onions with high solids content, 10–20%, are selected for dehydration. The bulbs should be large, preferably white and have a high pungency. Fully cured bulbs are required for dehydration. If they are not in this condition when delivered to the factory, they may be stored over perforated metal ducts through which warm air, at 32–38°C, flows to complete the curing. The onions are size-graded and screened to remove dirt, loose husks and other contaminants. The roots and tops are removed either by hand or by a machine featuring two rotating circular saws. The outer skin is removed by flame-peeling or high pressure water jets. The latter technique is more widely used in recent years. The bulbs are sliced at right angles to their vertical axis by high speed mechanical cutters. It is important to crush the tissue as little as possible so as to minimize the release of enzymes which might lead to discoloration of the flesh and/or a reduction in pungency. The onion slices, usually about 3 mm thick, may be dehydrated in cabinet, tunnel or conveyor driers. Two-stage concurrent-countercurrent tunnels may be used. Air inlet temperature in the first stage may be in the range 75–95°C while in the second stage lower inlet temperatures, 55–60°C, are used. Drying times to reach a moisture content of 5–7% may be 10–15 hours. The moisture content of the onion slices may be further reduced to 3–4% in bin driers, through which relatively dry air at 45–50°C flows. In recent years, multistage conveyor driers have been widely used for drying onion slices. Air inlet temperature in the early stages of drying is in the range 80–85°C and this is reduced in stages down to a final value of 55–60°C. The partially dehydrated slices are redistributed on each successive belt so that the bed depth may be 10 cm in the first stage and up to 150 cm on the final belt. About 6 hours is required to reach a moisture content of 6%. The onion slices may be equilibrated and further dried in bin driers. Dried onion is marketed in the form of slices, granules and powder. The dried product is hygroscopic and has to be handled and packed in a low humidity atmosphere. Dried onion products are marketed in consumer and catering packs. They are used as ingredients in dried soup and sauce mixes and complete meals. They are also used in tomato ketchup, chilli sauce, salad dressings, pickles and many other products (Van Arsdel et al, 1973b; Luh and Woodroof, 1975; Holdsworth, 1983; Salunkhe and Desai, 1984d). *See also* Cabinet drier, Conveyor drier, Tunnel drier.

Orange

Most dehydrated orange products are made from the juice of the fruit rather than the fruit in solid form. Woodroof and Luh (1975) report the drying of whole, seedless oranges down to a moisture content of 12–16% in a kiln drier, followed by secondary drying down to a moisture content of 3% in a bin drier. The dried product is milled into granular form. The granules have a very strong aroma. These granules are treated with an antioxidant, such as BHA, to prevent oxidation of the flavonoid compounds which are responsible for this aroma. This type of product may be used in spreads, jams, fillings and sugar confectionery products.

Orange peel may also be dried. The peel, a by-product from the juice production process, is mechanically treated to rupture some of the oil cells. A corn syrup-glycerol solution is added. This partly impregnates the peel and also acts as a heat transfer medium. The treated peel is then dehydrated under vacuum, using temperatures in the range 50–55°C, down to a moisture content of 16–8%. The corn syrup may be separated from the dried peel by screening. The dried peel may be milled if required. It is used in bakery and sugar confectionery products.

Orange juice may be dried by a number of different methods. In almost all cases the juice is preconcentrated prior to drying, in multiple effect evaporators. Evaporator designs which feature rapid heating and short processing times are preferred to minimize caramelization, oxidation and colour changes. Unless precautions are taken, the volatile compounds that contribute to the aroma of fresh orange juice will be lost during evaporation. These may be stripped from the juice prior to evaporation by partial vacuum distillation and added back to the concentrated juice or later to the dried product. Alternatively, these aroma compounds may be separated by condensation of the volatiles during the early stages of the evaporation process itself. This condensate will contain the oil and volatiles responsible for the aroma of the juice. The oil and volatiles are concentrated and removed form the condensate by distillation. They may be added back to the concentrate or later to the dried product.

Concentrated orange juice may be spray dried. However, the dry powder when hot is sticky and hygroscopic, tends to adhere to the wall of the drying chamber and is difficult to handle. Additives are used to overcome these problems. The most common additive used is liquid glucose with a dextrose equivalent in the range 15–30. This may be added in amounts up to 75% of the concentrate, calculated on a solids basis. Such a mixture may be spray dried satisfactorily, although the powder may be somewhat hygroscopic. Other additives, including spray-dried skim milk powder and carboxymethyl celluloses, have also been used but these limit the uses to which the dry powder can be put. Another important requirement for successful spray drying of juices containing insoluble solids is that the feed be very well homogenized prior to atomization. Both centrifugal and nozzle atomizers may be used. Tall chambers featuring downwards concurrent flow patterns are favoured. Chambers with cooled walls and/or equipped with devices to assist in product removal from the chamber are also used. The dried powder may be cooled in a fluidized bed to facilitate handling. Powders produced in this way have very little flavour. The flavour of such powders may be enhanced by the addition of volatiles and oils, separated from the juice before or during evaporation in suitable carriers, e.g. sugar alcohol, or microencapsulated orange peel oil.

Concentrated orange juice has been dried by foam-mat techniques. Methyl cellulose, glyceryl monostearate and soluble soya protein have been used as foaming agents. Two-stage drying has been used. In the first, concurrent stage an air inlet temperature of 75°C has been used while in the second, countercurrent stage the air inlet temperature was restricted to 50°C. Vacuum puff drying has also been applied to concentrated orange juice. In vacuum shelf driers shelf temperatures in the range 60–75°C have been used and drying times required to reach a moisture content of 3% ranged from 2 to 5 hours.

Freeze drying has only been used for concentrated orange juice to a very limited extent. Relatively low freezer and condenser temperatures are required. Only thin layers of the concentrate may be dried or very long drying times are required. Freezing the concentrate in layers and breaking up the frozen sheet, a technique used with coffee extract, can reduce drying time.

Concentrated orange juice may be drum dried but, because of the relatively high temperatures involved, changes in colour and development of cooked/burnt flavour does occur. This method of drying is little used for this product (Van Arsdel et al, 1973b; Woodroof and Luh, 1975; Nelson and Tressler, 1980; Masters, 1985; Brennan, 1991). *See also* Evaporation, Foam-mat drying, Spray drying, Vacuum puff drying.

P

Papaya
Dried papaya products are not produced commercially to any significant extent. Slices of this fruit may be sun-dried or dried in a hot-air drier. In the latter case, the sliced fruit is sulphured and dried in air at a temperature in the range 60–75°C. The dried product may be milled to a powder. Papaya slices may also be osmotically dried in 60% sugar solution for twelve hours and drying completed in a vacuum drier operated at 60°C. Freeze drying of papaya pieces has also been reported. Papaya fruit bars or 'leathers' are produced by adding 10% sugar and 500 ppm of sodium bisulphite to papaya purée, pouring the mixture on to trays and drying it in air at a temperature of about 85°C, down to a moisture content of 12–13%. The product has a chewy texture (Salunkhe and Desai, 1984b; Jagtiani et al, 1988). *See also* Freeze drying, Osmotic drying, Sun drying.

Paprika
Mild, sweet chilli peppers are diced and dried on conveyor driers, using air temperatures as high at 145°C in the early stages of drying. They are dried down to 6% moisture content, ground, including skin and seeds, and the powder rehydrated to about 12% moisture content to improve the colour of the powder. The powder may be stored under refrigeration (Van Arsdel et al, 1973b; Luh and Woodroof, 1975). *See* Conveyor drier.

Parsley
Cleaned and graded parsley may be dried down to 4% moisture content on a multistage conveyor drier in 30 minutes. After drying, the stems are removed by mechanical and/or air separation. The leaves may be marketed as flakes and stems milled to a powder. Alternatively, washed, dried parsley may be chopped, the stems removed and the leaves dried in a pneumatic drier of the airlift type (Van Arsdel et al, 1973b; Luh and Woodroof, 1975). *See* Conveyor drier, Pneumatic drier.

Passion fruit
No reports of commercial dehydration of the passion fruit itself were found. Passion fruit rinds have been hot-air dried and incorporated into feedstuff for dairy cattle. Passion fruit juice has been successfully freeze dried. In one report, the juice was dried in stainless steel trays. A thin layer of ice was precoated on to the bottom of the tray. The juice was added and frozen. A top layer of ice was applied. This technique prevented the formation of a gummy substance in the dried product. The trays of frozen juice were introduced into the freeze-drying chamber which was pumped down quickly to a pressure of 13.3 $N\,m^{-2}$ (0.1 torr) absolute. Drying proceeded for 24 hours with little heating. Residual moisture was removed by raising the shelf temperature to 20°C. Another technique that has been applied to the drying of passion fruit juice is known as vacuum puff freeze drying. The juice was placed under vacuum where it puffed up and froze evaporatively. The puffed juice was then freeze dried. The puffing reduced the drying time and the product had a porous structure which milled readily to a powder (Jagtiani, 1988). *See also* Freeze-drying.

Pasta
Pasta products are prepared by drying formed units of dough made from semolina, durum flour, farina (potato) flour or any combination of two or more of these materials, with water and with other optional ingredients, including whole egg, egg white, egg yolk, salt, onions, celery, garlic, bay leaf, disodium phosphate, gum, gluten and glycerol monostearate. There are two general classes of pasta products. Macaroni products may contain egg white but not whole egg or egg yolk and are tube shaped. Spaghetti is a macaroni product which may be tube or cord shaped. Noodle products must contain whole egg or egg yolk and may be ribbon, tube or cord shaped. Pasta is made by mixing the ingredients with water to produce a mix containing 30%

moisture, kneading this dough to form a homogeneous mass which is extruded through dies so as to produce a product of the desired shape. The extruded material is cut into suitable lengths depending on whether long or short cut products (or goods) are being made. The freshly made pasta contains approximately 30% moisture. The moisture content must be reduced to 12% or less to give a shelf-stable product. Therefore, the cut pieces are dried before being packaged.

If pasta is dried rapidly so that large moisture gradients develop within the pieces then checking or cracking is likely to occur. This results in a brittle product which may be mechanically damaged during packaging, storage and distribution. Consequently, hot-air drying is carried out in two or more stages, sometimes with a holding stage in between. Holding of the partially dried material, in a high humidity atmosphere, permits moisture to migrate within the pieces, reducing moisture gradients and relieving stresses which may have developed during the initial drying stage. If there is no separate holding stage, then secondary drying is done very slowly, over several hours, to avoid checking. Multi-tier, conveyor-type driers are most commonly used. In the case of long products such as spaghetti, the strands are draped over sticks or rods which carry them through the early stage of drying. The partially dried strands are then transferred to moving chain conveyors which carry them through the later stages. Air inlet temperature in the early stages of drying is about 44°C and in the latter stages 40°C, with a wet bulb depression of about 5°C. In the case of short cut products, perforated belt conveyors or vibrating screen conveyors are used to transport the pieces through the different stages of drying. Between stages, the pieces may be retained on a holding screen to permit equilibration of moisture content to occur. There are many variations on the above drying systems. In second-stage drying spaghetti strands may be placed in perforated cylinders which oscillate and rotate in the heated air stream. A Roto-therm drier has been described in which spaghetti strands are heated to a relatively high temperature by passing them between heated plates. Infrared, vibrating-deck driers have been used for second-stage drying. Microwave heating is also used. One system combines microwave heating at a frequency of 915 MHz with heated air, at a controlled humidity, to dry pasta in 8 hours. This system is said to offer substantial savings in energy and to provide a bacteriologically more acceptable product, as compared with conventional hot-air systems. Impingement drying would seem to be a suitable method for dehydrating pasta, but no record of its use commercially was found (Matz, 1970; Schiffmann, 1987). *See also* Conveyor drier, Impingement drying, Microwave drying.

Pea, green

Green peas for dehydration should be harvested when the tenderometer reading is in the range 100–105. Immature peas with a lower reading tend to wrinkle and the product is difficult to reconstitute. Older peas with a higher reading contain less sugar and tend to be starchy. After vining, the peas are dry cleaned, size-graded, washed in cold water and sorted. They are then blanched in near-boiling water or steam at atmospheric pressure for 1–4 minutes. The blanch water should contain sodium carbonate in solution so that its pH is in the range 9.5–10.0. This prevents the skins from hardening and improves the colour of the product. A 3 mm slit is made in each pea by a machine which may have a capacity of up to 3000 kg per hour. The peas are treated with a solution of sodium sulphite and bisulphite by dipping or spraying to give an SO_2 content of 300–500 ppm in the dried product. They have been dried in cabinet, tunnel and conveyor driers. The moisture content may be reduced to 8–15% in such equipment and further reduced to 4% in bin driers. Tunnel drying takes up to 8 hours. Multistage conveyor driers may perform the same task in 2–4 hours. In recent years, multistage fluidized-bed driers have been widely used for this application. Air inlet temperatures up to 95°C are used in the early stages of drying while the temperature range in the latter stages is 55–70°C. Drying times similar to those used for conveyor driers are required. Peas have been dehydrofrozen successfully and a high quality product may be produced by freeze drying. They may be puréed and the purée drum dried and milled to form a flaky product. Very well-milled (homogenized) peas may be spray dried to a powder. The powder tends to be very fine and difficult to handle. Dried peas are marketed in consumer and catering packs. They are also used as an ingredient in dried soup mixes, sauces and complete meals (Van Arsdel, 1973b; Luh and Woodroof, 1975; Holdsworth, 1983). *See also* Cabinet drier, Conveyor drier, Dehydrofreezing, Fluidized-bed drier, Tunnel drier.

Peach

Peaches are sun dried on a large scale. The ripe fruits are pitted and cut into halves. They may or may not be peeled. They are sulphured by exposure to sulphur fumes or treatment with bisulphite. The dried products may contain 2000–4000 ppm of SO_2. They are exposed to the sun on trays, with the cup upwards, for about four days. They are then stacked in well-ventilated sheds until a moisture content of about 25% is attained. The dried fruits are stored in wooden boxes or bins for some weeks to equilibrate to a uniform moisture content (sweating). Peeled, halved or sliced peaches may be dried in countercurrent tunnels, using air inlet temperatures of 65°C. Drying time to reach a moisture content of 25–30% may be 24–30 hours. Generally, tunnel dried peaches are of a lower quality than the sun dried product. Blanched peaches dry more rapidly, in 16–18 hours. Blanching may also improve the quality of the peaches. Dipping of peaches in an acid solution, consisting of 1% ascorbic acid and 0.25% malic acid, has been used to retard enzymic browning in the dried product. The moisture content of peaches may be further reduced to less than 5% by slicing or dicing and vacuum drying. Such low moisture peach products are used as filling for pies and tarts. The low moisture products may be milled to a powder which may be used in purées, spreads or glazes.

Peach purée has been drum dried for inclusion in dry baby foods. It can also be vacuum dried and freeze dried, but this is not practised on a commercial scale (Van Arsdel et al, 1973b; Woodroof and Luh, 1975; Salunkhe et al, 1991). *See* Solar drying, Tunnel drier, Vacuum drier.

Peanut protein
Peanuts are a useful source of protein. To obtain this protein, peanuts are sorted, roasted and have their skins and germ removed. They are then milled and extracted by a batch or continuous method. The extract is separated into three fractions, oil-rich, protein-rich and carbohydrate-rich, by centrifugation. The protein-rich fraction is further concentrated by centrifugation, neutralized and spray dried. The oil is cleaned and is used as an edible oil. The carbohydrate-rich fraction is washed, dried, milled and sold as carbohydrate meal. The peanut protein powder is used as an ingredient in soups, sausages, bread and cakes, infant and health foods (Masters, 1991). *See also* Spray drying.

Pear
Dehydrated pears are not widely produced commercially. Pears can be sun dried. The peeled and cored fruit may be treated with fumes of burning sulphur or sodium bisulphite to attain 1000–3000 ppm of SO_2 in the dried product. The fruit may then be exposed to the sun for ½–2 days. It is then placed in well-ventilated sheds for up to 5 weeks to reach a uniform moisture content of 24%. The dried fruit may then be sliced or diced. Alternatively, peeled, cored and sliced fruit may be sulphur treated and dried in a countercurrent tunnel drier. Typical air inlet and outlet temperatures would be 65 and 55°C, respectively, and drying time 8 hours to produce a product containing 24% moisture. This type of product has a limited shelf life unless refrigerated. It is used as a pie filler or for other catering purposes. The partially dried fruit may be further dried down to less than 3% moisture, usually in a vacuum drier (Van Arsdel et al, 1973b; Woodroof and Luh, 1975). *See also* Solar drying, Vacuum drier.

Pepper *see* Bell pepper, Chilli pepper, Pimiento.

Pimiento
These peppers are harvested when fully ripe. They are washed, sorted and peeled, by lye or flame techniques. They are cored by machine, halved and diced. They are not sulphited. They are dried in cabinet, tunnel or conveyor driers using temperatures of up to 75°C, although lower temperatures may result in better pungency in the product. The dried product may be ground and used as paprika. Like bell peppers, pimientos have been dehydrofrozen (Van Arsdel et al, 1973b; Luh and Woodroof, 1975). *See* Cabinet drier, Conveyor drier, Dehydrofreezing, Tunnel drier.

Pineapple
There is very little commercial drying of pineapple. Peeled and cored pineapples may be sliced, sulphured and dried down to 15–20% moisture in cabinet or tunnel driers. Relatively low air inlet temperature is used, about 65°C. Further drying may be accomplished in a vacuum drier. Pineapple pieces have been partially dried by osmosis in sugar solution, after which drying may be completed in a vacuum drier (osmovac process). The partially dried fruit may be frozen (dehydrofreezing) or canned (dehydrocanning) (Van Arsdel et al, 1973b; Woodroof and Luh, 1975). *See also* Dehydrocanning, Dehydrofreezing, Osmotic drying.

Pork *see* Meat.

Potato
A number of dried potato products are available, including sliced and diced potatoes, potato granules, potato flakes and potato flour. Potato varieties with high total solids and low reducing sugar contents are preferred for dehydration purposes. The potatoes are washed, peeled, trimmed and inspected prior to being sliced, diced or otherwise size reduced.

To produce a dehydrated *diced product*, the trimmed potatoes are diced using a mechanical dicer which can be adjusted to cut to a range of piece sizes. The knives should be kept in a clean and sharp condition to avoid damaging the potato pieces. The cutter is kept supplied with a stream of water to lubricate the blades and keep them clean. The potato pieces are then blanched to inactivate enzymes which might otherwise catalyse darkening of the cut potatoes and/or hydrolysis of lipids which could lead to the development of off-flavours during storage of the dried product. Blanching may be done in water or steam, at a temperature in the range 93–100°C for 2–12 minutes. Immediately after blanching the potato pieces should be rinsed with a water spray to remove gelatinized starch from their surfaces. They may then be treated with sulphite. SO_2 protects the product against non-enzymic browning and burning during drying. It thus permits higher temperatures to be used as compared with untreated potato. Sulphite is applied by spraying the potato dice with a 50:50 mixture of sodium sulphite and sodium bisulphite or by dipping them in a similar solution. The dried product may contain 200–600 ppm of sulphite, as SO_2. For some applications sulphiting is omitted. Treatment of the blanched potato dice with calcium salts can retard browning, prevent sloughing during cooking, improve the rate of reconstitution and increase the firmness of the reconstituted product. To effect this calcium salt treatment, calcium chloride is added to the sulphite solution discussed above.

Drying of the potato may be carried out in cabinet, tunnel or conveyor driers. The latter type of drier is most widely used. Two-stage conveyor drying is often practised with the partially dried potato pieces being transferred, and redistributed on to a second belt. Air inlet temperature at the wet end of the drier may be as high as 135°C while that entering at the dry end is in the range 60–73°C. The drying may be completed, down to a moisture content of 8% or lower, in a bin drier with air in the temperature range 38–60°C. Total drying time may be 10–14 hours. In modern driers, drying is completed in a final section of the conveyor drier, in 6–8 hours. Fluidized-bed driers have also been used to dry diced potato. After drying and before packaging, the dried potato pieces are size-graded by screening and discoloured pieces detected and removed by photoelectric colour sorters. Potato pieces have been subjected to explosion puffing, at an intermediate stage of drying, to accelerate the drying towards the end of the cycle. They have also been partially dried osmotically, using salt solutions, and then vacuum dried (osmovac process), frozen (dehydrofreezing) or canned (dehydrocanning), but only experimentally.

Potato flakes are produced by drum drying cooked, mashed potatoes. Sliced potatoes are cooked in two stages. Firstly they are cooked in water at 71°C for 20 minutes, then cooled to 24°C or lower for 20 minutes and finally cooked in steam at atmospheric pressure for 30–50 minutes. They are then mashed or riced. In one design of ricer, the potato slices are forced from the outside through the holes in a rotating, perforated cylinder by two solid rollers. The riced product is removed from inside the cylinder by a ribbon screw. A solution of sodium sulphite and bisulphite is mixed in to give concentrations of sulphite, as SO_2, of 400–600 ppm in the mash, or 150–200 ppm in the dried product. An emulsion is made up containing, typically, monoglyceride emulsifier (0.5% in dry product), sodium acid pyrophosphate as a chelating agent (0.1% in dry product), citric acid (75 ppm in dry product) and an antioxidant, BHA and/or BHT (15 ppm in dry product) and mixed into the mash. In some formulations milk powder is added. A single-drum drier, fitted with one feed roll and up to four applicator rolls, is usually used to dry the mash mixture. Steam at 520–560 kN m^{-2} is used to heat the drum and a drying time of 20 seconds is required to reduce the moisture content of the product down to 4.5–5.0%. The dried sheet is removed from the drum by a scraper knife. Two cutting rolls, set at right angles, convert the sheet into flakes, by pressing it against a rubber-covered roller. Potato flakes should reconstitute readily to form a smooth, homogeneous mashed product with good colour and flavour, suitable for domestic or catering applications.

Potato flour is made from raw potatoes not suitable for the manufacture of higher quality products. They are washed, peeled, trimmed, cooked in steam at atmospheric pressure for 45–60 minutes and mashed. They are then delivered to the top of a single-drum drier where they are distributed uniformly across the surface of the drum, by means of a screw conveyor. Each of three or four applicator rolls becomes coated with mash and in turn it deposits it on to the drum surface. A steam pressure of 700 kN m^{-2} is used and the mash dried down to 9% moisture content. The dried product is removed from the drum surface by a scraper knife and is conveyed to a milling room where it is ground and size-graded by means of screens. It is used to replace some of the wheat flour in potato bread and other baked products, in dried soup mixes, pie fillings and crusts and as a thickening agent in sauces and gravies.

Potato granules are made from cooked, mashed potato. The potatoes are sliced and then cooked. Cooking is usually carried out in steam at atmospheric pressure, with the potato pieces being conveyed through the steam chamber. Cooking times are in the range 30–40 minutes. In the 'add-back' process the cooked potatoes are then mashed in spiral ribbon or interrupted paddle mixers. At this stage, dry granules are added back into the hot, cooked potato pieces to attain a moisture content of 35–40%. The temperature of the mixture should be maintained above 60°C. The presence of the recycled dry granules helps to break down the potato pieces. A sodium sulphite-sodium bisulphite solution is added during mashing, enough to attain a sulphite content of 300–600 ppm, as SO_2, in the dried product. In a second mixing or granulation stage further breakdown of the potato flesh is brought about. It is important that cells are ruptured as little as possible at this stage and the product leaving the granulator should consist of single cells or very small groups of cells. Batch and continuous mixers of various types are used for granulation. After granulation, the mixture is cooled to 15.5–26.5°C by passing it over a vibrating screen through which cool air flows. It is then held at that temperature, or lower, for up to 60 minutes during which time retrogradation of the starch occurs, leading to further granulation and yielding a product which is easy to handle. After this 'conditioning' stage, there is a further gentle size reduction step to break up any large agglomerates which have formed. Any remaining large groups of granules and bruised pieces of potato flesh are removed and the granules dried. Usually drying takes place in two stages. They are first dried down to 12–13% moisture content in a pneumatic drier, of the vertical, airlift type. Then they are transferred to a fluidized-bed drier for final drying down to 6–7% moisture content. At least two other processes for the production of potato granules which do not involve adding back a proportion of the dried granules have been used. In both of these processes, the partially dried mash is frozen and thawed prior to granulation and final drying (Van Arsdel et al, 1973b; Talburt and Smith, 1975). *See also* Conveyor drier, Drum drier, Fluidized-bed drier, Pneumatic drier.

Potato fruitwater

In the manufacture of potato starch, large volumes of water containing some protein, carbohydrates, minerals and a little lipid material is produced. This is known as fruitwater and has to be disposed of and constitutes a difficult effluent problem. Protein may be recovered from this liquid. To do so, the fruitwater at 10°C is acidified to a pH of 4.0–5.5. It is then heated and pumped into coagulation tanks where the protein precipitates in the form of flakes. These are concentrated, cooled and spray dried. The powder has a moisture content of 10–12% and contains 70–80% crude protein, 6–12% carbohydrate, 1.5% ash and 0.5% lipid. The BOD of the fruitwater is reduced and so it is easier to dispose of (Masters, 1991). *See also* Potato starch, Spray drying.

Potato starch

Potatoes are a useful source of starch. The raw potatoes are washed and fed to a rasp or mill where the cells are ruptured and their contents, mainly starch, released. The slurry may be diluted with water and SO_2 added to inhibit enzymic browning. It is then passed through a stack of vibrating screens or a series of centrifugal sieves where the pulp is separated from the starch and water. The pulp may be fed to a second mill where it is ground again and more starch released. This suspension is screened again and the pulp is separated and removed. The starch suspension is pumped to a centrifugal separator where a more concentrated starch suspension is formed and an aqueous phase, consisting of wash water containing soluble materials, is separated off. The concentrated starch suspension is diluted with water and fed through a fine mesh screen to remove any residual pulp. The starch suspension which passes through this screen, is pumped to large tables where the starch settles and any remaining fibre and soluble solids are carried off in the aqueous stream. Large capacity centrifugal separators may be used instead of the tables. The starch cake from the tables or centrifuges is diluted with water and filtered through a rotary drum vacuum filter. The filter cake, now containing about 40% total solids, is dried in a conveyor drier, down to about 17% moisture content. As an alternative to a conveyor drier a multistage, pneumatic (flash) drier may be used. Potato starch is used in bakery products, as a thickener in soups and gravies, in many sugar confectionery products such as jellies and instant puddings, and as a dusting agent. It may also be hydrolysed for the production of glucose syrup. Other, non-food uses are in the manufacture of paper, textiles and adhesives (Van Arsdel et al, 1973b; Talburt and Smith, 1975). *See also* Conveyor drier, Pneumatic drier.

Prune

Prunes may be sun dried but are mostly dehydrated in countercurrent tunnel driers. The fruits are cleaned by an air blast and/or water sprays and then dipped in hot or cold water. Dipping in an alkaline solution containing 0.5% or less of sodium hydroxide or sodium carbonate can accelerate drying of the whole fruit by causing cracks to form in the skin. The fruits are spread in a single layer on trays and dried in the tunnel with an air inlet temperature of 75°C. Drying time may range from 24–36 hours and the final moisture content may be 16–19%. The moisture content of prunes may be further reduced to less than 4% by vacuum drying. Whole pitted fruits are perforated before drying. Prune flakes, granules and other-shaped fruit pieces are also dehydrated and are used as fillings and in spreads (Van Arsdel et al, 1973b; Woodroof and Luh, 1975; Salunkhe et al, 1991). *See also* Tunnel drier, Vacuum drying.

Pumpkin

Pumpkin flour has been produced by steaming pumpkin pieces and drying them down to about 6% moisture content in cabinet driers. The dried pieces were milled in a hammer mill, size-graded and packaged. However, dried pumpkin is usually marketed in the form of drum-dried flakes. The pumpkins are washed and their stems are cut off. They are cut into large pieces and the seeds removed through a rotating screen. The pieces are then steamed for 15–45 minutes, depending on their size, at 105–120°C. This process is known as wilting. The wilted pieces are pulped and remaining pieces of skin and seed removed by passing the pulp through a screen. Additives such as corn syrup or starch may be added to the pulp before it is dried on a double-drum drier (Van Arsdel et al, 1973b; Luh and Woodroof, 1975). *See* Drum drier.

R

Raisin *see* Grape.

Raspberry
Dehydration of raspberries is not commonly practised commercially. They have been dried in kiln and cabinet driers, using an air inlet temperature of 62.5°C or less. No pretreatment is required. Raspberries may also be freeze dried and raspberry purée or concentrate may be foam-mat dried or spray dried in a manner similar to that used for strawberry (Creuss, 1958; Woodroof and Luh, 1975; Masters, 1991). *See also* Strawberry.

Rhubarb
Rhubarb can be dried by trimming and washing the stalks, cutting them into short lengths and drying these in cabinet or tunnel driers using air temperatures of about 80°C. Very little rhubarb is dried commercially (Creuss, 1958). *See* Cabinet drier, Tunnel drier.

Rice flakes *see* Breakfast cereals.

Rice grain *see* Cereal grains.

S

Separated milk *see* Skim milk.

Single-cell protein (SCP)
Single-cell protein is produced by fermentation using a variety of substrates including derivatives of oil and gas such as *n*-paraffins, gas oil, methanol, ethanol, methane and various carbohydrates such as sugars and starches in molasses, sweet potato, cassava, rice and corn. The product is generally used as a source of protein in animal feed. The liquor from the fermenter typically contains 1–2% solids. This is concentrated up to 15–25% by centrifugation and spray dried. Relatively high inlet and outlet air temperatures are used, 350–650°C and 100–150°C respectively. The powder has a small particle size and a moisture content of 4%. Direct air heating is commonly used. There is a serious fire and explosion hazard when drying this material. Semi-closed-cycle (self-inertizing) spray-drying systems are used. The fine powder may be agglomerated to facilitate its handling. Other methods of drying, which produce a larger, granular product may also be used for this material. These include fluidized-bed and pneumatic driers. Back-mixing of some of the dried material with the feed is necessary to convert it into a state suitable for fluidization or pneumatic conveying (Masters, 1991). *See also* Fluidized-bed drier, Pneumatic drier, Safety aspects of drying, Semi-closed-cycle spray-drying system.

Skim (skimmed, separated) milk
This is produced when butterfat is removed from whole milk by skimming or centrifugal separation. Skim milk produced by centrifugation typically contains 9.15% total solids including 4.95% lactose, 2.65% casein, 0.52% albumin, 0.21% globulin, 0.05% fat and 0.72% ash. The composition of milk, and hence skim milk, depends on the breed of cattle and the season of the year.

Most dried skim milk powder is produced by spray drying. The raw skim milk is first heat treated. Powder produced from skim milk that receives a mild heat treatment, i.e. it is pasteurized, is known as low-heat powder. This type of powder contains a relatively high amount of soluble whey proteins (albumin and globulin). Skim milk powder intended for use in breadmaking is given a more severe heat treatment and is known as high-heat powder. It contains smaller amounts of soluble whey proteins as compared with low-heat powder. High-heat powder is preferred for breadmaking as it prevents dough slackening and yields loaves with an open texture and good volume. After being heat treated, the skim milk is usually concentrated to a total solids content of 40–50% by multiple-effect evaporation. Various designs of spray-drying chamber can be used for drying the concentrate. Both centrifugal and nozzle atomizers are used. Indirect heat is commonly used but direct heating with natural gas is also practised. Air inlet and outlet temperatures used are in the range 180–230°C and 85–100°C respectively. A number of different types of spray-dried skim milk powders are produced. Ordinary skim milk powder is produced by single-stage spray drying with no after treatment. Typically it has a moisture content in the range 3.5–4.0%, particle size 30–50 μm and bulk density 0.60–0.70 g ml^{-1}. It usually contains a proportion of very small particles which make it dusty and difficult to handle and package. It also exhibits poor reconstitution characteristics, forming lumps on the surface of the reconstituting liquid which are difficult to disperse. The amount of very small particles in the powder can be reduced by recycling the fines from the cyclone separators to the wet zone of the drying chamber. There they collide and form agglomerates with larger particles. The product produced in this way is sometimes known as dustless. Further improvements in reconstitution properties can be made by promoting agglomeration of the powder particles to produce an 'instant' product. A moderate degree of agglomeration can be achieved in a vibrating fluidized bed located downstream of the drying chamber. The powder from the drying chamber is cooled on this bed and some agglomeration occurs. The fines are also recycled to the drying chamber. A greater degree of agglomeration can be attained by discharging the powder from the drying chamber at a moisture content of 5–7% and completing the drying in a fluidized bed. Again the fines are recycled and a second fluidized bed may be used to cool the powder. To produce large agglomerates which reconstitute rapidly, spray-dried skim milk powder may be subjected to a rewetting process. This can produce a product with agglomerates in excess of 400 μm in diameter and a bulk density as low as 0.25 g ml^{-1}. This type of product is sometimes known as prime instant powder. Spray-dried skim milk powder may be reconstituted as a cold drink. It may be used as a whitener for coffee or tea. It is used as an ingredient in ice-cream mix, sausage meat and many bakery products. A relatively small amount of drum-dried skim milk is manufactured. Double-drum driers are usually used and are heated by steam at a pressure of 200–400 kN m^{-2}. Drum-dried milk is mainly used as animal feed (Van Arsdel et al, 1973b; Nielson, 1987; Masters, 1991; Brennan, 1992). *See also* Drum drier, Spray drying.

Skimmed milk *see* Skim milk.

Slaughterhouse by-products *see* Blood, Gelatine, Offal.

Sodium caseinate
Casein for sodium caseinate manufacture is produced from skim milk. The milk is heated to 40–45°C. Sufficient hydrochloric acid to reduce the pH of the skim milk to 4.6 is diluted with four parts of water and added to the skim milk with thorough mixing. The precipitated casein is separated from the whey by centrifugation. Water is added to adjust the solids content and the suspension is put through a colloid mill and mixer. Sodium hydroxide solution is fed to the mixer to adjust the pH back to 7.0, together with steam. This results in a viscous solution of sodium caseinate containing 20–25% total solids. This solution is spray dried, usually in two stages, the second stage in a vibrating fluidized-bed drier. A centrifugal atomizer is usually used. The dry product has a moisture content of 3.5–4.5% and a relatively low bulk density of 0.30 g ml^{-1}, due to occluded air in the particles. Sodium caseinate is used in whiteners, sausage meat, luncheon meat, bakery products and diet foods (Van Arsdel et al, 1973b; Masters, 1991). *See also* Spray drying.

Soya protein
Soya beans are rich in oil and protein. The oil is extracted and used in the manufacture of margarine, spreads and shortenings. Protein is recovered from the residue of soya bean meal. Two protein products may be produced, (i) concentrated soya protein and (ii) isolated soya protein.

(i) To produce concentrated soya protein, the bean meal is extracted by batch or continuous methods with a dilute hydrochloric acid solution at 50°C. The carbohydrate and inorganic salts go into solution and leave as extract. The solid residue in suspension is milled to a consistency suitable for atomization, pasteurized, adjusted to pH 7 and spray dried. A concurrent spray-drying chamber is used with either a centrifugal or jet atomizer. The dried product has a protein content of 60–70%.

(ii) To produce isolated soya protein, the bean meal is extracted by batch or continuous methods with an aqueous solution of sodium hydroxide at pH 9. In the batch process, extraction is carried out in an agitated vessel at 60°C. The soya bean protein, carbohydrates and inorganic salts dissolve to form solution containing 5–10% soluble solids and the insoluble cellulose material is removed by centrifugation. The protein is precipitated from the solution by adjustment of the pH to 4.3 with hydrochloric acid and removed by centrifugation. It is washed and pasteurized. The pH is adjusted to 7 by the addition of alkali and it is spray dried in a manner similar to that used for the concentrated protein. The dried product contains from 80 to 92% protein, depending on the extent of the washing of the precipitate. Continuous extraction is carried out in a countercurrent extractor with warm sodium hydroxide solution. The insoluble cellulose material leaves from one end of the extractor while the solution of protein, carbohydrates and organic salts leaves from the other end. The protein is precipitated from this solution and prepared for drying in a manner similar to that described above for the batch-extracted material. The dried product may contain from 88 to 92% protein, depending on the extent of the washing of the precipitate (Masters, 1991).

Spaghetti *see* Pasta products.

Spinach
The washed leaves of spinach are blanched in steam at atmospheric pressure for 3–5 minutes and dried in cabinet or tunnel driers using an air inlet temperature of 75°C (Cruess, 1958).

Strawberry
Dehydrated strawberries are not produced commercially to any significant extent. They have been dehydrated in cabinet driers using relatively low air inlet temperature, 37.5–50°C, to reduce the exudation of liquid from the fruit. Treatment with sulphur improves the colour of the dried product. Strawberries have been freeze dried. A slow freezing rate has been recommended. However, the texture of the rehydrated product is poor due to the cellular damage which occurs during freezing. Strawberry purée has been dried using the foam-mat technique. The product was said to have a good flavour, to be free flowing and not very hygroscopic. Very finely divided (homogenized) strawberry concentrate may be spray dried. Up to 75%, calculated on a solids basis, of a filler such as malto-dextrin, may have to be added to the concentrate. Spray driers equipped with devices to assist in the removal of product from the chamber wall may be used. Alternatively, a chamber fitted with a

cooling jacket may be employed (Cruess, 1958; Woodroof and Luh, 1975, Masters, 1991). *See also* Foam-mat drying, Spray drying.

Sugar-gelatine mixture

A mixture of sugar and gelatine, containing 30% solids, can be spray dried. A concurrent drying chamber with rotary atomizer is used. Air inlet and outlet temperatures are typically 230°C and 100°C, respectively. The powder is tacky when hot and so devices to assist in product removal from the chamber may be required. An after cooler, such as a fluidized bed, is used to facilitate handling of the powder (Masters, 1991). *See also* Spray drying, Spray-drying chamber.

Surimi

This is a concentrate of myofibrillar proteins of fish which posesses good gel-forming, water-holding and fat-binding properties making it suitable for conversion into a range of products including shellfish and meat analogues. It is usually made from underutilized species of fish such as pollock, whiting and hoki. It may be manufactured on factory ships or on shore. The fish are sorted on the basis of size, washed, the scales removed and filleted. It is important that the viscera and all black membranes be removed completely. The fillets are washed again and the flesh is separated from any remaining bones, fins or skins. This is done mechanically. The flesh is then soaked in water so that water soluble matter, blood and lipids leach out from it. The flesh is dewatered through a rotating screen. The leaching may be repeated one or more times. The flesh is then 'refined', i.e. connective tissue and any remaining skin or scales are removed. This is done in a cylindrical screen fitted internally with a rotating screw. The soft flesh is pushed through the screen and the harder material discharged at the end of the screen. The flesh is finally dewatered in a screw press to yield the product surimi. If the surimi is to be frozen or dried it is mixed with cryoprotecting agents to minimize denaturation of the protein during the processing. Sugar, sorbitol and phosphates are used as cryoprotectors. Typical quantities used are 4% sugar, 4–5% sorbitol and 0.2–0.3% polyphosphates. These are blended with the surimi in mixers or bowl choppers. The mixture may be packaged and frozen in blocks in a plate freezer. Alternatively it may be dehydrated.

Spray-dried surimi is produced by milling the mixture of protein and additives, usually in a colloid mill, straining the suspension to remove any remaining pieces of bone or scales and spray drying it using air inlet temperatures of 150–180°C and outlet temperatures of 50–80°C. The dried product contains typically 65% protein, 24% carbohydrate, 5% moisture, 4% fat and 2% ash. It is stable if packaged in a low water vapour and gas permeable material, preferably flushed with nitrogen gas. The powder may be reconstituted in the ratio 1 part powder to 4 parts water to give a product similar to thawed, frozen surimi but with not quite such good gelling properties.

Surimi mixed with cryoprotectors may be freeze-dried, using a plate temperature of 60°C or less to avoid structural collapse during drying. The dried product may be ground to a powder which is reasonably stable when appropriately packaged.

Thawed, frozen or rehydrated surimi may be mixed with other ingredients, such as starch, salt and flavouring materials, and converted into shellfish analogues by a process which involves extrusion, cooking, texturization, pasteurization, packaging and freezing.

A texturized, minced meat analogue based on surimi may be produced in dried form. The dewatered fish flesh from the screw press is comminuted with 1.0–1.5% salt and 0.5–1.0% sodium carbonate to give a mix with a pH of 7.4–7.8. This is then passed through an extruder and then mixed with ethanol at a temperature of 5–10°C in an extractor. The proteins are coagulated and the fats extracted into the solvent. The coagulated material is passed through a second extruder, is formed into pellets which are again mixed with ethanol. The hard surface coating formed during the first extraction is broken and further coagulation and defatting occurs. The ethanol is separated from the solid by centrifugation. The pellets are dried in air at a temperature of 70°C down to a moisture content of 10%. The exhaust air is sent to a condenser to recover the ethanol for reuse. The end product is in the form of granules, off-white or grey in colour with no discernible fishy odour or taste. When rehydrated it has a texture similar to boiled, minced beef or lamb. It has no gel-forming ability. It may be used as a replacement for meat in hamburgers, meat balls and many other savoury products (Lanier and Lee, 1992).

Swede

Diced swede may be dried in tunnel, conveyor and fluidized-bed driers under conditions similar to those used for drying diced carrot (Holdsworth, 1983). *See* Carrot.

Sweet potato

Diced sweet potato may be hot-air dried and a slurry of the cooked tubers may be drum dried to produce a flaked product. During harvesting and handling the periderm of the tuber may be damaged, which could lead to infection. If the tubers are stored at about 30°C and 85% relative humidity for 7–14 days the skin breaks should heal. During this 'curing' period, some starch is converted to sugar and there is an increase in the carotene content of the potatoes. Such cured tubers are suitable for dehydration. The potatoes are washed with high-pressure water sprays. Some experts recommend that they be heated in water at 55–70°C for 30 minutes to reduce enzymic discoloration. They are then peeled in steam or lye, washed to remove residual lye and held in dilute sodium chloride, citric acid or sulphite solution to prevent discoloration. If *dried, diced sweet potato* is being produced the tubers are diced, blanched in steam at atmospheric pressure for about 6 minutes and sprayed with or dipped in sodium sulphite and bisulphite solution long enough to give an SO_2 content of 200–500 ppm in the dried product. The diced tubers may be dried in cabinet, tunnel, conveyor or fluidized bed driers. A two-stage, concurrent-countercurrent tunnel may be used with inlet air temperature to the first stage in the range 80–100°C and 60–70°C to the second stage. A two- or three-stage conveyor drier is another option with an air inlet temperature in the early stages of drying as high as 115°C reducing to 60–70°C towards the end of drying. Another suitable drier would be a multistage fluidized-bed type with air inlet temperatures similar to those used with the conveyor drier. The potato dice may be dried down to a moisture content of about 11% in these continuous driers and further dried down to 4–5% in bin driers. Alternatively, dice may be reduced to 4% moisture content in the conveyor or fluidized-bed drier. To produce *sweet potato flakes* the tubers are sliced, blanched and cooked in steam at atmospheric pressure for about 30 minutes. They are then puréed and sufficient hot water added to give a slurry containing 20% solids. At this stage an amylolytic enzyme may be added to the cooled slurry which is held for a time to obtain an optimum starch/sugar ratio. This same effect can be brought about by heating the slurry to 70–80°C for 10–15 minutes. The amylase enzymes naturally present in the potato are activated by this heat treatment. Antioxidants BHA and BHT may be added up to the concentration permitted by law. About 0.02% of a mixture of sodium sulphite and bisulphite may also be added. As an alternative to the latter, a 3:1 mixture of sodium acid pyrophosphate and tetrasodium pyrophosphate may be added in an amount equivalent to 0.3–0.4% of the dry potato solids. These additives help to preserve the colour and flavour of the dried product. The slurry is dried on a double-drum drier. Steam pressure of about 500–600 kN m^{-2} is used within the drum and the moisture content of the product is 2–4%. The sheet of product leaving the drier is thermoplastic and needs a few seconds' cooling to become suitable for milling into flakes. Dried, diced and sweet potato flakes are used as ingredients for dried soup mixes, other dried vegetable mixes, canned soups and other similar applications (Cruess, 1958; Van Arsdel, 1973b; Luh and Woodroof, 1975; Salunkhe and Desai, 1984c). *See also* Cabinet drier, Conveyor drier, Drum drier, Fluidized-bed drier, Tunnel drier.

T

Tea extract
Dried, blended tea leaves are extracted with hot water and the extract dried to produce an instant tea powder. Extraction may be carried out in a static bed extractor comprised of 3–5 stages. After the extraction cells have been filled with the leaves, they may be evacuated and then the pressure restored to atmospheric with CO_2 gas. This procedure facilitates the flow of the water through the bed of leaves. The water temperature ranges from 70°C in the early stages of extraction to 90°C towards the end of the process. The soluble solids content of the leaves is c. 40%, but only 15–20% is extracted to produce a good quality end product. The solution leaving the battery of cells contains 2.5–5.0% solids. Some designs of continuous, moving-bed extractors are also used to extract tea leaves. The extract is concentrated by vacuum evaporation to 25–50% solids. Prior to the evaporation the volatile aroma compounds are stripped from the dilute extract, concentrated by distillation and added back to the concentrated extract before it is dried.

The concentrated extract may be spray-dried. A tall, cylindrical drying chamber with a conical base is used. The extract is introduced into the top of the chamber through a nozzle atomizer and follows a straight-line flow path, concurrent with the heated air, to the bottom of the chamber, from which the dry powder is removed. This flow pattern minimizes the chances of powder depositing on the wall of the chamber. Air inlet and outlet temperatures used are in the range 200–250°C and 90–100°C respectively. The fines from the cyclone may be recycled to the drying chamber. The powder, which contains 2.0–2.5% moisture, may be cooled on a vibrating fluidized bed. Two-stage spray drying, using an integrated fluidized bed, or a fluidized bed after the drying chamber, is also practised. The spray-dried powder may be instantized by the rewetting process.

Concentrated tea extract may also be dried in a vacuum band drier. A roller applies the extract to the band. After initial heating, some evaporation occurs at the belt surface which causes the layer of extract to puff up, creating a porous structure, which facilitates the drying. The dried material is scraped off the band by means of an oscillating knife and is removed from the chamber through a rotary valve or alternating air locks.

Dried, instant tea is also produced by freeze drying. The concentrated tea extract is frozen in trays or on belts and the frozen material is broken up into granules by methods similar to those used for coffee extract. The frozen granules are freeze dried in a cabinet, tunnel or continuous moving-bed freeze drier.

Instant tea is now marketed mixed with sugar, lemon flavouring and whitener (Van Arsdel et al, 1973b; Brennan et al, 1990; Masters, 1991). *See also* Coffee extract, Freeze drying, Spray drying, Vacuum band drier.

Tea whitener *see* Whiteners for coffee or tea.

Tomato
Sliced tomatoes are dehydrated commercially to a limited extent. Varieties that have a high solids and pectin content and a bright red colour are best suited to drying. The tomatoes are washed, inspected and sliced. Revolving stainless steel blades are used for slicing. It is important that these blades are kept very sharp so that the fruit is not crushed. The slices are placed on trays and sprayed with a sulphite/bisulphite solution. They are then dried in cabinet or two-stage, concurrent-countercurrent tunnel driers. The final moisture content is 4%. The dried slices are sticky, hygroscopic and difficult to handle. They are kibbled or milled into flakes or a powder for use in dried soup mixes, other dried vegetable mixtures and dried dishes such as chilli con carne and spaghetti bolognaise.

Tomato juice may be dehydrated by a variety of methods. It may be spray dried. To prepare tomatoes for creating dried juice, the fruit is crushed or broken by one of two methods. In the 'hot break' method, the fruit is broken down when cold and immediately heated up to 65–90°C. This rapid heating destroys the enzymes in the pulp which would otherwise cause the breakdown of the pectin. It also liberates the glutinous materials which surround the seeds in the tomatoes. These materials give consistency to the finished purée or paste, produced by concentrating the juice at a later stage. In the 'cold break' process, the tomatoes are pulped when cold and held in tank for a period of time to permit the enzymes, liberated during crushing, to catalyse the breakdown of pectins in the pulp. The juice produced by the cold break method has a relatively low viscosity and may be concentrated up to 38–48% total solids content prior to spray drying. However, the powder produced has rather poor reconstitution characteristics with a tendency for solids to settle quickly. A more viscous juice results from the hot break method which can only be concentrated to 26–32% total solids content for spray drying. However, the powder thus produced reconstitutes well, forming a homogeneous suspension with little tendency to settle. After pulping, the juice is separated from the pulp and concentrated in a multiple (usually double) effect evaporator to the appropriate concentration of total solids. When hot, the tomato powder is thermoplastic, sticky and hygroscopic. Consequently, the powder tends to deposit on the walls of the drying chamber. Various mechanical devices, hammers, chains and brushes, have been used to assist in releasing the powder from the chamber wall, but with limited success. A better solution is to use a double-walled drying chamber and to draw cool air through this jacket to maintain the wall temperature in the range 38–50°C. Under these conditions, powder will build up on the chamber wall, but will fall off periodically in large lumps. These lumps fall on to a fluidized bed where they are broken up and cooled. The air leaving the cooling jacket may be introduced into the drying chamber as part of the drying air, thus conserving some of the heat extracted from the chamber to cool the wall. Centrifugal atomization is usually used and air inlet temperature to the drying chamber is in the range 140–150°C. The powder, containing 3.5% moisture or less, is hygroscopic, with a strong tendency to cake. It needs to be packaged in a low humidity atmosphere. Anti-caking agents may be dry mixed into the powder. Another design of spray drier suited to this application features a tall chamber in which the atomized liquid follows a straight-line flowpath from top to bottom, with the drying air flowing parallel to it. The cone-shaped base of the chamber may be fitted with a cooling jacket. The BIRS process featured a very tall chamber in which relatively cool (50°C), dehumidified air flowed upwards, countercurrent to the atomized liquid. This process was used to produce tomato powder. Tomato paste containing 30% total solids has been dried by the foam-mat technique. Glyceryl monostearate was used as a foaming agent and the foamed paste was extruded on to a moving belt in the form of strips. The belt carried these strips through a two-stage, concurrent-countercurrent drying tunnel. The air inlet temperature to the concurrent stage was as high as 95°C, while that entering the countercurrent stage was 55°C. Tomato paste has also been dried using the vacuum puffing technique and by freeze drying. Drum dried tomato flakes are also produced. However, considerable changes in the colour and flavour occur due to the relatively high temperature to which the paste is exposed, up to 160°C for 2–30 seconds. Double-drum driers are used. Problems can occur in removal of the dried product from the drum surface. The product is thermoplastic and hygroscopic. It is usually converted into flakes for inclusion in dried soup mixes and baby foods (Goose and Binsted, 1973; Van Arsdel et al, 1973b; Luh and Woodroof, 1975; Masters, 1991). *See also* Drum drying, Foam-mat drying, Spray drying, Vacuum puff drying.

Trehalose *see* entry in Part 2.

V

Vitamins

Many foods are rich in vitamins and dried preparations may be used in tablets for humans or in animal fed. Spray dried products are most commonly used. Spray dried brewers' yeast and liver extract are rich in vitamin B. Ascorbic acid, mixed with carbohydrate and binder may be spray dried and tableted. Oil soluble vitamins A and D are spray-dried in the form of oil-in-water emulsions (Masters, 1991). *See also* Yeast, Spray drying.

W

Wheat gluten
Gluten separated from wheat grain may be spray-dried. A concurrent drying chamber, equipped with a pressure nozzle or centrifugal atomizer, may be used. The dried product is used as a protein additive in animal feed (Masters, 1991). *See also* Corn gluten, Spray drying.

Wheat flakes *see* Breakfast cereals.

Wheat grain *see* Cereal grains.

Whey
Whey is a by-product of cheesemaking or the manufacture of casein. Typically, it has a total solids content of 6.85% including 5.00% lactose, 0% casein, 0.52% albumin, 0.20% globulin, 0.30% fat and 0.54% ash. Sweet whey, also known as cheese whey, is produced in cheesemaking when rennet is used to produce the curd. It has a pH between 5.2 and 6.7. Sour or acid whey has a pH between 3.8 and 4.6 and is produced in casein manufacture or in making quarg or cottage cheese. Both sweet and sour whey can be spray dried. Ordinary whey powder is produced by single-stage spray drying. The whey is heat treated and concentrated to 40–50% total solids by multiple-effect evaporation before spray drying. Equipment similar to that used for drying skim milk is used. Inlet and outlet temperatures used are in the range 180–200°C and 90–100°C respectively. Ordinary whey powder has a moisture content of 3.0–4.0% and a bulk density of 0.60–0.70 g ml^{-1}. It is very hygroscopic and difficult to handle. This is mainly due to the fact that the lactose in this type of whey is in an amorphous state. This lactose can be converted into the more stable α-lactose monohydrate by a crystallization process before and/or after spray drying. Non-caking whey powder may be produced in two ways. (i) The whey is further concentrated to 50–60% total solids, seeded with lactose crystals and held for a period of up to 24 hours while being agitated. This pre-crystallized whey is then dried in two stages, the first stage in a spray drier, the second in a vibrating fluidized-bed drier. It is then cooled in a second fluidized-bed. This results in a powder made up of small agglomerates which is free flowing. (i) Whey containing 50% total solids is partly crystallized before drying and dried to c. 14% moisture in a spray drier. It is held on a conveyor belt for a period of time to allow further crystallization to occur. The drying is completed in a vibrating fluidized bed and the dry powder cooled in a second fluidized bed. The powder produced in this way consists of large agglomerates. These are not very hygroscopic and their bulk density is 0.4–0.5 g ml^{-1}.

Whey powder has many uses. It is used in alcoholic and non-alcoholic beverages, bakery and sugar confectionery products, for the manufacture of whey cheese and recovery of fat.

A small amount of whey is drum dried. It tends to roll up on the knives removing it from the drum surface. Additives such as skim milk or cereals can reduce this problem. Drum-dried whey is used mainly for animal feed (Van Arsdel et al, 1973b; Masters, 1985; Brennan, 1993). *See also* Drum drier, Skim milk, Spray drying.

Whiteners for coffee or tea
These are formulated oil-in-water emulsions typically containing sodium caseinate, sugar, vegetable oil, emulsifying agents, anti-caking agents such as potassium phosphate, fine synthetic silicates or metallic oxides. The ingredients are mixed at optimum temperature and homogenized, usually in a pressure homogenizer. The emulsion can then be spray dried. Chambers equipped with integrated fluidized beds may be used or secondary drying may take place in a vibrating fluidized bed. Nozzle atomization is usually employed. The spray-dried powder may be agglomerated by the rewetting process (Masters, 1991). *See also* Homogenization, Spray drying.

Whole egg *see* Egg.

Whole-fish meal *see* Fish meal.

Whole milk

Whole bovine milk typically contains 12.65% total solids including 3.75% fat, 4.85% lactose, 2.60% casein, 0.50% albumin, 0.20% globulin and 0.72% ash. Its composition depends on the breed of cattle and the season of the year. Most dried whole milk is produced by spray drying. The milk is heat treated, concentrated to 45–50% total solids by multiple-effect evaporation and homogenized prior to drying. The drying equipment used is similar to that employed in the manufacture of skim milk powder. Air inlet and outlet temperatures used are in the range 175–200°C and 75–95°C respectively. Whole milk powder tends to be rather sticky when hot and can build up as a deposit on the drying chamber wall. Hammers which tap the wall of the chamber at intervals may be employed to assist in powder removal. Ordinary whole milk powder has a moisture content of 2.5–3.0% and a bulk density of 0.55–0.65 g ml^{-1}. Whole milk powders similar to skim milk powders and agglomerated to varying degrees are produced. Agglomeration may not improve the wettability of whole milk powder to the same extent as skim milk powder. Fat may migrate to the surface of the powder particles, making them hydrophobic (water repellent). The addition of small quantities of lecithin, a surface active agent, to whole milk powder improves its wettability so that it reconstitutes rapidly in cold liquid. This process is known as lecithinization (Van Arsdel et al, 1973b; Masters, 1991; Brennan, 1993). *See also* Skim milk, Spray drying.

Y

Yam

Yam flour is produced from sun-dried yam slices. The yams are sliced and the peel removed from the slices, which are sun dried down to 5–10% moisture content. In some cases, the pieces may be partially cooked prior to drying. The pieces are ground to a flour. Yam pieces may also be dehydrated in cabinet or tunnel driers. Ayernor and Brennan (1973) produced a flaked yam product by drum-drying a precooked yam slurry. The product retrograded very rapidly when reconstituted to a dough (Ayernor and Brennan, 1973; Salunkhe and Desai, 1984d). *See* Drum drier, Solar drying.

Yeast

Yeast is a useful source of protein in areas where protein shortage occurs. Yeast may be produced by fermentation of the sugar and organic acids found in molasses. The yeast produced in this way is rich in organic and mineral products. The addition of nitrogen and phosphorus in the form of salts increases its nutritive value. *Torulopsis utilis* strains are mostly used as food yeast. The molasses is diluted with water and clarified by centrifugation. The nutritive salts are added and the molasses fermented aerobically in large vessels. Heat generated during fermentation is removed by cooling coils. Fermentation takes place at 27–30°C and the pH value is controlled at 3.9–4.4, using ammonia. After fermentation, air is removed in a separation tank and this is followed by a series of separations and washings in a number of centrifuges. The yeast is then concentrated to 22–24% solids content by evaporation. It is then spray dried, using relatively high air inlet and outlet temperatures of 330°C and 100°C, respectively. The moisture content of the dry material is about 8%. Because of the high temperatures used, there is a danger of fire and precautions have to be taken, as discussed in this text under Safety aspects of drying. This dried yeast is mainly used as animal feed and is also known as fodder yeast.

Brewers' yeast consists of strains of *Saccharomyces cerevisiae*. It can be liquefied by treatment with acid or heat, concentrated to 50% total solids and spray dried under conditions similar to those used for fodder yeast. The dried product is hygroscopic and needs to be handled and packed in a low humidity environment. This product is used in soups, sauces and gravies (Masters, 1991). *See also* Safety aspects of drying, Spray drying.

Part 4

Bibliography

Bibliography

AITKEN, A., MACKIE, I.M., MERRITT, J.H. and WINDSOR, M.L. (1981), *Fish Processing and Handling*, 2nd ed., Edinburgh: Her Majesty's Stationery Office.

ALVAREZ, P.I. and LEGUES, P. (1986), A semi-theoretical model for the drying of Thompson seedless grapes, *Drying Technology*, **4**, 1–17.

ANDERSON, T.D.C. (1962), Dehydration of food products by the BIRS process, *Food Trade Review*, **33**, 12, 52–53.

ANON (1985), Pump is gentle with delicate foods, *Food Processing*, **54**, 24.

ANON (1992), Dryers: Technology and Engineering, In *Encyclopedia of Food Science and Technology*, Vol. **1**, Y.H. Hui (ed.), pp. 619–656, Chichester: John Wiley & Sons, Inc.

ARTHEY, D. and DENNIS, C. (1991), *Vegetable Processing*, Glasgow: Blackie and Son Ltd.

BAKKER, M. (ed.) (1986), *The Wiley Encyclopedia of Packaging Technology*, New York: John Wiley & Sons.

BARNES, D. and WILSON, F. (1983), *Chemistry and Unit Operations in Water Treatment*, London: Applied Science Publishers.

BARTHOLOMAI, G.B., BRENNAN, J.G. and JOWITT, R. (1975a), Mechanisms of volatile retention in freeze-dried food liquids, *Lebensm.-Wiss.+Technol*, **8**, 25–28.

BARTHOLOMAI, G.B., BRENNAN, J.G. and JOWITT, R. (1975b), Influence of cellulose and starch on the retention of volatiles during freeze-drying of a liquid extract of mushroom, *Lebensm.-Wiss.+Technol.*, **8**, 174–176.

BARTHOLOMAI, G.B., BRENNAN, J.G. and JOWITT, R. (1976), Mechanisms of volatile retention in freeze-drying, *Proceedings IV International Congress Food Science and Technology*, Madrid, 1974, Vol. **II**, pp. 169–174, Institute Nacional de Cienca y Technologia de Alimentos, Consejo Superior de Investigaciones Cientificas, Spain.

BEEVER, P.F. and CROWHURST, D. (1989), Fire and explosion hazards associated with milk spray drying operations, *J. Society of Dairy Technology*, **42**, 65–70.

BERQUIST, D. (1987), Dried Whole Egg Plant, *In Food Factories-Processes, Equipment, Costs*, A. Bartholomai (ed.), pp. 104–109, Weinheim, Germany: VCH.

BIZOT, H. (1983), Using the G.A.B. Model to Construct Sorption Isotherms, In *Physical Properties of Foods*, R. Jowitt F. Escher, B. Hallstrom, H.F.T. Meffert, W.E.L. Spiess and G. Voss (eds.), pp. 43–54, London: Applied Science Publishers.

BOERSEN, S. (1988), The Stork shaking fluidbed with well mixed section, *North European Food and Dairy J.*, **6**, 216–221.

BOLIN, H.R., FULLER, G. and POWERS, J. (1973), Apricot concentrate product development, *Food Product Development*, **7**, 30

BOLIN, H.R. and SALUNKHE, D.K. (1982), Fluid dehydration by solar energy, *CRC Critical Reviews in Food Science and Technology*, **16**, 327–354.

BRENNAN, J.G. (1971), Packaging biological materials, *Process Biochemistry*, **6**, 47–50.

BRENNAN, J.G. (1978), Developments in food dehydration, *Process Chemical Engineering*, **31**, 9, 11–12, 14, 16–17, 20, 24.

BRENNAN, J.G. (1989), Dehydration of Foodstuffs, In *Water in Foods*, T.M. Hardman (ed.), pp. 33–70, London: Elsevier Applied Science.

BRENNAN, J.G. (1992), Developments in Drying, In *Food Technology International Europe 1992*, A. Turner (ed.), pp. 77–80, London: Sterling Publications.

BRENNAN, J.G. (1993a), Fluidized Bed Drying, In *Encyclopedia of Food Science, Food Technology and Nutrition*, R. Macrae, R. Robinson and M. Sadler (eds.), London: Academic Press.

BRENNAN, J.G. (1993b), Spray Drying, In *Encyclopedia of Food Science, Food Technology and Nutrition*, R. Macrae, R. Robinson and M. Sadler (eds.), London: Academic Press.

BRENNAN, J.G., BUTTERS, J.R., COWELL, N.D. and LILLY, A.E.V. (1990), *Food Engineering Operations*, 3rd ed., London: Elsevier Applied Science Ltd.

BRENNAN, J.G., HERRERA, J. and JOWITT, R. (1971), A study of some of the factors affecting the spray-drying of orange juice, on a laboratory scale, *J. Food Technology*, **6**, 295–307.

BRENNAN, J.G. and PRIESTLY, R.J. (1972), Foam-spray drying using a centrifugal atomizer, *Process Biochemistry*, **7**, 25–26.

BRENNAN, J.G. and SODAH-AYERNOR, S. (1973), A study of the kinetics of retrogradation in a starch-based dough made from dehydrated yam, *Die Starke*, **25**, 276–279.

BRUNAUER, S., EMMETT, P.H. and TELLER, E. (1938), Adsorption of gases in multimolecular layers, *J. American Chemical Society*, **60**, 309–319.

BURROWS, I.E. and BARKER, D. (1976), Intermediate Moisture Petfoods, In *Intermediate Moisture Foods*, R. Davies, G.G. Birch and K.J. Parker (eds.), pp. 43–53, London: Applied Science Publishers.

BURT, J.R. (ed.) (1988), *Fish Smoking and Drying – The Effect of Smoking and Drying on the Nutritional Properties of Fish*, London: Elsevier Applied Science.

CAURIE, M. (1981) Derivation of Full Range Moisture Sorption Isotherms, In *Water Activity: Influences on Food Quality*, L.B. Rockland and G.F. Stewart (eds.), pp. 63–87, London: Academic Press.

CHARM, S.E. (1981), *Fundamentals of Food Engineering*, 3rd ed., 2nd Printing, pp. 298–433, Westport Connecticut: AVI Publishing Company, Inc.

CHIRIFFE, J. (1983), A Survey of Existing Sorption Data, In *Physical Properties of Foods*, R. Jowitt, F. Escher, B. Hallstrom, H.F.Th. Meffert, W.E.L. Speiss and G. Vos (eds.), pp. 55–64, London: Applied Science Publishers.

CHIRIFFE, J. and IGLESIAS, H.A. (1978), Equations for fitting sorption isotherms of foods, *J. Food Technology*, **13**, 159–174.

CLARKE, R.J. and MACRAE, R. (1987), *Coffee*, **Volume 2**: *Technology*, London: Elsevier Applied Science.

COOK, E.M. and LANG, R.W. (1979), Optimum design of multi-stage drying systems, *Chem Eng. Progr.*, **75**, (4), 37–39.

COTSON, S. and SMITH, D.B. (1963), *Freeze-drying of Foodstuffs*, London: Columbine Press.

COULSON, J.M. and RICHARDSON, J.F. WITH BACKHURST, J.R. and HARKER, J.H. (1990), Fluid Flow, Heat Transfer and Mass Transfer, 4th ed., *Chemical Engineering*, **Vol. 1**, Oxford: Pergamon Press.

COULSON, J.M. and RICHARDSON, J.F. WITH BACKHURST, J.R. and HARKER, J.H. (1978), Unit Operations, 3rd ed., *Chemical Engineering*, **Vol. 2**, Oxford: Pergamon Press.

CRUESS, W.V. (1958), *Commercial Fruit and Vegetable Products*, 4th ed., New York: McGraw-Hill Book Company Inc.

DALGLEISH, J.McN. (1990), *Freeze-drying for the Food Industries*, London: Elsevier Applied Science.

DUCKWORTH, R.B. (1981), Solute Mobility in Relation to Water Content and Water Activity, In *Water Activity: Influences on Food Quality*, L.B. Rockland and G.F. Stewart (eds.), pp. 295–317, London: Academic Press.

DUCKWORTH, R.B., ALLISON, J.Y. and CLAPPERTON, H.A.A. (1976), The Aqueous Environment for Chemical Change in Intermediate Moisture Foods, In *Intermediate Moisture Foods*, R. Davies, G.G. Birch and K.J. Parker (eds.), pp. 89–99, London: Applied Science Publishers Ltd.

DZIEZAK, J.D. (1988), Microencapsulation and encapsulated ingredients, *Food Technology*, **42**, (4), 136–148.

EICHNER, K. (1975), The Influence of Water Content on Non-Enzymic Browning Reactions in Dehydrated Foods and Model Systems and the Inhibition of Fat Oxidation by Browning Intermediates, In *Water Relations of Foods*, R.B. Duckworth (ed.), pp. 417–434, London: Academic Press.

FAST, R.B. (1990), Manufacturing Technology of Ready-To-Eat Cereals, In *Breakfast Cereals and How They Are Made*, R.B. Fast and E.F. Caldwell (eds.), pp. 15–42, Minnesota: American Association of Cereal Chemists.

FELLOWS, P. (1988), *Food Processing Technology*, Chichester: Ellis Horwood Ltd.

FERGUSSON, P.H. (1989), Developments in the evaporation and drying of dairy products, *J. Society of Dairy Technology*, **42**, 94–101.

FILKOVA, I. and MUJUMDAR, A.S. (1987), Industrial Spray Drying Systems, In *Handbook of Industrial Drying*, A.S. Mujumdar (ed.), pp. 243–293, New York: Marcel Dekker Inc.

FLINK, J.M. (1977a), Intermediate moisture food products in the American marketplace, *J. Food Processing and Preservation*, **1**, 324–339.

FLINK, J.M. (1977b), Energy analysis in dehydration processes, *Food Technology*, March, 77–87.

FOSTER, G.H. (1964), *Dryeration – a corn drying process*, USDA Agr. Marketing Serv. AMS 532.

GOLDBLITH, S.A., REY, L. and ROTHMAYR, W.W. (eds.), (1975), *Freeze Drying and Advanced Food Technology*, London: Academic Press.

GOOSE, P.G. and BINSTEAD, R. (1973), *Tomato Paste and Other Tomato Products*, London: Food Trade Press.

GORLING, P. (1958), Physical phenomena during the drying of foodstuffs. In *Conference on the Fundamental Aspects of the Dehydration of Foodstuffs*, Aberdeen, March 1958, pp. 42–53, London: SCI.

GRAVES, B. and WEISS, H. (1992), Encapsulation Techniques, In *Encyclopedia of Food Science and Technology*, **Vol. 2**, Y.H. Hui (ed.), pp. 697–703, Chichester: John Wiley & Sons, Inc.

GREEN, D.W. (1984), *Perry's Chemical Engineers' Handbook*, 4th ed., New York: McGraw Hill.

GRIKITIS, K. (1988), Crispy fillings, *Food Processing*, **57**, 12–13.

GUISE, W. (1989), Taking on the tamperers, *Food Processing*, **58**, 28–30, 33.

GUTCHO, M.H. (1977), *Freeze Drying Processes for the Food Industry*, New Jersey: Noyes Data Corporation.

GUTIERREZ-LOPEZ, G.F. (1989), *Convective drying and solid-moisture interactions*, PhD Thesis, Department of Food Science and Technology, University of Reading, UK.

HALL, C.W. (1979), *Dictionary of Drying*, New York: Marcel Dekker Inc.

HALL, C.W., FARRALL, A.W. and RIPPEN, A.L. (eds.) (1986), *Encyclopedia of Food Engineering*, **2nd ed.**, pp. 820–822, Westport, Connecticut: AVI Publishing Company, Inc.

HALL, C.W. and HEDRICK, T.I. (1966), *Drying Milk and Milk Products*, Westport, Connecticut: AVI Publishing Company Inc.

HALL, C.W. and UPADHYAYA, R.L. (1986), Roller drum dryers – a bibliography, *Drying Technology*, **4**, 477–489.

HANG, Y.D. (1992), Waste Management and Food Processing, In *Encyclopedia of Food Science and Technology*, **Vol. 4**, Y.H. Hui (ed.), New York: John Wiley & Sons Inc.

HANSEN, R. (1988), The first Filtermat in Denmark, *North European Food and Dairy Journal*, **6**, 190–196.

HAYASHI, H. (1989), Drying technologies of foods – their history and future, *Drying Technology*, **7**, 315–369.

HAYWOOD, R.W. (1968), *Thermodynamic Tables in SI (metric)*

Units, Cambridge: Cambridge University Press.

HELDMAN, D.R. and LUND, D.B. (eds) (1992), *Handbook of Food Engineering*, pp. 437–562, New York: Marcel Dekker Inc.

HELDMAN, D.R. and SINGH, R.P. (1981), *Food Process Engineering*, Westport Connecticut: AVI Publishers, Inc.

HOLDSWORTH, S.J. (1969), Recent developments in dehydration and canning, *Food Manufacture*, November, 44–49.

HOLDSWORTH, S.D. (1983), *The Preservation of Fruit and Vegetable Products*, London: Macmillan Press.

HOLMES, Z.A. and WOODBURN, M. (1981), Heat transfer and temperature of foods during processing, *Critical Reviews in Food Science and Nutrition*, **14**, (3), 231–294.

HOVMAND, S. (1987), Fluidized Bed Drying, In *Handbook of Industrial Drying*, A.S. Mujumdar (ed.), pp. 165–225, New York: Marcel Dekker.

IGLESIAS, H.A. and CHIRIFFE, J. (1982), *Handbook of Food Isotherms,* London: Academic Press.

IMRIE, L.L. (1987), Solar Drying, In *Handbook of Industrial Drying*, A.S. Mujumdar (ed.), pp. 357–417, New York: Marcel Dekker.

JAGTIANI, J., CHAN, H.T. Jr. and SAKAI, W.S. (1988), *Tropical Fruit Processing*, London: Academic Press.

JANSEN, L.A. and ELGERSMA, R.H.C. (1985), Direct heating of drying air with natural gas in the preparation of milk powder, *J. Society of Dairy Technology*, **38**, 4, 134–139.

JASON, A.C. (1958), A study of evaporation and diffusion processes in the drying of fish muscle, In *Conference on the Fundamental Aspects of the Dehydration of Foodstuffs*, Aberdeen, March 1958, pp. 103–1034, London: SCI.

JOHN, V.B. (1993), *Introduction to Engineering Materials*, 3rd ed., p. 242, London: Macmillan Press Ltd.

JOWITT, R. (ed.) (1980), *Hygienic Design and Operation of Food Plant*, Chichester: Ellis Horwood Ltd.

KAPSALIS, J.G. (1981), Moisture Sorption Hysteresis, In *Water Activity: Influences on Food Quality*, L.B. Rockland and G.F. Stewart (eds.), pp. 143–177, London: Academic Press.

KAREL, M. (1974), Fundamentals of Dehydration Processes, In *Advances in the Preconcentration and Dehydration of Foods*, A. Spicer (ed.), pp. 45–94, London: Applied Science Publishers.

KAREL, M. (1976), Technology and Application of New Intermediate Moisture Foods, In *Intermediate Moisture Foods*, R. Davies, G.G. Birch and K.J. Parker (eds), pp. 4–31, London: Applied Science Publishers.

KELLY, J.J. (1987), Rotary Drying, In *Handbook of Industrial Drying*, A.S. Mujumdar (ed.), pp. 133–154, New York: Marcel Dekker.

KELLY, P.M. and SLATTERY, J. (1985), Direct gas-fired heating of a pilot-scale spray drier using natural gas for milk powder manufacture, *J. Society of Dairy Technology*, **38**, 2, 55–59.

KELLY, P.M., GRAY, J.I. and SLATTERY, J. (1989), Direct 'low-NOX' gas combustion heating of a spray drier during milk powder manufacture, *J. Society of Dairy Technology*, **42**, 1, 14–18.

KERKHOF, P.J.A.M. and SCHOEBER, W.J.A.H. (1974), Theoretical Modelling of the Drying Behaviour of Droplets in Spray Driers, In *Advances in the Preconcentration and Dehydration of Foods*, A. Spicer (ed.), pp. 349–397, London: Applied Science Publishers.

KESSLER, H.G. (1986), Energy Aspects of Food Preconcentration, In *Concentration and Drying of Foods*, D. MacCarthy (ed.), pp. 147–163, London: Elsevier Applied Science.

KING, C.J. (1971), *Freeze-drying of Foods*, CRC Monoscience Series, London: Butterworths.

KING, C.J. (1973), *Application of freeze-drying to food products*, International Course, Burgenstock.

KIRK, J.R. (1981), Influence of Water Activity on Stability of Vitamins in Dehydrated Foods, In *Water Activity: Influences on Food Quality*, L.B. Rockland and G.F. Stewart (eds), pp. 531–566, London: Academic Press.

KISAKUREK, B. (1987), Flash Drying, In *Handbook of Industrial Drying*, A.S. Mujumdar (ed.), pp. 475–499, New York: Marcel Dekker.

KRAUS, M.N. (1980), *Pneumatic Conveying of Bulk Materials*, 2nd ed., New York: McGraw-Hill Publications Co.

LABUZA, T.P. (1968), Sorption phenomena in foods, *Food Technology*, **22**, 263.

LABUZA, T.P. (1975), Oxidative Changes in Foods at Low and Intermediate Moisture Levels, In *Water Relations of Foods*, R.B. Duckworth (ed.), pp. 455–474, London: Academic Press.

LABUZA, T.P. (1977), The properties of water in relationship to water binding in foods: a review, *J. Food Processing and Preservation*, **1**, 167.

LABUZA, T.P. and SALTMARCH, M. (1981), The Non-Enzymatic Browning Reactions as Affected by Water in Foods, In *Water Activity: Influences on Food Quality*, L.B. Rockland and G.F. Stewart (eds), pp. 605–650, London: Academic Press.

LAMBERT, D.J. (1989), The use of pumps in food processing, *Food Technology International Europe*, pp. 112–113, London: Sterling Publications Limited.

LANIER, T.C. and LEE, C.M. (1992), *Surimi Technology*, New York: Marcel Dekker Inc.

LERICI, C.R., PINNAVAIA, G., DALLA ROSA, M. and BARTOLUCCI, L. (1985), Osmotic dehydration of fruit: Influence of osmotic agents on drying behaviour and product quality, *J. Food Science*, **50**, 1217–1219.

LEWIS, M.J. (1987), *Physical Properties of Foods and Food Processing Systems*, Chichester: Ellis Horwood.

LEWIS, W.K. (1921), The rate of drying of solid materials, *J. Industrial and Engineering Chemistry*, **13**, 427–432.

LIAPIS, A.I. (1987), Freeze Drying, In *Handbook of Industrial Drying*, A.S. Mujumdar (ed.), pp. 295–326, New York: Marcel Dekker.

LIAPIS, A.I. and LITCHFIELD, R.J. (1979), *Chemical Engineering Science*, **34**, 975.

LITCHFIELD, R.J. and LIAPIS, A.I. (1982), *Chemical Engineering Science*, **37**, 45.

LONCIN, M. and MERSON, R.L. (1979), *Food Engineering*, pp. 195–202, London: Academic Press.

LORENTZEN, J. (1974), New Directions in Freeze-drying, In *Advances in Preconcentration and Dehydration of Foods*, A. Spicer (ed.), pp. 413–433, London: Applied Science Publishers Ltd.

LORENTZEN, J. (1975), Industrial Freeze-drying Plants for Foods, In *Freeze Drying and Advanced Food Technology*, S.A. Goldblith, L. Rey and W.W. Rothmayr (eds), pp. 429–44[illegible], London: Academic Press.

LUH, B.S. and WOODROOF, J.G. (1975), *Commercial Vegetable*

Processing, Westport, Connecticut: AVI Publishing Company, Inc.

McCABE, W.L., SMITH, J.C. and HARRIOTT, P. (1985), *Unit Operations of Chemical Engineering*, 4th ed., New York: McGraw Hill.

MAROUSIS, S.N. and SARAVACOS, G.D. (1990), Density and porosity in drying starch materials, *J. Food Science*, **55**, 1367–1372.

MASTERS, K. (1991), *Spray Drying Handbook*, 5th ed., New York: Longman Scientific & Technical.

MASTERS, K. and PISECKY, J. (1983), New spray drier concept for the food industry, In *Proceedings of the Sixth International Congress in Food Science and Technology*, Dublin, Vol. 5, J.V. McLoughlin and B.M. McKenna (eds.), pp. 379–386, Dublin: Boole Press.

MATZ, S.A. (1970), *Cereal Technology*, Westport, Connecticut: AVI Publishing Company, Inc.

MILES, C.A., VAN BEEK, G. and VEERKAMP, C.H. (1983), Calculation of the Thermophysical Properties of Foods, In *Physical Properties of Foods*, R. Jowitt, F. Escher, B. Hallstrom, H.F.T. Meffert, W.E.L. Spiess and G. Vos (eds), pp. 269–312, London: Applied Science Publishers.

MOORE, J.G. (1987), Drum Dryers, In *Handbook of Industrial Drying*, A.S. Mujumdar (ed.), pp. 227–242, New York: Marcel Dekker.

MORSE, R.E. (1974), Antioxidants, In *Encyclopedia of Food Technology*, A.H. Johnson and M.S. Peterson (eds.), pp. 43–49, Westport, Connecticut: AVI Publishing Company, Inc.

MUJUNDAR, A.S. (1987), Impingement Drying, In *Handbook of Industrial Drying*, A.S. Mujumdar (ed.), pp. 461–474, New York: Marcel Dekker.

NARAYAN, S.B. and MUJUMDAR, A.S. (1987), Fire and Explosion Hazards in Drying Plants, In *Handbook of Industrial Drying*, A.S. Mujumdar (ed.), pp. 863–869, New York: Marcel Dekker.

NELSON, P.E. and TRESSLER, D.K. (1980), *Fruit & Vegetable Juice Processing Technology*, 3rd ed., Westport, Connecticut: AVI Publishing Company, Inc.

NIELSON, A. (1987), Powder Milk Plant, In *Food Factories – Processes, Equipment, Costs*, A. Bartholomai (ed.), pp. 101–104, Weinheim, Germany: VCH.

NONHEBEL, G. and MOSS, A.A.H. (1971), *Drying of Solids in the Chemical Industry*, London: Butterworths.

OSORIO REVILLA, G. (1990), *Spouted Bed Drying of Liquids*, PhD Thesis, Department of Food Science and Technology, University of Reading, UK.

PAINE, F.A. (ed.) (1990), *The Packaging User's Handbook*, Glasgow: Blackie & Son.

PAINE, F.A. and PAINE, H.Y. (1992), *A Handbook of Food Packaging*, 2nd ed., Glasgow: Blackie Academic and Professional.

PAKOWSKI, Z. (1987), Drying of Pharmaceuticals, In *Handbook of Industrial Drying*, A.S. Mujumdar (ed.), pp. 605–641, New York: Marcel Dekker.

PALLAI, E., NEMETH, J. and MUJUMDAR, A.S. (1987), Spouted Bed Drying, In *Handbook of Industrial Drying*, A.S. Mujumdar (ed.), pp. 419–460, New York: Marcel Dekker.

PEARSON, A.M. and TAUBER, F.W. (1984), *Processed Meats*, 2nd ed., Westport, Connecticut: AVI Publishing Company Inc.

PISECKY, J. (1987), Evaporation and Spray Drying in the Dairy Industry, In *Handbook of Industrial Drying*, A.S. Mujumdar (ed.), pp. 571–603, New York: Marcel Dekker, Inc.

POLLEY, S.L., SNYDER, O.P. and KOTNOUR, P. (1980), A compilation of thermal properties of foods, *Food Technology*, **34**, (11), 76–92.

PRICE, J.F. and SCHWEIGERT, B.S. (1971), *The Science of Meat and Meat Products*, 2nd ed., San Francisco: W.H. Freeman and Company.

RAJAH, K.K. and BLENFORD, D.E. (1988), *The ALM Guide to Lactose Properties and Uses*, The Hague: Association of Lactose Manufacturers.

RHEINLANDER, P.M. (1982), Filtermat – the 3-stage dryer from DEC, In *Proceedings of the Third International Drying Symposium*, Birmingham, **Vol. 1**, J.C. Ashworth (ed.), pp. 538–534, Wolverhampton: Drying Research Limited.

ROBERTS, A.C. and McWEENY, D.J. (1972), The uses of sulfur dioxide in the food industry. A review, *J. Food Technology*, **7**, 221–228.

ROBERTSON, G.L. (1993), *Food Packaging – Principles and Practice*, New York: Marcel Dekker Inc.

ROTHWELL, J. (1984), *Ice Cream Making*, Reading: J. Rothwell.

ROOS, Y.H. (1992), Phase Transitions and Transformations in Food Systems, In *Handbook of Food Engineering*, D.R. Heldman and D.B. Lund (eds.), pp. 145–197, New York: Marcel Dekker, Inc.

ROSER, B. (1991), Trehalose a new approach to premium dried foods, *Trends in Food Science and Technology*, July, 166.

ROSER, B. and COLACO, C. (1993), A sweeter way to fresher food, *New Scientist*, **138**, (1873), 24–28.

ROSIN, RAMMLER and INTELMANN (1932), *Z., Ver deut. Ing.*, **76**, 433–437.

SALUNKHE, D.K., BOLIN, H.R. and REDDY, N.R. (1991), *Storage, Processing and Nutritional Quality of Fruits and Vegetables*, **Vol. II**, 2nd ed., Boca Raton, Florida: CRC Press Inc.

SALUNKHE, D.K. and DESAI, B.B. (1984a), *Postharvest Biotechnology of Fruits*, **Vol. I**, Boca Raton, Florida: CRC Press Inc.

SALUNKHE, D.K. and DESAI, B.B. (1984b), *Postharvest Biotechnology of Fruits*, **Vol. II**, Boca Raton, Florida: CRC Press Inc.

SALUNKHE, D.K. and DESAI, B.B. (1984c), *Postharvest Biotechnology of Vegetables*, **Vol. I**, Boca Raton, Florida: CRC Press Inc.

SALUNKHE, D.K. and DESAI, B.B. (1984d), *Postharvest Biotechnology of Vegetables*, **Vol. II**, Boca Raton, Florida: CRC Press Inc.

SCHAR, W. and RUEGG, M. (1985), The evaluation of G.A.B. constants from water vapour sorption data, *Lebensm.-Wiss. u.-Technol.*, **18**, 225–229.

SCHIFFMANN, R.F. (1987), Microwave and Dielectric Drying, In *Handbook of Industrial Drying*, A.S. Mujumdar (ed.), pp. 327–356, New York: Marcel Dekker, Inc.

SCHOFIELD, C. (1984), *Guide to Dust Explosion Prevention Part 1 – Venting*, Rugby: Institution of Chemical Engineers.

SCHOFIELD, C. and ABBOTT, J.A. (1988), *Guide to Dust Explosion Prevention Part 2 – Ignition Prevention, Containment, Inerting, Suppression and Isolation*, Rugby: Institution of Chemical Engineers.

SHARAF-ELDEN, I., HAMDY, M.T. and BLAISDELL, J.L. (1979), *Falling rate drying of fully exposed biological materials: a review of mathematical models*, American Society of Agricultural Engineers, USA, Paper No. 79-6522.

SIMATOS, D., LE LESTE, M., PETROFF, D. and HALPHEN, B. (1981). Use of Electron Spin Resonance for the Study of Solute Mobility in Relation to Moisture Content in Model Food Systems. In *Water Activity: Influences on Food Quality*, L.B. Rockland and G.F. Stewart (eds.), pp. 319–346, London: Academic Press.

SINGH, R.P. (1992), Heating and Cooling Processes for Foods, In *Handbook of Food Engineering*, D.R. Heldman and D.B. Lund (eds), pp. 247–276, New York: Marcel Dekker, Inc.

SINGH, R.P. and HELDMAN, D.R. (1993), *Introduction to Food Engineering*, 2nd ed., New York: Academic Press, Inc.

SIVETZ, M. and DESROSIER, N.W. (1979), *Coffee Technology*, Westport Connecticut: AVI Publishing Company Inc.

SLADE, L. and LEVINE, H. (1991), Beyond water activity: recent advances based on alternative approach to the assessment of food quality and safety, *Critical Reviews in Food Science and Nutrition*, **30**, 115–360.

SPICER, A. (ed.) (1974), *Advances in Preconcentration and Dehydration of Foods*, London: Applied Science Publishers Ltd.

STADELMAN, W.J. and COTTERILL, O.J. (1986), *Egg Science and Technology*, 3rd ed., Westport, Connecticut: AVI Publishing Company Inc.

STRUMILLO, C. (1987), Problems of energy savings in drying of food and agricultural products, *Acta Alimentaria Polonica*, **13**, 3, 175–189.

STRUMILLO, C. and LOPEZ-CACICEDO, C. (1977), Energy Aspects in Drying. In *Handbook of Industrial Drying*, A.S. Mujumdar (ed.), pp. 823–862, New York: Marcel Dekker.

STRUMILLO, C., MARKOWSKI, A. and KAMINSKI, W. (1983), Modern Developments in Drying of Pastelike Materials, In *Advances in Drying*, A.S. Mujumdar (ed.), p. 193, Washington: Hemisphere.

SULLIVAN, J.F. and CRAIG, J.C.Jr. (1984), The development of explosion puffing, *Food Technology*, **38**, 52–55, 131.

TALBURT, W.F. and SMITH, O. (1975), *Potato Processing*, 3rd ed., Westport, Connecticut: AVI Publishing Company Inc.

TOLEDO, R.T. (1991), *Fundamentals of Food Process Engineering*, 2nd ed., New York: Van Nostrand Reinhold.

TROLLER, J.A. (1989), Water Activity and Food Quality, In *Water and Food Quality*, T.M. Hardman (ed.), pp. 1–31, London: Elsevier Applied Science,

TROLLER, J.A. and CHRISTIAN, J.H.B. (1978), *Water Activity and Food*, London: Academic Press.

VACCAREZZA, L.M. and CHIRIFE, J. (1978), On the application of Fick's law for the kinetic analysis of air drying of foods, *J. Food Science*, **43**, 236–238.

VAN ARSDEL, W.B., COPLEY, M.J. and MORGAN, A.I. Jr. (1973a), *Food Dehydration*, 2nd ed., **Vol. I**, Drying Methods and Phenomena, Westport Connecticut: AVI Publishing Company, Inc.

VAN ARSDEL, W.B., COPLEY, M.J. and MORGAN, A.I. Jr. (1973b) *Food Dehydration*, 2nd ed., **Vol. II**, Practices and Applications, Westport, Connecticut: AVI Publishing Company, Inc.

VAN DEN BERG, C. and BRUIN, S. (1981), Water Activity and its Estimation in Food Systems: Theoretical Aspects. In *Water Activity: Influences on Food Quality*, L.B. Rockland and G.F. Stewart (eds), pp. 1–61, London: Academic Press.

VAN DEN BOS, M.J. (1982), Backgrounds of technologies used for production of lactose, *Bulletin of the International Dairy Federation*, **212**, 99–102.

VERSIC, J.R. (1988), Flavor Encapsulation, An Overview. In ACS Symposium *Flavor Encapsulation*, Series 370, Chapter 1, p. 5, Washington, ACS.

WANG, N. (1992), *Convective Drying – Modelling and Water Solid Interactions*, PhD Thesis, Department of Food Science and Technology, University of Reading.

WANG, N. and BRENNAN, J.G. (1991), Moisture sorption isotherm characteristics of potatoes at four temperatures, *J. Food Engineering*, **14**, 269–287.

WANG, N. and BRENNAN, J.G. (1992), Effect of water binding on the drying behaviour of potato, *Proceedings 8th International Drying Symposium*, Montreal, In *Drying '92*, Part B, A.S. Mujumdar, (ed.), pp. 1350–1359, New York: Elsevier.

WANG, N. and BRENNAN, J.G. (1992), Thermal conductivity of potato as a function of moisture content, *J. Food Engineering*, **17**, 153–160.

WANG, N. and BRENNAN, J.G. (1993), The influence of moisture content and temperature on the specific heat of potato measured by differential scanning calorimetry, *J. Food Engineering*, **19**, 303–310.

WANG, N. and BRENNAN, J.G. (1994b), Changes in structure, density and porosity of potato during dehydration, *J. Food Engineering* (in press).

WANG, N. and BRENNAN, J.G. (1994c), A mathematical model of simultaneous heat and moisture transfer during drying of potato, *J. Food Engineering* (in press).

WEDZICHA, B.L. (1986), Interactions Involving Sulphur Dioxide in Foods, In *Interactions of Food Components*, G.G. Birch and M.G. Lindley (eds), pp. 99–116, London: Elsevier Applied Science Publishers.

WILLIAMS, J.C. (1976), Chemical and Non-Enzymatic Changes in Intermediate Moisture Foods, In *Intermediate Moisture Foods*, R. Davies, G.G. Birch and K.J. Parker (eds), pp. 100–119, London: Applied Science Publishers Ltd.

WILLIAMS-GARDNER, A. (1976), *Industrial Drying*, 2nd Issue, London: George Godwin Ltd.

WINDSOR, M. and BARLOW, S. (1981), *Introduction to Fishery By-products*, Farnham, Surrey: Fishing News Books Ltd.

WOLF, W., SPEISS, W.E.L. and JUNG, G. (1984), *Sorption Isotherms and Water Activity of Food Materials*, New York: Elsevier.

WOLF, M., WALKER, J.E. and KAPSALIS, J.G. (1972), Water vapor sorption hysteresis in dehydrated foods, *J. Agriculture and Food Chemistry*, **20**, 1073.

WOODROOF, J.G. AND LUH, B.S. (1975), *Commercial Fruit Processing*, Westport, Connecticut: AVI Publishing Company, Inc.

ZEMELMAN, V.B. and KETTUNEN, D.M. (1992), Agglomeration and Agglomerator Systems, In *Encyclopedia of Food Science and Technology*, Vol. 1, Y.H. Hui (ed.), pp. 11–17, Chichester: John Wiley & Sons Inc.

ZIEMBA, J.V. (1962), Now drying without heat, *Food Engineering*, **34**, (7), 84–85.